T0177543

Mormonism, Medicine, and Bioethics

Mormonism, Medicine, and Bioethics

COURTNEY S. CAMPBELL

OXFORD
UNIVERSITY PRESS

OXFORD
UNIVERSITY PRESS

Oxford University Press is a department of the University of Oxford. It furthers
the University's objective of excellence in research, scholarship, and education
by publishing worldwide. Oxford is a registered trade mark of Oxford University
Press in the UK and certain other countries.

Published in the United States of America by Oxford University Press
198 Madison Avenue, New York, NY 10016, United States of America.

Library of Congress Cataloging-in-Publication Data
Names: Campbell, Courtney S., 1956– author.
Title: Mormonism, Medicine, and Bioethics / Courtney S. Campbell.
Description: New York, NY, United States of America :
Oxford University Press, 2021. |
Includes bibliographical references and index.
Identifiers: LCCN 2020030310 (print) | LCCN 2020030311 (ebook) |
ISBN 9780197538524 (hardback) | ISBN 9780197538548 (epub)
Subjects: LCSH: Medical ethics—Moral and ethical aspects. | Ethics. |
Medicine—Religious aspects. | Mormons—Attitudes.
Classification: LCC R724 .C3334 2021 (print) | LCC R724 (ebook) |
DDC 174.2—dc23
LC record available at https://lccn.loc.gov/2020030310
LC ebook record available at https://lccn.loc.gov/2020030311

DOI: 10.1093/oso/9780197538524.001.0001

1 3 5 7 9 8 6 4 2

Printed by Sheridan Books, Inc., United States of America

For Juliette, Jason, Scott, and Cassandra . . . the oceans, the sunsets,
the volcanoes, the mountains, the waterfalls, the flowers, the forests reveal
the divine in ways that no words can.

Contents

Preface

Books have their origins in conversations and seek to extend and expand those conversations over time and with different audiences. The conversations that have culminated in this book were initially stimulated through a research project at the Hastings Center on the role of religious voices in the professional fields of bioethical inquiry. Those professional conversations have continued throughout my academic career as a member of various institutional ethics committees, organizational ethics task forces, and in local, state, and national public policy settings.

The professional context of bioethics conversations can sometimes miss the richness of conversations that occur in the classroom and with various communities, including family members, friends, and religious and civic communities. These conversations provide an experiential depth, a groundedness in the lives and stories of persons, that augments and corrects the professionalized perspectives.

I have been particularly fortunate and appreciative of opportunities to bridge the academic and professional with the personalized and communal through conversations about the ethical commitments and moral culture cultivated by The Church of Jesus Christ of Latter-day Saints (LDS or Mormon). I was invited to develop an overview essay on "Bioethics in Mormonism" for the professional reference work *Encyclopedia of Bioethics* (3rd ed., 2004), and some years later received my first invitation to make a presentation on "LDS Ethics" in an academic setting at the University of Virginia. This book is the outgrowth of these many conversations and seeks to advance my calling for professional and communal ethical dialogue.

My aim in this book is to begin bridging these various intersections between the LDS religious community and its moral culture, the professional fields of bioethics, and practical decision-making. This work seeks to be a catalyst for expanding discourse within the interdisciplinary field of Mormon studies to include ethics and bioethics. Ethics has not been a well-developed area in Mormon studies, in contrast to studies in LDS history, theology, or literature. To remedy this oversight, I present a substantive interpretation of the sources, theological background, and moral principles of LDS ethics. The historical narratives and conceptual intertwining I offer of both bioethics and of LDS moral culture is intended to complement and expand the realm of Mormon studies.

A further objective is to create opportunities for reciprocal dialogues between the bioethics community and LDS scholarship. This conversation has yet to occur within academic disciplines or professional communities, or in public policy deliberations. My exposition, analysis, and critiques intertwine and contextualize LDS moral values and health care practices within the ethical inquiry undertaken in the broader professional scholarship of bioethics. My arguments disclose some points of common ground as well as areas of divergence toward the end of establishing the LDS faith tradition as a community of moral discourse for the bioethics field and the healing professions it informs (medicine, nursing, pharmacy, etc.). My claim is that given its emerging cultural prominence, LDS ethical scholarship should engage in bioethical literacy and bioethics should be LDS-literate.

I am also engaged in an effort to initiate more reflective dialogues regarding LDS ethics and moral culture among LDS scholars, LDS health care professionals, and the interested general LDS reader. The focus of the book on the interrelationship of religion, ethics, medicine, and health care should present for these various audiences new opportunities for mindful reflection and creative scholarship on the ethical implications of faith commitments, the responsibilities of the healing professions, and religious dimensions of public policy and public bioethics. A religious community that is formed through narratives and practices of covenantal commitments of love of neighbor needs to have a robust discourse about its ethical character.

I have understood my scholarship in biomedical ethics and in religious ethics through a linking metaphor of my moral culture, medicine, and the law, of *bearing witness*. The witness offers moral realities and moral truths about the way things are, vocalizes and embodies moral experience, and prophetically critiques the hypocrisies of the powerful and their oppression of the vulnerable by offering a new story, a re-storying, of tradition and conventional practice.

Acknowledgments

Any book, while exhibiting the flaws and frailties of its author, is always an exercise in moral community. I have been encouraged by many intellectual mentors and colleagues in these aspirations, including James F. Childress of the University of Virginia, Daniel Callahan of the Hastings Center, Richard Miller of the Divinity School at the University of Chicago, and Thomas Cole of the McGovern Center for Humanities and Ethics at the University of Texas Health Science Center at Houston. I express my appreciation to Oregon State University and the Hundere Endowment in Religion and Culture for providing an academic leave that enabled me to initiate conversations, learn from others, engage in scholarly research, and transmit the smallest part of that life wisdom into writing. I am also appreciative of my students who are my best teachers. My scholarship in this book has been stimulated by many persons who have helped me with their insights, narratives, and comments. I especially wish to thank Steve Athay, Denise and Ronald Bjornasen, Curtis and Kristin Black, Cassandra Brown, Emily Campbell, Juliette Douglas, Andrew Gessel, John and Britni Lefevre, Neil Olsen, Kara Ritzheimer, Julia and Patrick Salemme, Samantha Shivji, and Christin Sterling. My editors at Oxford University Press, Cynthia Read, Drew Anderla, and Rajakamuri Ganessin, have been patient and accessible, and their work will be appreciated by readers.

Note on Sources

The scriptural canon of The Church of Jesus Christ of Latter-day Saints consists of the Holy Bible, the Book of Mormon, the Doctrine and Covenants, and the Pearl of Great Price. My use of teaching or narratives from the latter three distinctive books of scripture cites the text as a whole (e.g., *The Book of Mormon*), then the specific book within the text (e.g., *Alma*), followed by chapter or section and verses.

Note on Sources

Introduction

Moral Realities of Latter-day Saint Ethics

Consider the ethical choices and moral convictions about medicine and health care embedded in the following vignettes:

- A recently married couple, unable to conceive naturally, considers using in vitro fertilization to become parents.
- A couple that has completed their family contemplates donating their excess embryos for biomedical research.
- A family seeks guidance about the best forms of nutrition and disease-prevention compatible with the health teachings of their faith community.
- Parents discuss with their adolescent children receiving the HPV vaccine to protect against a common sexually transmitted disease.
- A patient requests a blessing of faith healing to supplement the expertise of physicians.
- A family history of an inherited disease prompts a discussion among family members about undergoing genetic testing to determine susceptibility to the genetic mutation.
- Family members express reservations about an organizational request to retrieve organs and tissues from the body of their deceased loved one.
- Parents seek out medical advice and ecclesiastical counsel about discontinuing an aggressive course of medical interventions for their gravely ill newborn.
- A family member appointed as a health care proxy struggles with a medical recommendation about cessation of medical life support for their relative.
- A physician receives a request from a longtime patient for a prescription to hasten the patient's death.
- A couple resists a physician's recommendation to have a selective reduction of pregnancy.

Mormonism, Medicine, and Bioethics. Courtney S. Campbell, Oxford University Press (2021). © Oxford University Press.
DOI: 10.1093/oso/9780197538524.003.0001

- A family contemplates postponing annual physical exams and dental care because of inadequate health insurance.
- Citizens are perplexed about whether new laws permitting medical marijuana are compatible with the values of their faith tradition.

These scenarios reflect the personal, professional, and philosophical terrain of bioethics, an emerging field of academic study, clinical practice, and public policy on medicine and health care of the last half-century. While professional medical ethics has existed from antiquity in diverse cultures such as Greece and India, the advent of bioethics in the mid-1960s marks profound shifts in professional and cultural approaches to moral problems in medicine generated by technological advances, recognition of patients and their families as decision-makers, and the necessity of policy guidance on personal ethical choices at the beginnings and endings of life. The field of bioethics has sought to bring moral wisdom to bear on everyday ethical interactions between persons and healing professionals in medicine and nursing, in biomedical research on innovative medical interventions, and in technological advancements that can define, create, continue, enhance, and end human life. The realms of bioethics are discerned in the profound practical choices in medicine and health care made by persons, families, professionals, communities, and citizens that enact deep sources of human meaning, choices that inescapably express convictions—religious or nonreligious—about human purposiveness, the point to pain and suffering, and human mortality and destiny. The issues pursued and the choices made in bioethical decision-making reflect and inform answers to a common question of moral life: *how ought I to live?*

The scenarios presented earlier have a particular ethical valence in that they reflect personal, familial, professional, or civic circumstances encountered by members of The Church of Jesus Christ of Latter-day Saints (LDS) religious tradition.[1] The objective of this book is to integrate the professional and moral wisdom of bioethical inquiry with the unique spiritual insights, moral culture, and lived experience of this religious community. The ubiquitous experience of ethical questions in health care and medicine means that bioethical inquiry can be a meeting ground for reflective dialogue and meaningful practice among many communities, including patients and families, professionals, faith traditions, scholars, citizens, and policy-makers. My contention is that LDS communal and professional engagement with bioethical discourse holds the potential to reconceptualize LDS moral culture and ethical reflection, and advance intellectual intertwining between the broader academic field of bioethics, the specific field of Mormon studies, and Latter-day Saint scholars and practitioners.

The Church of Jesus Christ of Latter-day Saints was established in 1830 following a series of visions and revelations during the preceding decade to a prophetic leader, Joseph Smith (1805–1844).[2] Over the subsequent two centuries, the LDS Church has become the fourth-largest religious denomination in the United States. These developments have not gone unnoticed by various luminaries historically and have led some social commentators recently to draw attention to the "Mormon moment" in American political culture.[3] At the same time, an aggressive program of worldwide evangelization has led to a greater proportion of Latter-day Saints living outside the United States.

The LDS religious community manifests a robust moral culture and profound commitments to various moral practices, particularly as enacted within the family, through responsibilities (or callings) within ecclesiastical congregations, in voluntary community programs, and in global humanitarian assistance initiatives. A 2013 study showed that Latter-day Saints in the United States give an average of 336.5 hours of volunteer service per year, roughly nine times more than the average American volunteer.[4] However, continual ecclesiastical growth has not been accompanied by a parallel development in ethical reflection, including discussions regarding ethics and medicine. Three decades ago, a prominent LDS physician and scholar, Lester E. Bush, Jr., observed, "One way in which individual Mormons, at any level, have *not* contributed appreciably to the development of church teachings is through an influential body of reflective medical-ethical literature. Very little has been attempted to date comparable in any sense to the theological discussions found in most other faiths."[5] The rather sparse landscape in LDS medical ethics prevails even as Mormon studies has emerged as a thriving field in academic discussion. Significantly, recent definitive scholarship on LDS culture and on Mormon studies makes no mention of ethics or medicine and ethics.[6] The ethics of medicine and of bioethics is substantially underdeveloped within both LDS culture and Mormon studies relative to interpretations of Mormon history, theology, and scriptural studies.

This is a problematic situation as Latter-day Saints make decisions about bioethical issues—as individuals, as parents, as patients, as professionals, and as citizens—on a regular if not everyday basis. LDS members bring their faith convictions to bear when they address choices pertaining to procreation and contraception in creating life, as they encounter choices to use medical care for their newborn child, and subsequently whether or not to vaccinate their children for childhood and for sexually transmitted diseases. They make lifestyle choices that possess preventive medical implications, such as observance of a health covenant known as "The Word of Wisdom." Familial commitments that manifest ecclesiastical emphasis on researching family history and genealogical lineage can open difficult conversations about undergoing tests for a genetic predisposition to a disease. LDS moral culture is continually displayed in voluntary coordination of communal blood drives and a donative emphasis for the common good that carries potent symbols of love of neighbor embedded in organ and tissue donation. Invariably, faith convictions are challenged by the perplexities experienced in deciding on appropriate care for a family member at the end of life.

LDS health care professionals, including physicians, nurses, and pharmacists, encounter ethical choices and moral distress in providing care laden with ethical ambiguities for both faith and professionalism. They are in the forefront of conversations with families and parents about immunizations and vaccines, even when some professional recommendations may convey medical messages about sexuality different from those reflected in ecclesiastical moral teaching on the "law of chastity."[7] They have professional responsibilities for providing care and appropriate medical treatments at the edges of life, whether these involve aggressive life-sustaining interventions in a neonatal intensive care unit, curative treatments for a young adult afflicted with cancer, or palliative measures for persons at the end of their mortal life span. LDS physicians provide professional counsel to patients seeking relief of pain through medicinal marijuana. In the wake of legalized physician-aided death in many states with large LDS populations, LDS physicians receive requests from their terminally ill patients to write a life-ending prescription. The health professions are situated by ecclesiastical teaching and the moral culture within an overriding objective of healing, but this commitment has yet to generate an LDS ethic of professionalism.

Members of the LDS Church may also encounter bioethical questions as citizens enacting a civic responsibility to be "anxiously engaged in a good cause,"[8] including participating in public policy processes. Civic engagement may be shaped by a shared concern about a corrosion of civility in public and political discourse, by a perceived diminishing standard of ethical accountability in governmental and public life,[9] and by ecclesiastical statements directed to public policy issues on legalized abortion, physician-assisted death, and medical marijuana; public health concerns such as vaccinations; and the societal question of health care reform.[10] The global humanitarian health initiatives of the LDS Church offer opportunities for individual members and congregations to express their moral commitment of love of neighbor by contributing to essential health care services for diverse cultures outside the religious community.[11]

These practices, choices, and deliberative reflection constitute a rich, if incomplete, LDS moral culture. This moral culture is shaped through *narratives*—scriptural, individual, and communal—that esteem personal and familial commitments to cultivating virtues and character; community-defining *values* such as love of neighbor and covenant relationships; moral *practices* of service to benefit others within the religious community; responsible exercise of *moral agency* in professional, societal, and civic realms beyond the religious community; and a *progressive soteriology* that interprets lived experience as an occasion for acquiring embodied knowledge and as a pathway to eternal life. This means that a way of life lived with ethical integrity is, for a Latter-day Saint, a necessarily demanding quest: there are no moral holidays. Ecclesiastical leaders often contrast the moral culture of the religious community with the thin, minimalist ethic of secular moral culture,[12] which is premised on three values: freedom of choice, refraining from harming others, and advancing individual self-interest in a manner more conducive to moral atomism than community. This trinity of values promotes the virtual moral equivalence of all choices in secular culture, but such an equivalence is inadequate for the bioethical choices that confront Latter-day Saints.

Moral Realities

An ethic capable of providing meaningful moral guidance to address and resolve the bioethical questions confronted by Latter-day Saints as patients, families, professionals, and citizens comprises three necessary elements that collectively constitute a *moral reality*. First, the ethic must offer a method of reflective deliberation that enables persons to generate *practical ethical judgments* in concrete situations of moral choice (such as those presented at the beginning of this chapter). Second, the ethic should furnish *normative ethical principles*, *values*, and *virtues* derived from various sources of moral wisdom that both guide deliberation and decision-making and serve as standards for defending or critiquing moral choices. Third, these normative values must be interrelated with *broad pre-moral convictions* and interpretations of God, the world, human nature and destiny, knowledge, and ultimate meaning. These convictions, often narratively constructed, provide the context and character for a moral culture.

The remainder of this introduction initiates a project of constructing and interpreting the moral reality of LDS bioethics by (1) identifying critical and distinctive pre-moral convictions that manifest the *revealed realities* of the LDS religious worldview and (2) providing a substantive exposition of formative principles, values, and virtues of LDS ethics that constitute a *restored morality*. The subsequent chapters display the relevance of this bivalent moral reality in the context of practical moral choices about health, health care, and medicine.

Revealed Realities

The moral culture of Latter-day Saint communities is embedded in what ecclesiastical leader Dallin H. Oaks has designated as a "gospel culture," which refers to "a distinctive way of life, a set of values and expectations and practices common to all members."[13] This "distinctive" culture, including commitments to an examined moral life and communally shared practices and values of family, calling, and service, is formatively shaped by the character and content of the LDS restoration narrative. Since its founding in the 1820s and 1830s, Mormonism has affirmed what can be designated as a *revealed reality*. The visions, revelations, and prophetic teachings of founder Joseph Smith and his apostolic successors over nearly two centuries have articulated theological affirmations about the character of God and the nature of human beings, a narrative of salvation, sacred texts, and the delegation of divine authorization to perform salvific ordinances and rituals. The LDS origins narrative claims that essential religious truths on these matters embodied in Jesus Christ, his sacrificial atonement, and the early Christian church had been lost to the world over time and culture; the LDS revelatory reality restores such truths and re-stories the purpose of human life in light of teachings that establish saving knowledge, authority, and a covenantal relationship with God. This revealed reality necessarily confers ecclesiastical primacy on the epistemic category of "truth" rather than the ethical category of "right"; notably, a generalized dispersion of knowledge about nature, the human body, human communities, and the world, including scientific know-ledge that informs biomedical practices and technologies, accompanies the restoration of religious and spiritual truths.

The witness of these pre-moral convictions about God, the human condi-tion, and the natural and social worlds offers a set of affirmations addressing the pervasive question of the ethical skeptic, "Why be moral, anyway?" The LDS revealed reality offers a distinctive set of exemplary narratives and motivations for adhering to ethical standards, caring about ethics as a serious endeavor, and constructing a vibrant and lived moral culture and experience. The truth claims invite a way of seeing or envisioning the world that enables discernment of correct ethical principles and influences how situations of moral choice are constructed and understood. These convictions further-more present interpretive realities about the nature of persons as moral agents and their processes of moral decision-making, the form and identity of human relationships (including the moral identity of a church commu-nity), the sources of authoritative knowledge (including sources of moral wisdom within and beyond the religious community), and constructions of science, technology, and the healing professions.

This understanding that Mormonism's religious truths constitute a re-vealed reality about human nature, science, health, and the human body that shape the character and content of morality can be illustrated through an overview of the initial encounter of bioethics discourse with LDS moral cul-ture that transpired in research on the totally implantable artificial heart in human beings conducted at the University of Utah in 1982.[14] LDS ecclesi-astical teaching did not address the research or ethical questions about the implantable artificial heart in any direct way. An ecclesiastical document compiled in 1974 provided some very general-principled and procedural direction on human research studies: "The Church recognizes the need for carefully conducted and controlled experimentation to substantiate the effi-cacy of medicine and procedures. We believe, however, that the free agency of the individual must be protected by informed consent and that a qualified group of peers should review all research to ascertain that it is needed, is ap-propriately designed and not harmful to the person involved."[15]

The formulation of ecclesiastical policy on medical experimentation occurred in the context of significant national controversy about abuses in medical research and the necessity for ethical oversight. In the wake of disclosures of unethical research in the professional literature,[16] in 1974 the U.S. Congress established the first public bioethics commission, with the unwieldy name of the National Commission for the Protection of Human Subjects of Biomedical and Behavioral Research. Among many important matters, the National Commission endorsed the use of institutional review boards (IRBs) to ensure respect for the rights and welfare of research subjects in the pursuit of scientific knowledge, a procedural safeguard intended to end determinations of what constituted justifiable research based solely on the good intentions of the medical researcher. As important, the National Commission issued a foundational document, the "Belmont Report," which effectively became a kind of "constitution" for bioethical inquiry.[17] The report established three core principles—respect for persons, beneficence (promoting welfare), and justice—as an ethical framework for human research. These review processes and principles addressed the substantive concerns about human research in the LDS ecclesiastical statement. The implementation of these policies implied that ethical issues specific to the artificial heart study were settled from the ecclesiastical perspective and that the ethical integrity of the research relied on the professional expertise, knowledge, and practical wisdom of researchers and the commitments of healing professionals to protect the rights and welfare of research subjects embedded in procedural review.

However, in a critical sociocultural analysis of the artificial heart study, pioneering bioethicists Renee Fox and Judith Swazey maintained that "Mormonism . . . added certain dimensions of meaning and mission to the implantation of the artificial heart in Barney Clark."[18] While ambivalence if not outright tension has frequently characterized the relationship between religious values and scientific and biomedical innovations in American culture, Fox and Swazey argued that the distinctive world view, moral culture, and way of life embodied in Mormonism shaped the medical, institutional, and public discourse in which the narrative of a machine supplanting the human heart to sustain life was enacted. In a separate analysis, Fox asserted that "the artificial heart implant dramatically exemplified, in a morality play-like way, fundamental principles of Mormonism. . . ."[19] This "morality play" is illustrative of how pre-moral convictions of the LDS revealed reality about human nature, salvation, knowledge, health, embodiment, and family converged with scientific aspirations for biomedical progress.

Narratives of Progress. LDS teachings on personal and communal identity, and reflection on the ultimate human questions of human origins; human identity and purpose; meanings of pain, suffering, and mortality; and human destiny are situated within a distinctive teleological and soteriological narrative of divine and human purpose. This narrative of eternal human progress in the imitation of God and Christ and ultimate destiny in the presence of God invites persons to discover their nature and identity in a story and purpose greater than self. The narrative, whose meaning is dramatized and ritualized in the LDS temple liturgy, affirms that a divine plan and purpose for human mortal life were established prior to the creation of the earth. This "plan of salvation" was ratified by persons as unembodied spirit beings in a pre-mortal life.[20] Mortal life is a passage during which all persons who approved the plan experience embodied life, receive opportunities to acquire embodied knowledge in relationship with family and community, and make choices, religious and moral, necessary for their ongoing eternal progress. Birth, the afflictions and suffering experienced in mortality, and death are all necessary passages in the progressive salvation narrative. The eschatological culmination of the narrative is an eternal relationship with family in the presence of God.[21]

In this context, the artificial heart study represented a unique unification of two morally potent narratives of progress: the scientific narrative of perpetual progress through technological advances, such as life prolongation through mechanical assistance of heart functions, converged with a religious narrative of eternal human progress. An implicit assumption of these converging narratives is that of an ultimate harmony between the revealed realities of religion and the discovered realities of scientific inquiry.[22] A progressivist biomedical science could not have constructed a more compatible moral anthropology than the LDS narrative of eternal human progress.

Knowledge as Quest. A eulogy at Mr. Clark's funeral by an LDS ecclesiastical leader emphasized the interrelationships of human nature as progress in the imitation of God and knowledge generated through scientific research: "True scientific research is but the carefully ordered expression of that mortal hunger and drive and quest to know more of what God knows."[23] An eternal quest for knowledge is a second pre-moral conviction that bears great importance for LDS ethics and moral culture. This quest for knowledge and wisdom assumes soteriological significance in the teachings of Joseph Smith: "A man is saved no faster than he gets knowledge."[24] This interrelationship of progress and knowledge is reinforced in LDS scriptural teachings that explain that "Whatever principle of intelligence we attain unto in this life, it will rise with us in the resurrection"; the early Mormon community also received various scriptural admonitions to promote broad cultural, scientific, civic, and moral literacy that remain relevant in contemporary communal discourse.[25] The salvific status of knowledge has profound implications for biomedical research: biological knowledge acquired through scientific study about the human body—the functioning of its organs, tissues, cells, and DNA—and the advancement of technological devices that assist bodily functions possesses practical, instrumental value and symbolic, religious meaning. These epistemic convictions provide distinctive motivations to undertake scientific inquiry, to pursue new knowledge through biomedical research, and to find a life calling through the healing professions.

The theological embrace of empirical knowledge as part of the human journey of eternal progress underlies communal interpretations of advancements in scientific study, biomedical research, and applied medical technologies as symbols of divine inspiration. The scientific unveiling of knowledge about the human body that heretofore resided in the realm of the unknown and mysterious does not portend a "god of the gaps" theology in which "god" signifies what science has not yet explained. Rather, the presumed ultimate harmony between revealed religious realities and discovered scientific truths constructs an ecclesiastical witness both that God is the source of all human learning and that science is a supportive epistemic equivalent of religion.

In contrast to the receding presence of the sacred embedded in "god of the gaps" theologies, LDS history and moral culture share with the biomedical sciences an ethos of exploring frontiers and expanding the boundaries of the known material world. A pioneering identity was narratively formed through the migrations of the first generation of the LDS community from the familiar regions of New York and Ohio to build new frontier settlements in Missouri and Illinois, and received its definitive expression in an 1846 exodus from Illinois due to religious and political persecution to the relatively unknown and unpopulated landscape of the Great Salt Lake Valley (which at the time was part of Mexico). The pioneering, exploratory ethos provides a culture of moral support for new developments and progressive advances in the biomedical sciences. The perseverance required to push back the frontiers of human limitations and boundaries in the quest for knowledge while shadowed by uncertainty, human fallibility, and prospective failures is illustrative of what Fox and Swazey refer to generally in biomedicine as "the courage to fail."[26]

Embodiment and Health. These pre-moral convictions also inform communal constructions about the human body and about health that are integral to the identity of the healing professions. The LDS salvation narrative affirms that a person's journey of eternal progress necessarily requires the experience and wisdom conferred only through embodiment in mortal life. Consequently, the human body holds a sacred status in LDS teaching, even as it becomes an object of scientific study and medical intervention. The body is not fundamentally a prison for the human essence that deprives us of knowledge, or an inert machine that is the source of continual illusions, or private property over which a person has ownership, but is ultimately conceived as a "temple" or "tabernacle" that commands awe and reverential respect.[27] The body reveals human creation in the image of God (*imago Dei*) and thereby is an intrinsic, rather than an accidental or contingent, feature of the human self. The fullness of human flourishing requires an integration of eternal spirit and physical body; this integration constitutes the "soul."[28] Persons are then embodied souls and ensouled bodies. The valuation of embodiment as essential in eternal progress means that the existential experiences of adversity, pain, suffering, and mortality that confer a point and purpose to medicine can be re-storied and transformed from ordeals imposed by an arbitrary, capricious, and indifferent cosmos to a school for character that inevitably bestows new knowledge of self and cultivates personal capacities for empathic understanding of suffering others.[29]

The theology of embodiment bears important implications for communal constructions of health, personal responsibility, and medical interventions to promote or restore health. Health assumes soteriological significance and bridges embodied experience and the acquisition of knowledge. Latter-day Saints who adhere to the proscriptions and admonitions of the "Word of Wisdom" receive promises not only of "health in their navel and marrow to their bones" but also of "wisdom and great treasures of knowledge."[30] The domain of health, and thereby the scope and responsibilities of the healing professions, extends beyond the realm of physical well-being and is intertwined with progress and the quest for knowledge. What the theology of embodiment and the value of health call for from the healing professions is a holistic medicine, encompassing physical, emotional, cognitive, relational, and spiritual health and well-being.

Precisely because the body has sacred value as an intrinsic element of personal identity, it exerts an initial moral claim of resistance against scientific and biomedical reductionistic tendencies to treat the body merely as an object for study, manipulation, or intervention. The sanctuary that is the ensouled body commands respect in a manner not present in the prison, mechanistic, or property philosophies of the body, which have shorn off the self from the body. In LDS teaching, the totality of the bodily organism is clearly much more than the sum of its organic functions, and this reality places constraints on the nature of scientific investigation or research experimentation in the absence of good cause, clear design, and authorization. The reason the bioethical assessments of the Utah artificial heart study provide such an illuminating illustration of the interrelationship of revealed religious realities with medical judgment and ethical deliberation is that, in the "morality play" analysis provided by Fox and Swazey, the narratives of progress in both biomedical science and LDS culture supported a technological aspiration rather than an ethical responsibility: the body was displaced literally by technology. The absence from current LDS ecclesiastical policies on medical and health practices of *any* statement regarding ethical standards for research with human beings might perpetuate this posture.

Familialism. The process and professional responsibilities in the artificial heart study were also influenced by a moral culture of "familialism." This relational value reflects LDS pre-moral convictions that the family is foundational to societal cohesion and community identity, and as indicated previously, is of eternal soteriological significance. The communal stress on family is often emphasized in worship services that cite a prophetic maxim that "no other success can compensate for failure in the home."[31] The theology of family was formally institutionalized in the 1995 promulgation of "The Family: A Proclamation to the World," issued by the presiding LDS ecclesiastical leadership bodies, the First Presidency and the Quorum of the Twelve Apostles. The proclamation affirmed that the family is "ordained of God," is "central to the Creator's plan for the eternal destiny of His children," and is "the fundamental unit of society." Furthermore, the family is the core relationship of salvation: "the divine plan of happiness enables family relationships to be perpetuated beyond the grave . . . [and] for families to be united eternally."[32] It is thus a meaningful symbol of community identity that local congregations are constructed as "ward families" and that members refer to each other as "Sister" or "Brother" rather than by first names.

Scholars have recognized the LDS emphasis on family as one of the appealing features of Mormonism within contemporary American society.[33] However, Fox and Swazey argued that the familial feel of the community had problematic implications in biomedical research. The concept of family was extended beyond a congregational setting into the professional setting to such a degree that the boundaries between family ethics and professional ethics dissolved: "The Mormon emphasis on the family as the basic unit of society . . . shaped and reinforced these medical team attitudes and behaviors [on group solidarity and extended kinship]. They would be unlikely to occur in this form . . . in comparable hospital settings."[54] This attribution of familial "shaping" implies that LDS culture conferred a markedly uncritical endorsement of innovative research that involved more complex professional relationships than treating another family member.

A responsible ethic must refrain from simply underwriting any and all prospective medical developments as a form of progress and also from delegating moral assessments about vital matters of human and patient welfare to professional and policy elites. Insofar as scientific, medical, and religious narratives about progress, knowledge, family, the body, and health do converge, the capacity for a morally *critical* religious witness is diminished. The revealed realities of LDS teaching thereby give rise to and require supplementation by a framework of normative ethical principles, values, and virtues that provide not only moral guidance for adherents as patients, families, and professionals, but also grounds for prophetic moral critique of specific medical practices and health care policies. This necessitates a focal change from the revealed reality to the *restored morality*.

Restoring a Prophetic Morality

The formative prophetic vision of Joseph Smith, the ecclesiastical articulation of a revealed reality, and early LDS communal experience generated a moral culture better characterized as a restored or recovered morality than as a revealed, discovered, or invented morality. It is illustrative that only the concluding statement in the thirteen articles composing the principal LDS creedal statement, the Articles of Faith, addresses values and virtues required for moral life, having been preceded by twelve statements affirming the nature of the revealed reality. This concluding expression of moral commitment voices community aspirations to honesty, integrity, fidelity, chastity, virtue, and benevolence and essentially reiterates a New Testament or Pauline ethic of virtue.[35] The revealed reality does not also reveal new ethical precepts or principles but furnishes a morality that appears largely derivative from a predecessor Christian ethos. Two claims require brief exposition: how LDS morality is a "restored" morality, and how it is also a "re-storied" morality.

The concept of a restored morality draws on a background narrative of generalized moral decline largely constructed as a divergence from a broader

Judeo-Christian ethos in Western cultures. The formative ethical standards of Jewish and Christian communities may remain relatively intact, but their relevance for societal interactions beyond specific faith communities are deemed to be substantially receding and the target of societal critiques. The LDS narrative of moral decline as articulated during the mid-20th century through the 21st century comprises several features, including eroding societal support for the family and an increased prevalence of divorce and cohabitation; libertine attitudes of sexual permissiveness, and related moral evils such as pornographic consumption and abortion; political corruption and a generalized diminished trustworthiness, honesty, and promise-keeping in everyday interactions; challenges to core freedoms, especially religious liberty, as part of the ongoing secularization of law; and, misuse of drugs, or use of illicit drugs, that spawn in their wake public health crises such as the current opioid epidemic and drug-related violence.[36]

The narrative of moral decline is a significant aspect of American LDS culture. A recent study indicated that, for LDS members from the silent and baby boomer generations and the succeeding GenX generation, "moral or religious decline" is the single most important issue facing America. Among the emerging Millennial population, concerns of "moral or religious decline" are the third-most-important concern, topped by concerns about poverty and homelessness and about terrorism,[37] issues with their own moral valence even if they are not featured in the moral declension narrative. The implication of this narrative is that LDS teaching both reveals religious truths and restores ethical *standards* as well as *moral motivations* for adhering to those standards.

The Jewish and Christian moral legacy is, however, situated differently within LDS theological claims and narratives about God, humanity, salvation, and the world. This restored morality is thereby *re-storied* and so furnishes distinctive motivations for living with ethical integrity. A morally paradigmatic biblical parable is illustrative of how a morality can be restored, re-storied, and prophetic. In a New Testament narrative, Jesus is confronted with a question about the greatest commandment in Jewish law. His formulation of loving God and loving one's neighbor as intertwined requirements of *Torah* is explicitly not a new or revealed morality but a recovery of the community's moral tradition. However, a subsequent question about the identity of the neighbor who is to be loved becomes the occasion for Jesus to relate the memorable parable of the Good Samaritan that exemplifies the meaning of love.[38] The ethic of love is retrieved from religious law but then re-storied with a deeper and expansive meaning: the exemplar of love, a Samaritan, is a person from an ethnicity outside the covenant community of Israel. The restored and re-storied morality thereby constitutes a prophetic form of moral criticism, exposing the community's failure to live by its historically formative values. Retrieved from religious law, love is re-storied as a demanding prophetic ethic, altruistic in nature, and universal in application.[39]

The restored, re-storied, and prophetic morality expressed in LDS ethical teaching and enacted in LDS moral culture can be represented through a normative framework of five principles—love, hospitality to strangers, covenant, justice, and moral agency. These principles and values are embedded in five principal sources of LDS ethical teaching, including scripture, ecclesiastical teaching, personal reflection, communal experience, and non-LDS moral wisdom. These principles and values are presupposed in LDS ecclesiastical guidelines on moral issues and health policies and in the community moral culture. This ethical framework does not underwrite either a secular libertarian ethos in which all choices are morally equivalent or a progressivist ideology in which all biomedical advances are immune from moral critique. The framework rather establishes normative moral guidance that, precisely because it possesses continuities with the moral legacy of predecessor religious traditions and the moral wisdom of other cultures, creates a common or shared set of criteria for ethical accountability.

Love. A restored and re-storied ethic positions neighbor-love as the moral essence of Christian life. This ethic emerges not as a matter of willed choice, calculative reciprocity, or even divine commands, but rather from grateful *responsiveness* to divine blessings and human gifts bestowed on each person. Responsiveness to the human experience of giftedness is expressed in generative moral responsibilities of solidarity, empathy, and compassionate caring for equally vulnerable persons.[40] An ethic of neighbor-love is informed by the moral virtues of charity, mercy, and compassion and is the organizing value of personal integrity. As a normative action-guide, love of neighbor encompasses the minimal secular moral constraint of refraining from inflicting harm on other persons, but it also expands the scope of moral concern to removing and preventing harm and promoting the basic welfare and flourishing of other persons.

The re-storying of love displayed in the Good Samaritan parable illustrates two important features by which an ethic of love includes but expands beyond a minimalist secular ethic of beneficence in mutual relationships.

Jesus's concluding admonition to "do likewise" as the Samaritan necessitates reflection on what the Samaritan actually *did*. The Samaritan enacted care for a needy and wounded person, disclosing that love involves a self-sacrificial feature of setting aside personal projects and plans to care for the basic needs of vulnerable others.[41] The Samaritan's moral capacity for self-sacrificial love presupposed an initial exercise of empathic imagination to apprehend the embodied needs of the injured traveler. While religious and professional communities often draw analogies for moral action through identifying with the Samaritan,[42] what the Samaritan *did* was identify with the bodily woundedness of the injured traveler.

Re-storied love is also a principle of prophetic moral critique: the ostensibly religious figures in the Good Samaritan parable are indifferent to the plight of the traveler while love is enacted by a person from a despised ethnicity. Jesus's audience could not have heard the parable as a heartwarming, feel-good story about rescuing a wounded person but rather as a prophetic indictment of an ossified social and religious structure that oppressed the socially vulnerable. The love ethic exposes the moral hypocrisy of elites to self-justify, rationalize, or exempt themselves from certain actions because the other person does not conform to socially constructed classifications. The attitude of "otherizing" that allows persons to dismiss any claim of moral responsibility is contrary to love.

LDS moral culture enacts the moral commitment of love in numerous ways, including a culture of service, care for the poor, bearing the burdens of the bereft, personal and communal contributions to welfare programs, and ongoing partnerships and participation in interfaith and relief organizations to alleviate poverty, prevent disease and provide needed assistance to the vulnerable. These practices reflect habits of virtuous character interrelated with love, including solidarity, cooperation, gratitude, selflessness, a mutual covenant of care, and compassion. These values reflect communal constructions of the neighbor as the moral equivalent of self: "A foundational premise of Mormon community is that there is no separation between you and your neighbor. Your well-being is tied to your neighbor's well-being."[43]

Hospitality to Strangers. Though implicit in the love ethic, the principle of hospitality to the stranger broadens the scope of moral community. The "stranger" is a moral identity in the biblical texts that pertains primarily to persons outside the covenant people of Israel, yet Israel is repeatedly reminded to care for the stranger as though they were full members of the community. The moral character of this inclusionary community relies on an analogy with Israel's divine deliverance from oppression and slavery: "Love . . . the stranger: for you were strangers in the land of Egypt."[44] As with love, an experience of giftedness (deliverance from slavery) precedes the call to care for the stranger.

The hospitality principle reflects a core claim about the equality of persons: as all persons bear the image of God, loving regard for neighbors within the religious community and care for the stranger should not display a moral difference. The narrative re-storying of hospitality is ethically expressed in a discourse related in *The Book of Mormon* by the prophetic leader Benjamin. Benjamin's sermon at one point addresses a perennial question for any moral community: the scope of responsibility to care for needy persons, community exiles, or "the beggar." Benjamin imaginatively re-stories the communal question by drawing an analogy from the human-divine relationship, a relationship suffused by the revealed reality of radical human dependency on divine gifts: "Are we not all beggars? Do we not all depend upon the same Being, even God, for all the substance which we have, for both food and raiment . . . ?" Upon understanding community relationship to needy persons, exiles, and strangers through the prior relationship with the divine source of all goodness, moral identity should shift in a radically empathic manner: a communal responsibility of solidarity and care for vulnerable persons manifests as an imitation of divine love. "Now, if God, who has created you, on whom you are dependent for your lives and for all that ye have and are, doth grant unto you whatsoever you ask that is right . . . then, how you ought to impart of the substance that ye have one to another."[45] Matters of lineage, ethnicity, ecclesiastical affiliation, or geography should make no ethical difference in the extension of love, empathy, and hospitality.[46]

The principle of hospitality for vulnerable strangers manifests not only a re-storied ethic but also a prophetic ethical path. Moral actions and social policies that work in intent or outcome to isolate strangers (including the ill and medically indigent) and are catalysts for conflict rather than agents of healing are susceptible to moral critique. This prophetic dimension of hospitality is displayed in LDS ecclesiastical statements on polarizing controversies about immigration to the United States. The statements assert that immigration reaches a "bedrock moral issue," namely, "how we treat each other as children of God," and situate LDS moral commitments on care for strangers within the communal narrative legacy of migration and immigration.[47] That is, immigration is re-storied as a "moral" issue rather than a political question, and the ethic is then re-storied through the historical experience of the LDS community as refugees from religious persecution. The hospitality principle transforms the moral identity of the stranger from outcast to kinship with a more intimate moral community.

Covenant. The ethical principles of love and hospitality invite persons into a moral community marked by covenantal relationships. The dynamic and complex ethical structure of covenant comprises four interrelated features: an initial experience of *giftedness, relational responsiveness,* a *promise* or *witness* to assume moral responsibilities, and a *calling* to a distinctive way of life. A covenantal relationship is thereby always characterized by a moral structure of grace-response, relationship-gratitude, witness-responsibility, and call-fidelity that issues in a transformed moral identity and re-storied way of life rather than an episodic ethics.

The moral dynamic of a covenantal call to ethical responsibility is exemplified in a community-forming and -transforming narrative related in *The Book of Mormon.* A community of exiles, having received a gift of divine deliverance from tyranny and oppression, is invited to baptism by their prophetic teacher, Alma, to confirm their ongoing relational responsiveness to God. This invitation reflects an assumption of intertwining covenantal responsibilities in which a moral covenant with others—"bear one another's burdens, that they may be light; . . . mourn with those that mourn; . . . comfort those that stand in need of comfort, . . ."—precedes and may be a prerequisite for the life-demanding religious covenant with God: "stand[ing] as witnesses of God at all times and in all things, even until death."[48] The covenantal responsibility of moral presence to the needy, afflicted, wounded, and vulnerable is inseparably intertwined with a witness of religious fidelity. The intertwining of these covenants implies the desirability of integrating revealed realities with restored and re-storying moralities and to enact knowing truth with living truthfully. The consummation of the covenantal promise in baptism exemplifies a transformation of identity of the exiles from strangers and slaves into a binding moral community.

The moral logic of covenant bears a prophetically critical dimension that is as meaningful within the religious community as it is for mediating relationships with persons outside the community. LDS ecclesiastical discourse and scholarly commentary has tended to (a) focus almost exclusively on the religious or vertical covenantal relation, to the neglect of the moral or horizontal covenant relationship, and (b) subsequently conflate covenant with the minimalism of contractual agreements.[49] A covenant is always conveyed through a relationship-initiating gift that situates persons within a narrative of gift-and-call that is absent from the mutual promotion of self-interest embedded in contractual interactions. The gift relationship of covenant symbolizes giving rather than merit, grace and mercy rather than negotiation and entitlement, other-regard rather than personal advantage. A covenantal ethical principle presumes trust in the covenanted partners to freely determine how to live with fidelity to their binding promise while a contract is constructed through negotiation that delineates the specific responsibilities that must be performed by both parties for compliance.

The covenantal principle offers a basis for critiquing the reduction of ethics to only those legally enforceable duties specified in a contract. A covenantal ethic will be more demanding than legal requirements, and its extensive responsibilities can serve as a ground for moral critique and reform of the law. Covenantal ethics express an ongoing resilience and bindingness of the partners to a shared purpose and ongoing relationship rather than organizing ethics through contractual interactions that terminate upon completion of a specific action. As exemplified in a marriage, an ethic oriented by covenant is laden with moral potency for transforming identities, while an ethic oriented by a contract, as illustrated in everyday commerce, is composed of a transactional exchange of goods extraneous to self-identity. The moral logic of covenant thereby provides an edge for prophetic moral criticism that is absent from either religious or secular contractual ethics.[50]

The primary LDS relational locus for covenantal expression is the family organization, which includes not only the covenant of marriage, but also parental callings, responsibilities to lineage, and covenants to posterity. Marriage as ordained of God witnesses to the mutual sharing of gifts, responsiveness, and promises of responsibility by each of the partners. Marriage is also the covenantal context for sexual expression and procreation, with parents assuming a sacred stewardship to raise "children in love and righteousness, to provide for their physical and spiritual needs, and to teach them to love and serve one another, observe the commandments of God, and be law-abiding citizens wherever they live."[51] The covenantal family is the bulwark against the narrative of societal moral decline. The family is also the relationship through which the cultural, religious, and moral legacies of past generations and lineal predecessors are honored and embodied in the present and bequeathed to future descendants and generations. The family covenant binds both persons and past, present, and future in an eternal relationality.

Justice. Although issues of social justice have been characterized as a "blind spot" in LDS moral culture,[52] the sources and practices of LDS ethics make it both possible and necessary to construct a robust and prophetic principle of justice. In its basic formulation, justice requires treating similar cases in a similar way and giving to other persons what they deserve. The revealed reality and the salvation narrative re-stories this construction of justice through articulating (a) the grounds for just treatment of fellow human beings, (b) the basis for the moral equality of persons, (c) the interrelationships of love and fairness within the "aspirations [of] a just and caring society,"[53] and (d) the symbol of an ideal community of justice designated as "Zion."

The core concept of justice is re-storied in the LDS salvation narrative through a theological affirmation that, as "man was also in the beginning with God,"[54] all persons have an eternal identity and possess *inherent value and dignity.* Justice thereby requires recognizing the eternal dignity of each person, which is not effaced or diminished by the accidents and contingencies of embodied mortal life. The moral responsibility to render just treatment to others is reinforced through the biblical creation narrative that confers on all persons the unique status among creation as bearers of the image of God (*imago Dei*).[55] This revealed reality provides a basis for the *moral equality* of all persons irrespective of whether or not legal equality is acknowledged or enforced by laws in civic society. It becomes possible on the basis of human dignity and equality to prophetically differentiate between just and unjust laws.

The moral equality of persons means that justice can be constructed as the social extension of love in responsibility to broader political and societal organizations. The ethical interrelationship of justice and love within a society can assume various forms, including distributive, restorative, and preferential justice. Distributive justice specifies a social commitment to fairness in distributing basic benefits (security, education, health care) and burdens (service, taxation) as part of civic membership and participation. The inherent dignity and moral equality of persons entail that there is a moral floor requiring communal provision for the basic needs of others below which no person should fall. Each person as a bearer of the image of God has a moral claim of justice to a decent minimum of the liberties and material goods sufficient for the basic needs of living and well-being.

The moral threshold for justice as socially extended love and caring is manifested in how a society treats its most vulnerable persons. The entire canon of LDS scripture presents a common witness to the moral reality that infidelity to the principles of love, hospitality, and covenant as displayed primarily in neglect of the poor, the vulnerable, the wounded, and the stranger is necessarily a violation of justice. The societal propensity to injustice reflects ingratitude, callousness, and moral myopia. In an unjust culture, persons with privilege beyond common liberties and excess material goods have failed to remember their own radical vulnerability, the fundamental experience of giftedness, and their dependency on God and others. This failure of moral memory diminishes capacities to empathically identify with vulnerable persons and to act in a manner relationally responsive to how one has received from others. Indifference to the basic needs of others generates a moral myopia in which persons who experience such deprivation are no longer even noticed. A prophetic ethic of justice will embody the *moral memory* of a community or society when the community betrays its constitutive values.

Justice implies a profound, indissoluble covenantal solidarity and an empathic identification with other persons in virtue of their humanity, dignity, and embodied witness of the image of God. Insofar as certain persons or communities have historically been excluded from equality and fair opportunity to receive common liberties and societal benefits, or have experienced the imposition of a disproportionate share of social burdens, justice can require rectification for the poor, the vulnerable, and the marginalized. A preferential form of justice is necessitated by historical patterns of injustice so deep-rooted they now constitute systemic injustice or structural violence that erodes a moral culture. Institutionalized oppression and systemic violence designate cultural conditions that severely constrain moral agency and degrade the dignity and sanctity of persons who bear the image of God. The shared scriptural witness to God's ongoing care for the powerless and victimized is characteristically situated within a prophetic indictment of the moral hypocrisy of social elites and the structures of unauthorized power that elites establish to retain and perpetuate unrighteous dominion over others.[56]

The ethical principle of justice is also re-storied in LDS moral culture through a powerful symbol of the ideal moral community, "Zion."[57] Morally, Zion is multivalent: it conveys a conception of justice as righteousness, wholeness, and integrity as members of the community embody the virtue of purity of heart.[58] Zion manifests ideals of human equality and unity in common vision and purpose: a Zion community aspires to remedy divisive worldly tendencies of power, classism, and unrighteous dominion manifest in social and economic stratification such that there are neither social divisions nor poor persons in Zion. The ethical fabric of Zion illuminates the vibrancy of intermediating communities of moral culture, including

families, friendships, particular religious communities, professions, educational institutions, and civic associations and communities, which are
transformed into critical schools for moral experience in which covenantal
care can be enacted. The scripturally based imperative of LDS moral culture
to establish Zion in the world draws on distinctive narratives and furnishes
distinctive motivations for enacting justice in the world.

The re-storied moral witness to a Zion community is especially important
in settings of liberal democratic cultures that are agnostic about the questions
of the good society, the good community, or the good life for human beings
in relationship. The Zion community, albeit an aspiration, provides grounds
for a justice-based prophetic moral critique of this philosophical agnosticism. The morally critical standard embedded in the LDS construction of
Zion is analogous to the interpretation of the love ethic of Jesus developed by
Protestant theologian Reinhold Niebuhr.[59] The moral ideal of Zion presents
an *indiscriminate* standard of moral criticism that discloses the proximate
and limited nature of all human social achievements. At the same time, Zion
is a *discriminating* standard that permits assessments of particular moral
cultures as more or less just, loving, hospitable, covenantal, egalitarian, and
respectful of human moral agency.

Moral Agency. I situate moral agency last to emphasize that any account of
LDS ethics that neglects this principle is incomplete. Moral agency comprises
our capacities for discernment and deliberation and our existential personal
liberty to make responsible choices, a capacity so expansive that it is invio-
lable even by God. The ethic of agency is instantiated in LDS moral culture
by a profound commitment regarding the authority of personal experience.[60] The appeal to moral agency is a mediating bridge between generaliz-
able precepts of scripture and ecclesiastical teaching and personal or familial
decision-making in concrete situations of moral choice. Significantly, LDS
ecclesiastical teaching on virtually all bioethical issues ultimately defers
to a decision-making procedure in which the moral authority of personal
agency, consultation with health care professionals, and reliance on spiritual
practices such as prayer are determinative for practical choice.

The culture of agency differentiates the moral reality of LDS ethics from
an ecclesiastical authoritarian ethics and a libertarian ethic that avoids accountability. Oaks re-stories the concept of agency through the soteriological narrative as an experience of giftedness: "moral agency—the right to
choose—is a fundamental condition of mortal life. Without this precious gift
of God, the purpose of mortal life could not be realized." This intertwining

of revealed reality and re-storied morality provides a critical distinction between choice as a *method* and as a *goal*. Persons created in the image of God
assume ethical responsibility for exercising their capacity for agency as a
method of moral deliberation and judgment in a world of choices between
both good and evil and between good and good.[61] This gift-based interpretation of agency contrasts with political accounts of personal autonomy that
construe freedom of choice as an entitlement-based goal and risk a "slip into
the position of trying to justify any choice that is made."[62] The philosophical defense of autonomy as an end entails that there are no criteria external
to choice by which to differentiate justifiable from unjustifiable choices and
actions. Autonomy as a goal levels all moral distinctions. The ethic of moral
agency and the secular principle of autonomy thus overlap but are not morally equivalent.

The principle of moral agency does not imply endorsing the moral indifference reflected in the relativistic sentiment that "each person chooses
what's right according to their own values." Participation in a vibrant, living
moral culture involves openness to persuasive arguments and narrative
witnesses of others. While authoritarian methods of coercion, compulsion,
and unrighteous dominion clearly violate personal agency, it is equally disre-
spectful to isolate a moral agent from community in the name of protecting
freedom and confer immunity from moral suasion. Respecting moral agency
seeks a middle path between paternalistic pretensions to make choices for
others and moral apathy regarding another's ethical choices. The distinction between agency as a method for making choices and autonomy as a goal
confers a prophetic moral edge to agency that manifests in a sharp critique of
ethical subjectivism and cultural moral relativism. A legal (and theological)
right to *choice* does not itself ensure *right* choices.

The prophetic critique of autonomy reflects a claim that moral agency
comprises freedom of choice and *accountability* for actions.[63] A religious conception of stewardship is embedded in moral agency: Persons are entrusted
with and have accountability for physical well-being, responsible care of
their bodies, care for the earth, and wise disposition of material resources.
A moral agent has a responsibility of answering to several moral audiences,
including conscience and community, about the agent's use of the gifts received rather than presuming immunity from accountability.[64] Agency as
accountability presumes persons possess a minimal knowledge of right and
wrong or a moral conscience independent of socialization processes (family,
education, peers, religious teaching) for acquiring values. The moral reality

of agency resides in part in agents responsibly incorporating foundational moral knowledge to make discerning assessments of right, permitted, or wrong actions in concrete situations.

I have presented five basic principles—love, hospitality, covenant, justice, and moral agency—that are the normative element of the moral reality of LDS ethics. These principles give rise to a cluster of interrelated values and virtues, including solidarity, empathy, care and compassion, gift relationships, and stewardship, to integrate character with ethical deliberation. My claim is that these principles constitute a *restored* morality recovered or retrieved from predecessor religious and moral cultures, a *re-storied* morality insofar as these principles and values are situated within narratives and scripture unique to LDS moral culture, and a *prophetic* morality that exposes and critiques moral decline and failures in contemporary society to live according to its organizing and formative values. This moral reality provides both normative action-guides for participants in LDS moral culture and a moral memory for the broader culture.

Moral Continuities

A morality that is restored, re-storied, and prophetic necessarily possesses continuities of moral wisdom with predecessor moral cultures of religious traditions and with contemporaneous secular ethics. The LDS community in general has a profound indebtedness relative to the intellectual, moral, and political legacy of other moral cultures and traditions. The principles, values, and practices composing LDS moral culture have been articulated and practiced in other moral cultures in the absence of the revealed reality. As one illustration of both claims, for theological reasons (protecting moral agency) and political reasons (the historical legacy of religious persecution and exile), the LDS Church is a perennial advocate of religious liberty. The principles of religious toleration, religious freedom, and the moral equality of all religions has an intellectual and cultural history that long precedes the 19th-century LDS restoration events, including compelling arguments, centuries of debate, and deep cultural chasms occasioned by religious wars.[65] LDS scriptural and ecclesiastical teaching ratifies, confirms, and re-stories the moral wisdom of pre-restoration traditions of religious liberty.[66] It follows from my distinctions in the moral reality of LDS ethics between the revealed reality and the restored morality and from my claims about the scope and sources of the restored morality that a person can embody ethical goodness without affirming a specific faith conviction or even any belief system. This is a deeply contested question culturally and globally.[67] My claim is that while morality may be psychologically, motivationally, or culturally dependent on religious belief, it is not logically dependent: a person can be morally good and embody ethical integrity without religious convictions.

The substantive moral continuities between the LDS moral reality and other traditions and cultures is supported by scripture, ecclesiastical teaching, and practices of the moral culture.[68] Each person as an image-bearer of the divine is bestowed with a conscience, designated as the "light of Christ," which provides basic insights about good and evil.[69] The moral conscience provides a source for any individual to make judgments about right and wrong choices irrespective of their religious affiliation or cultural or historical situatedness. The innate capacity of conscience is a necessary condition for the exercise of moral agency. Intimations of a common moral core also emerge in LDS scriptural texts that indicate all peoples and cultures have received some portion of divine wisdom.[70] The LDS First Presidency specified these scriptural intimations as "moral truths" bestowed by God upon culturally transformative religious leaders and philosophers, including Confucius, the Buddha, Muhammad, and Socrates. Such individuals are perceived as recipients of "moral truths . . . given to them by God to enlighten whole nations and to bring a higher level of understanding to individuals."[71] The moral wisdom of these diverse cultures provides insight into questions essential to ethical deliberation. The scope of moral traditions and exemplars that provides resources for and can be integrated with a restored and re-storied morality is exceptionally broad-based. Furthermore, the substantial deference in LDS ecclesiastical teaching on medical, moral, and health practices to the integrity of the healing professions presumes a profound moral and practical continuity between the sacred, the professional, and the secular.

These continuities in moral wisdom do not imply an absence of distinctiveness to the LDS moral reality and moral culture. The ethical principles and moral practices are distinctive relative to the ethics of both non-LDS religious and secular systems as well as professional morality because they are embedded in distinctive LDS *sources* and *narratives*. The revealed religious reality both retrieves and re-stories the principles of love, hospitality, covenant, justice, and moral agency for distinctive forms of ethical *reflection* and *moral practice*. The moral realities of LDS ethics furnish distinctive *motivations* and *exemplars* for ethical conduct. The distinctiveness resides less in the *what* of morality than the *why*. This is how an LDS ethic can be a recovered morality, a restoring of the moral culture embedded in predecessor traditions and cultures, a re-storied morality attributable to narratives unique to LDS experience, and a prophetic morality that embodies the moral memory of a broader culture susceptible to moral critique when it betrays its constitutive values.

Overview: The Path Ahead

My convictions on moral continuities inform the core purpose of this book: to initiate dialogues on the intersections of religion, ethics, and medicine in the context of moral choices in health care. These dialogues seek to expand the realm of Mormon studies to include ethics and bioethics, provide for a bioethics-literate LDS moral culture and an LDS-literate bioethics, and provide a critically reflective ethic for members of the LDS community, including LDS health care professionals. My analysis and argumentation is organized into three major areas. This Introduction and Chapter 1 provide expositions of the LDS moral reality and its framing ethical principles that provide a conceptual framework for the prospective dialogues. The ethical principles of love, hospitality, covenant, justice, and agency can guide ethical deliberation in concrete circumstances of moral choice; mediate relationships between families, patients, and the healing professions; and present ethical criteria for assessing health care public policy. The following chapter expands on the theme of moral continuity by presenting an exposition of healing as a core feature of convergence between LDS moral culture and the healing professions. The overlapping commitment to healing provides a framework for interpreting the healing professions as moral callings that embody vocational responsibilities of ethical integrity.

Chapters 2 through 7 apply the foundational framework of LDS ethical principles and the healing aspirations of medical professionalism to an array of specific bioethical topics: reproductive technologies, stem cell research, preventive medicine, vaccinations, genetic screening and editing, biomedical enhancements, organ and tissue donation, and life endings in neonatal intensive care, withdrawing medical treatments, and medical-assisted death. The conceptual trajectory of these chapters as a whole evolves from personal and familial realms of ethics to broader communal and professional concerns.

These chapters intertwine personal and communal experience and ecclesiastical policy positions with perspectives in bioethical discourse. I seek to
honor the moral authority of personal experience by introducing many of
the topical chapters with an illustrative narrative. These narratives present
stories of bioethical choices encountered by LDS adherents as patients, family
members, members of a religious community, and professionals: couples
assessing options regarding infertility, parents contemplating having their
preadolescent children vaccinated against sexually transmitted disease; persons who receive a physician's recommendation for genetic testing based on
lineage and family history, proxy decisions regarding organ donation, families and professionals deliberating on treatment decisions about seriously
impaired infants, moral anguish over stopping life-sustaining medical treatment, and physician responses to patient requests for hastened death. These
narratives are the grounding "data" for LDS bioethics. The moral authenticity
of the stories witnesses to the complexity of living the truth of personal and
communal convictions.

The concluding portion of Chapter 7 on medical-assisted death, the subsequent chapter on public bioethics, and the Epilogue shift the ethical context from personal, familial, communal, and clinical settings to public policy
issues. Although the LDS Church expresses great reticence about ecclesiastical involvement in the political culture as such, many important bioethical
issues—medical-assisted death, legalized abortion, and legalized medical
marijuana—are deemed to possess such moral import as to warrant an ecclesiastical policy witness for a public broader than the LDS community. I expand on this focus in the Epilogue, making a case for health care reform and
universal access to basic health care rooted in both the moral reality and the
moral culture of LDS humanitarian initiatives.

I engage in various ethical tasks of critical analysis throughout the
book—including descriptions of practices, interpretations of experience,
expositions of scriptural and ecclesiastical teaching, construction of theological or ecclesiastical rationales, and normative argumentation on specific
issues—in such a manner that the LDS moral reality and culture and the ethical wisdom of bioethics are in dialogue. Ultimately, however, I understand
my role as one of *witnessing*, a metaphor that connects LDS ethics and medical experience. I aim to disclose truths I have seen, experienced, and learned.
Witnessing is a practice of public discourse, and the intelligibility of such an
action presupposes community and relationship. The public bearing of witness transforms the private into the communal and invites different moral

communities into a shared narrative. Witnessing is a moral act laden with ethical presuppositions of truthful dialogue and solidarity in community; courage and honesty in speaking truth about moral hypocrisy to persons who possess social, professional, or ecclesiastical power; and public accountability for enacting values.

Notes

1. Throughout the remainder of the book, I use the shorthand designators of "LDS" or "Mormon" when referring to the ecclesiastical teaching or religious moral culture of The Church of Jesus Christ of Latter-day Saints.
2. Matthew J. Grow et al., *Saints: The Story of the Church of Jesus Christ of Latter-day Saints,* vol. 1, *The Standard of Truth: 1815–1846* (Salt Lake City: The Church of Jesus Christ of Latter-day Saints, 2018), provides the most current ecclesiastical interpretation of the LDS founding and its subsequent developments.
3. J. B. Haws, *The Mormon Image in the American Mind: Fifty Years of Public Perception* (New York: Oxford University Press, 2013); Mitchell Landsberg, "Romney's Conservative Roots Lie in Mormon Faith," *Los Angeles Times,* September 21, 2012, https://www.latimes.com/politics/la-xpm-2012-sep-21-la-na-mormon-conservatism-20120921-story.html; Ross Douthat, "Romney's Mormon Story," *New York Times,* August 11, 2012, https://www.nytimes.com/2012/08/12/opinion/sunday/douthat-romneys-mormon-story.html; Pew Research Center, "Reactions to the 'Mormon Moment,'" January 12, 2012, https://www.pewforum.org/2012/01/12/mormons-in-america-mormon-moment/; Leland A. Fetzer, "Tolstoy and Mormonism," *Dialogue: A Journal of Mormon Thought* 6 (1972), 13–29, https://www.dialoguejournal.com/wp-content/uploads/2010/05/Dialogue_V06N01_15.pdf
4. Van Evans, Daniel W. Curtis, and Ram A. Cnaan, "Volunteering Among Latter-day Saints," *Journal for the Scientific Study of Religion* 52:4 (December 2013), 827–841.
5. Lester E. Bush, Jr., *Health and Medicine among the Latter-day Saints: Science, Sense, and Scripture* (New York: Crossroad, 1993), 201.
6. See Patrick Q. Mason, ed., *Directions for Mormon Studies in the Twenty-First Century* (Salt Lake City: University of Utah Press, 2016); Quincy D. Newell, Eric F. Mason, *New Perspectives in Mormon Studies: Creating and Crossing Boundaries* (Norman: University of Oklahoma Press, 2013); Terryl L. Givens, Philip L. Barlow, *The Oxford Handbook of Mormonism* (New York: Oxford University Press, 2015).
7. This moral law requires abstinence from sexual relations prior to marriage and fidelity within marriage, understood as a covenantal partnership between a man and a woman.
8. *Doctrine and Covenants* 58:27–28.
9. Mormon Women for Ethical Government, *The Little Purple Book: MWEG Essentials* (Salt Lake City: By Common Consent Press, 2018).

10. Appendix A provides a comprehensive representation of important LDS ecclesiastical statements pertaining to medical and health ethics.

11. Russell M. Nelson, "The Second Great Commandment," LDS General Conference, October 2019, https://www.churchofjesuschrist.org/study/general-conference/2019/10/46nelson?lang=eng

12. Jeffrey R. Holland, "The Inconvenient Messiah," *The Ensign,* February 1984, https://www.lds.org/ensign/1984/02/the-inconvenient-messiah?lang=eng&_r=1

13. Dallin H. Oaks, "The Gospel Culture," *The Ensign,* March 2012, https://www.lds.org/ensign/2012/03/the-gospel-culture?lang=eng&_r=1

14. Bush, *Health and Medicine Among the Latter-day Saints,* 106–107; Renee C. Fox, Judith P. Swazey, *Spare Parts: Organ Replacement in American Society* (New York: Oxford University Press, 1992), 162–166.

15. "Attitudes of the Church of Jesus Christ of Latter-day Saints Towards Certain Medical Problems," in "Mormon Medical Ethical Guidelines," ed. Lester E. Bush, Jr., *Dialogue* 12 (Fall 1979), 98, 100.

16. James H. Jones, *Bad Blood: The Tuskegee Syphilis Experiment,* 2nd ed. (New York: Free Press, 1993); Susan M. Reverby, "More Than Fact and Fiction: Cultural Memory and the Tuskegee Syphilis Study," *Hastings Center Report* 31 (September–October 2001), 22–28; Henry K. Beecher, "Ethics and Clinical Research," *New England Journal of Medicine* 274 (June 14, 1966), 1354–1360; Paul Ramsey, *The Patient as Person: Explorations in Medical Ethics* (New Haven, CT: Yale University Press, 1970), 47–54; "Willowbrook Hepatitis Experiments," https://science.education.nih.gov/supplements/webversions/bioethics/guide/pdf/Master_5-4.pdf

17. National Commission for the Protection of Human Subjects of Biomedical and Behavioral Research, "The Belmont Report," April 1979, https://www.hhs.gov/ohrp/regulations-and-policy/belmont-report/read-the-belmont-report/index.html; James F. Childress, Eric M. Meslin, Harold T. Shapiro, eds., *Belmont Revisited: Ethical Principles for Research with Human Subjects* (Washington, DC: Georgetown University Press, 2005).

18. Renee C. Fox, *Essays in Medical Sociology: Journeys in the Field,* 2nd ed. (New Brunswick, NJ: Transaction Books, 1988), 607; cf. Fox, Swazey, *Spare Parts,* 162–166.

19. Renee C. Fox, "'It's the Same, but Different': A Sociological Perspective on the Case of the Utah Artificial Heart," in *After Barney Clark: Reflections on the Utah Artificial Heart Program,* ed. Margery W. Shaw (Austin: University of Texas Press, 1984), 84.

20. LDS scriptural texts refer to this salvation story as "the plan" of "salvation," "happiness," "mercy," "redemption," "restoration," and "deliverance." See Dallin H. Oaks, "The Great Plan of Happiness," LDS General Conference, October 1993, https://www.churchofjesuschrist.org/study/general-conference/1993/10/the-great-plan-of-happiness?lang=eng

21. A couplet attributed to LDS leader Lorenzo B. Snow crystallizes the nature of eternal progression: "As man is now, God once was; as God is now, man may become." Lester E. Bush, Jr., "The Mormon Tradition," in *Caring and Curing: Health and Medicine in the Western Religious Traditions,* ed. Ronald L. Numbers, Darrel W. Amundsen (Baltimore, MD: Johns Hopkins University Press, 1986), 402.

22. David H. Bailey, Jeffrey M. Bradshaw, "Science and Mormonism," *Interpreter: A Journal of Latter-day Saint Faith and Scholarship* 19 (2016), 17–37; Gene A. Sessions, Craig J. Oberg, eds., *The Search for Harmony: Essays on Science and Mormonism* (Salt Lake City: Signature Books, 1993); Erich Robert Paul, "Science and Religion," in *The Encyclopedia of Mormonism*, ed. Daniel H. Ludlow, 1992, 1270–1272, https://contentdm.lib.byu.edu/digital/collection/EoM/id/4169

23. Neal A. Maxwell, as quoted in Fox and Swazey, *Spare Parts*, 164.

24. Joseph Fielding Smith, ed., *Teachings of the Prophet Joseph Smith* (Salt Lake City: Deseret Book Company, 1977), 217; Richard N. Williams, "Knowledge," in *The Encyclopedia of Mormonism*, ed. Daniel H. Ludlow, 1992, 799–800, https://contentdm.lib.byu.edu/digital/collection/EoM/id/3856

25. *Doctrine and Covenants* 130:18–19; 88:78–80, 118.

26. Renee C. Fox, Judith P. Swazey, *The Courage to Fail: A Social View of Organ Transplants and Dialysis*, 2nd ed. (New York: Routledge, 2002).

27. For a discussion of the historical evolution of embodiment in LDS thought, see Benjamin E. Park, "Salvation Through a Tabernacle: Joseph Smith, Parley P. Pratt, and Early Mormon Theologies of Embodiment," *Dialogue: A Journal of Mormon Thought* 43 (2010), 1–44.

28. *Doctrine and Covenants* 88:15; 138:50; Barbara Lockhart, "The Body: A Burden or a Blessing?," *The Ensign*, February 1985, https://www.lds.org/ensign/1985/02/the-body-a-burden-or-a-blessing?lang=eng

29. A Christological claim in LDS scripture (*Alma* 7:13) is that Christ's embodiment and suffering were essential to his perfection: "he will take upon him their infirmities, that his bowels may be filled with mercy, according to the flesh, that he may know according to the flesh how to succor his people according to their infirmities."

30. *Doctrine and Covenants* 89. See the fuller discussion of these requirements and promises in Chapter 3.

31. The maxim entered LDS discourse through 1935 and 1964 sermons of David O. McKay, ninth LDS president, who was quoting from J. E. McCulloch, *Home: The Savior of Civilization* (Washington, DC: Southern Cooperative League, 1924), 42. See Russell Arben Fox, "About the McKay Quote," *Times and Seasons*, March 11, 2004, https://www.timesandseasons.org/harchive/2004/03/about-the-mckay-quote/

32. The Church of Jesus Christ of Latter-Day Saints, "The Family: A Proclamation to the World," 1995, https://www.churchofjesuschrist.org/study/manual/the-family-a-proclamation-to-the-world/the-family-a-proclamation-to-the-world?lang=eng. The concept of "family" is always a vexed issue within LDS tradition given the 19th-century experience of plural marriage.

33. Jana Riess, *The Next Mormons: How Millennials Are Changing the LDS Church* (New York: Oxford University Press, 2019), 72–76.

34. Fox, *Essays in Medical Sociology*, 603–604; Fox, " 'It's the Same, but Different,'" 82.

35. *The Pearl of Great Price*, 13th Article of Faith: "We believe in being honest, true, chaste, benevolent, virtuous, and in doing good to all men; indeed, we may say that we follow the admonition of Paul—We believe all things, we hope all things, we have

endured many things, and hope to be able to endure all things. If there is anything virtuous, lovely, or of good report or praiseworthy, we seek after these things."

36. Gordon B. Hinckley, et al., *Speaking Out on Moral Issues* (West Valley City, UT: Bookcraft, 1998).

37. Riess, *The Next Mormons*, 175–182.

38. *Luke* 10:25–37.

39. See Courtney S. Campbell, *Bearing Witness: Religious Meanings in Bioethics* (Eugene, OR: Cascade Books, 2019), 98–124.

40. The experience of giftedness as a ground for ethics invites a gift-call or grace-response ethical dynamic rather than the command-obedience dynamic common in theistic religions and commonly dismissed by ethical theorists.

41. Thomas S. Monson, "Your Jericho Road," *The Ensign,* May 1977, https://www.lds.org/general-conference/1977/04/your-jericho-road?lang=eng

42. Albert R. Jonsen, *The New Medicine and the Old Ethics* (Cambridge, MA: Harvard University Press, 1990), 38–60; Larry R. Churchill, *Rationing Health Care in America: Perceptions and Principles of Justice* (Notre Dame, IN: University of Notre Dame Press, 1987), 33–40.

43. Newsroom, "The Mormon Ethic of Community," October 16, 2012, https://newsroom.churchofjesuschrist.org/article/the-mormon-ethic-of-community

44. *Deuteronomy* 10:19; cf. *Exodus* 22:21; *Doctrine and Covenants* 121:45.

45. The Book of Mormon, *Mosiah* 4:19–21.

46. Matthew Soerens, Jenny Yang, Leith Anderson, *Welcoming the Stranger: Justice, Compassion, and Truth in the Immigration Debate* (Downers Grove, IL: IVP Books, 2016).

47. Newsroom, "Deferred Action for Childhood Arrivals (DACA) Statement," January 26, 2018, https://newsroom.churchofjesuschrist.org/article/daca-statement-january-2018

    The DACA statement observes, "Most of our early Church members emigrated from foreign lands to live, work, and worship, blessed by the freedoms and opportunities offered in this great nation." See also Walker Wright, "LDS Moral Commitments on Immigration Are Grounded in History and Scripture," *Deseret News*, June 19, 2018, https://www.deseretnews.com/article/900022114/op-ed-lds-moral-commitments-on-immigration-are-grounded-in-history-and-scripture.html

48. The Book of Mormon, *Mosiah* 18:8–10.

49. Wouter van Beek, "Covenants," in *The Encyclopedia of Mormonism*, ed. Daniel H. Ludlow, 1992, 331–333, https://contentdm.lib.byu.edu/digital/collection/EoM/id/5640

50. My covenantal critique of contractual ethics is indebted to the analysis of William F. May, *The Physician's Covenant: Images of the Healer in Medical Ethics* (Louisville, KY: Westminster John Knox Press, 2001), 112–154.

51. The Church of Jesus Christ of Latter-Day Saints, "The Family," 1995.

52. Patrick Q. Mason, "The Possibilities of Mormon Peacebuilding," *Dialogue: A Journal of Mormon Thought* 37 (2004), 31.

53. Newsroom, "Church Supports Principles of *Utah Compact* on Immigration," November 11, 2010, https://newsroom.churchofjesuschrist.org/article/church-supports-principles-of-utah-compact-on-immigration

54. *Doctrine and Covenants* 93:29.

55. Campbell, *Bearing Witness*, 60–97.

56. Michael L. Hadley, ed., *The Spiritual Roots of Restorative Justice* (New York: SUNY Press, 2001); *Doctrine and Covenants* 121:39.

57. Stephen C. Taysom, *Shakers, Mormons, and Religious Worlds: Conflicting Visions, Contested Boundaries* (Bloomington: Indiana University Press, 2017), 51–98; Thomas Carter, *Building Zion: The Material World of Mormon Settlement* (Minneapolis: University of Minnesota Press, 2015); A. D. Sorenson, "Zion," in *Encyclopedia of Mormonism*, ed. Daniel H. Ludlow (New York: Macmillan, 1992), 1625–1626, https://contentdm.lib.byu.edu/digital/collection/EoM/id/4373

58. *Doctrine and Covenants* 97:21; The Pearl of Great Price, *Moses* 7:18; The Book of Mormon, *4 Nephi* 16–17.

59. Reinhold Niebuhr, *An Interpretation of Christian Ethics* (New York: Seabury, 1979), 62–83.

60. A recent study indicated that the two most important authorities for Latter-day Saints in making moral decisions are conscience and spiritual impressions, both of which presume subjective and personal experience. See Riess, *The Next Mormons*, 193–195.

61. Dallin H. Oaks, "Good, Better, Best," October 2007, The Church of Jesus Christ of Latter-day Saints, https://www.churchofjesuschrist.org/study/general-conference/2007/10/good-better-best?lang=eng

62. Dallin H. Oaks, "Weightier Matters," *The Ensign*, January 2001, https://www.lds.org/ensign/2001/01/weightier-matters?lang=eng

63. C. Terry Warner, "Agency," in *The Encyclopedia of Mormonism*, ed. Daniel H. Ludlow, 1992, 26–27, https://contentdm.lib.byu.edu/digital/collection/EoM/id/5455

64. Campbell, *Bearing Witness*, 90–97; Courtney S. Campbell, "The Gift and the Market: Cultural Symbolic Perspectives," in *Transplanting Human Tissue: Ethics, Policy, and Practice*, ed. Stuart J. Youngner, Martha W. Anderson, Renie Schapiro (New York: Oxford University Press, 2004), 139–159.

65. Daniel L. Dreisbach, Mark David Hall, eds., *The Sacred Rights of Conscience: Selected Readings on Religious Liberty and Church-State Relations in the American Founding* (Indianapolis: Liberty Fund, 2009).

66. *Doctrine and Covenants* 134; "Religious Freedom," The Church of Jesus Christ of Latter-day Saints, https://www.lds.org/religious-freedom?lang=eng; Quentin L. Cook, "Restoring Morality and Religious Freedom," *The Ensign*, September 2012, https://www.lds.org/ensign/2012/09/restoring-morality-and-religious-freedom?lang=eng

67. A 2017 study by the Pew Research Center indicates that 56% of Americans believe it is not necessary to believe in God to be moral and have good values. Gregory A. Smith, "A Growing Share of Americans Say It's Not Necessary to Believe in God to Be Moral," Pew Research Center, October 16, 2017, http://www.pewresearch.org/fact-tank/2017/10/16/a-growing-share-of-americans-say-its-not-necessary-to-believe-in-god-to-be-moral/

68. Brian D. Birch, "A Portion of God's Light: Mormonism and Religious Pluralism," *Dialogue: A Journal of Mormon Thought* 51:2 (2018), 85–102.
69. The Book of Mormon, *Moroni* 7:16–19.
70. The Book of Mormon, *Mosiah* 3:13; *Alma* 13:22; 26:37; 29:5, 8; *Moroni* 7:16–19.
71. "Statement of the First Presidency Regarding God's Love for All Mankind," February 15, 1978, as cited in R. Lanier Britsch, "What Is the Relationship of The Church of Jesus Christ of Latter-day Saints to the Non-Christian Religions of the World?," *The Ensign*, January 1988, https://www.lds.org/ensign/1988/01/i-have-a-question/what-is-the-relationship-of-the-church-of-jesus-christ-of-latter-day-saints-to-the-non-christian-religions-of-the-world?lang=eng; Carlos E. Asay, "God's Love for All Mankind," in *Mormons and Muslims: Spiritual Foundations and Modern Manifestations*, 2nd ed., ed. Spencer J. Palmer (Provo, UT: Brigham Young University, 2002), 54–55.

1

Faith, Medicine, and Healing

The status of contemporary medicine in the 21st century moral culture formed by The Church of Jesus Christ of Latter-day Saints (LDS) is instructively displayed by the fact that the current prophetic leader and Church president, Russell M. Nelson, initially received worldwide acclaim as a pioneering heart surgeon: Nelson was part of a medical team that designed the first-ever heart-lung machine. When appointed to the Quorum of Twelve Apostles in 1984, Nelson was the first physician to receive such a calling; his prominence in his professional vocation has been a source of inspiration within the LDS community. Nelson's apostolic ascendancy to the church presidency might seem an anomaly from the perspective of 19th-century Mormons who relied on faith healing, herbal and botanic medicine, and folk remedies to combat sickness and disease. Living primarily on the American frontier, the early Latter-day Saint communities experienced the ravages of infectious diseases like smallpox, malaria, and tuberculosis against which the medicine of the 1800s was largely impotent as well as the ravages of medical quackery. It is understandable that the community expressed pronounced aversions to physicians trained in the heroic medical practice of the era.

In this context of nonscientific medicine, a pervasive communal practice to prevent, remove, or cure disease was faith healing. The restoration narrative encompassed a recovery of numerous spiritual gifts portrayed in biblical texts, including reliance on faith and prayer for healing from diseases and impairments. The spiritual gift of healing as a sign of divine presence was enunciated in the first year of the church organization: "And whoso shall ask it in my [Christ's] name in faith, they shall cast out devils; they shall heal the sick; they shall cause the blind to receive their sight, and the deaf to hear, and the dumb to speak, and the lame to walk."[1] The moral realities of disease and healing could be re-storied through the healing ministry of Jesus and his disciples and interpreted as a witness of authenticity of the LDS Church. Healing through faith, including anointing a sick person with consecrated oil and providing a blessing, has evolved over nearly two centuries of communal

Mormonism, Medicine, and Bioethics. Courtney S. Campbell, Oxford University Press (2021). © Oxford University Press.
DOI: 10.1093/oso/9780197538524.003.0002

experience but continues to be a meaningful ritual in the 21st-century LDS community.

This historical faith healing experience of the community and its presumptive reliance on providential care raises a fundamental issue about the rationale and justification for the profession of medicine and for treatments oriented by scientific understandings of the body and various pathologies. The faith healing approach presupposes a revealed reality about divine interventions in disease processes and was an indispensable religious practice during a prescientific medical era when rudimentary medical knowledge about diagnosis, prognosis, and treatments frequently made the ill person worse to the point of death. A communal shift in the late 19th century to increasing acceptance of medical expertise, informed by recognition that medicine had developed a scientific basis integrating the germ theory of disease, opened the community to new possibilities of treatment and cure while necessitating reconceptualizing the relationship of faith and medicine. The emergence of an effective, science-based medicine inevitably raised questions about the value of faith healing practices for treating serious disease and preventing death. As medicine and the healing professions have empirically demonstrated their efficacy, LDS communal attitudes have evolved from critiquing and disparaging medicine to encouraging adherents to make full use of its various preventive, diagnostic, and curative powers.

The contemporary LDS understanding that faith and medicine are complementary paths of healing offers a middle ground between an extreme of relying exclusively on faith and prayer, as retained by some smaller Christian evangelical communities,[2] and an alternative extreme of relying exclusively on secular medical methods immune from faith or spirituality. Advances in scientific knowledge, technical expertise, and treatments characteristic of modern medicine have been re storicd as signs of, and ultimately derived from, the divine source of healing most fully exemplified in the ministry of Jesus. The medical profession is communally constructed as a divinely inspired human endeavor of healing and a form of professional calling.[3] This approach to the healing professions can be situated within the tradition's more general claims about an ultimate harmony of religious and scientific truths and of the complementarity of both faith (healing) and works (medicine).

The articulation of a theological and moral rationale for the healing professions is of great salience for the moral reality of LDS bioethics. An administrative guidebook for lay leaders providing short policy statements on various medical, health, and moral issues includes an admonition situated under the general topic of "medical and health practices": "Members should not use medical or health practices that are ethically or legally questionable."[4] The policy presumes the relevance of ethics and law pertaining to decisions about health care and medicine but also a general communal deference to professionally determined standards of medical practice. This implies a commitment of trust in the ethical integrity of medicine and the healing professions. However, the admonition does not offer any exposition of the values that might inform judgments about the general acceptability and trustworthiness of the healing professions or about the ethically or legally problematic nature of a proposed medical or health intervention.

My claim is that *healing* provides moral convergence and continuity between communal rituals and practices oriented by faith convictions and reliance on medical interventions. The moral reality of healing is characterized by several features: (a) restoring *wholeness* to persons, (b) *relationships* enacting covenantal commitments to bear the burdens of others, (c) bearing witness to the *patient's narrative* of illness experience, (d) the potency of *touch*, and (e) an *empathetic solidarity* with vulnerable persons.[5] Insofar as modern medicine articulates its fundamental self-identity as a healing profession,[6] the practice patterns of contemporary medical professionals can be shaped by aspects of the religious roots of healing. While healing is not the exclusive province of medicine and health care, healing is the culminating aspiration of medical and nursing interventions oriented by concrete ends of preventing illness, relieving pain and suffering, promoting health, and curing disease.

In this chapter, I develop an ethic of healing by presenting a typology of three broad patterns of relationship between faith convictions and medical practice that emerged historically in LDS culture: Faith *against* medicine, faith *and* medicine, and faith *in* medicine. The pattern of faith and healing against medicine is represented in the early LDS historical recourse to faith healing and herbal and botanical remedies in caring for adherents who were invariably afflicted with untreatable conditions and outbreaks of contagious disease in frontier settings. The pattern of compatibility of faith and medicine emerged in late-19th-century transitions in both LDS and medical cultures that allowed a re-storying of the paradigmatic features of modern medicine and its underlying scientific method as part of the ministry and calling of healing. Current LDS culture exhibits a faith in medicine as adherents are encouraged to utilize the remarkable professional and technological resources of contemporary medicine, interpret biomedical knowledge as ultimately sourced in divine wisdom, and retain the community ritual of administering healing blessings to persons who suffer from disease and other afflictions.[7] This healing ethos provides grounds for a prophetic critique of the corporatization and commodification of medicine.

Faith and Healing Against Medicine

The LDS Church was organized in 1830 during an era when infectious diseases were rampant and medicine was still premodern, not yet an organized profession nor structured by scientific knowledge, largely ineffective in treating disease, and lacking established methods for alleviating pain. Wisdom about health, the body, nutrition, and remedies for disease required resources internal to the community. One of the earliest scriptural teachings regarding sickness reflected a practical experience of community members being "healed" by faith or "nourished" through tenderness, herbs, and mild food;[8] medical treatment was not a viable option. Historian Jonathan Stapley contends that the early Mormon "healing ritual" situated itself against two currents in the broader culture, the concept of "cessationism" in American religiosity that held that spiritual gifts including healing had ended with the biblical era, and "heroic allopathic medicine," the orthodox medicine of the day that relied on practices such as bloodletting and consumption of toxic agents such as calomel.[9] The affirmation of faith healing as a prophetic critique of cessationism and herbs as a critique of allopathic medicine led some ecclesiastical leaders to compare botanic medicine (or "Thomsonianism") to the LDS restoration of truth and orthodox medicine to a fallen Christianity.[10] These theological convictions and practical considerations in caring for the ill in the early LDS community have been viewed by other scholars as bearing similarities with contemporary naturopathic medicine.[11]

Faith in either faith healing or botanical medicine in turn relied on communal storytelling of individual healings. The re-storying significance of communal depictions of faith healing are illustrated through a narrative of "a day of God's power" that is ubiquitous in contemporary LDS literature. Forcibly exiled from Missouri in 1838, the LDS community had taken refuge in mosquito-infested swamplands adjacent to the Mississippi River that provided the ecological conditions for an outbreak of malaria in the summer of 1839. An apostle, Wilford Woodruff, related that the prophetic leader Joseph

Smith emulated the healing ministry of Jesus: "[Joseph Smith] called upon the Lord in prayer, and the power of God rested upon him mightily; and as Jesus healed all the sick around Him in His day, so Joseph, the Prophet of God, healed all around on this occasion."[12] Smith bestowed blessings of healing on many sick persons within and adjacent to his own house, as well as in many homes near the river, alleviating symptoms at least temporarily for numerous persons who were near death.[13] This manifestation of healing presence in the community was mediated entirely by faith and prayer.

Smith's teachings on healing re-storied both the illness experience and healing by situating any circumstance of sickness within a narrative of divine providential care, Jesus as both exemplary sufferer and exemplary healer, and the human experience of giftedness. Insofar as the prospect of healing relies on divine powers that persons may access but do not control, the implication is that giftedness precedes healing and invites "prayer and thanksgiving for the manifestation of divine power already granted and realized."[14] Gratitude for providential care and the gift of healing should be expressed even when a person does not recover from their affliction. Significantly, the healing ministry was radically democratized rather than manifested through a single charismatic leader. Smith taught that any faithful elder could invoke the divine blessings to heal others, as "the actual administration by the anointing with oil and by the imposition of hands by those who hold the proper office in the priesthood is an authoritative ordinance."[15] The rituals of anointing and laying of hands reflect the healing potency of touch, especially as within the early Mormon community, the anointing could apply to the afflicted portion of the body.

Moreover, the gift of healing was not coextensive with male priesthood holders but was both exclusive of some males who lacked faith or worthiness to heal and inclusive of women.[16] Women were called to minister to the needs of the sick through the establishment of the female Relief Society in 1842, and LDS literature for the subsequent century contains both numerous examples of healing through the faith and healing gifts of women and ecclesiastical sanction for women to administer to sick persons. Women also assumed religious callings as midwives through pregnancy and the birthing process, a vital role given cultural concerns about modesty and the presence of a male doctor during childbirth.[17] This democratization of healing also extended to parents who as persons uniquely positioned to know the medical needs of their children possessed a symbolic healing power. Brigham Young, the prophetic successor to Joseph Smith, commented that God willed that "every father and mother should know just what to do for their children when they are sick. Instead of calling for a doctor, you should administer to them mild foods, herbs, and medicine that you understand."[10]

The parental stewardship to use "medicine that you understand" rather than "calling for a doctor" reflected communal disparagement of allopathic medical practices. In contrast to botanical doctors, physicians trained in the conventional medical education of the prescientific era were in general considered to be ignorant of human bodily systems, incompetent in their skills, and otherwise callous individuals who were indifferent as to whether they helped or harmed a patient so long as the physician could profit from the patient's illness. The presence of a physician at a person's house could symbolize to the community that faith in God had been supplanted by trust in fallible human understanding. The frequent migrations of the early LDS community to different frontier areas to avoid religious persecution also meant the community lacked the long-term relationships and resources to differentiate between a qualified, credible physician and pervasive medical quackery. The primary exception to this communal aversion to physicians concerned what were designated as "mechanical" skills or procedures, such as setting broken bones, amputating infected or wounded limbs, and dentistry.[19] A surgeon or dentist could be legitimately relied on to remedy fairly evident matters of anomalous anatomy and physiology, making technical repairs peripheral to the mysterious work of healing and reliance on providential care. However, as medicine became increasingly professionalized, science-based, and effective in the later 19th century, the LDS community inescapably reflected on the compatibility of faith healing practices with resort to medical practices.

Faith and Medicine: Communal Transitions

Although the prevalence of faith healing persisted throughout the 19th century, a communal transition to increased trust in allopathic physicians and more frequent utilization of medical interventions was initiated in second-generation Mormonism in the late 1860s. Brigham Young encouraged communal appropriation of an increasingly scientific medicine—transformed by research on the germ theory of disease, the development of the stethoscope, and the introduction of anesthesia in surgery—by instructing several LDS young men and women to attend medical schools in the eastern United

States. The presence of LDS women in medical education occurred just as obstetrics and gynecology were becoming recognizable fields and reinforced the importance of medical knowledge in matters specific to women's health, including pregnancy and childbirth.[20] Young continued to express ambivalence about fully embracing the evolving scientific medicine, claiming that a successful treatment might be attributable to random luck: "Doctors make experiments, and if they find a medicine that will have the desired effect on one person, they set it down that it is good for everybody, but it is not so, for upon the second person the medicine is administered to, seemingly with the same disease, might produce death."[21] Still, Young's view seems to endorse a rationalized scientific method for acquiring evidence of medical efficacy through an experimental process accounting for both individual variations and generalized replicability of results. Despite personal skepticism regarding much of the allopathic medicine of his era, Young encouraged both practicality and providential care, admonishing openness to "every remedy that comes within the range of my knowledge, and to ask my Father in heaven, in the name of Jesus Christ, to sanctify that application to the healing of my body."[22]

The communal transition from aversion to ambivalence to cautious acceptance of medicine paralleled the organization of professional associations of medicine, such as the American Medical Association, which promulgated a code of ethics at its founding in 1848.[23] Professionalized medicine aimed to generate social credibility by establishing generalized standards and procedures for the safety and efficacy of medical interventions, diminishing the risks of causing harm or hastening death in treating patients, and so eradicate both randomness and quackery from "medical" practice. In addition, following a quarter-century period of hesitancy and internal debate, medical practitioners increasingly accepted innovative medical methods such as the stethoscope to improve diagnosis and the use of analgesics to relieve pain and anesthesia to control pain following an invasive procedure.[24] These innovations meant that the Hippocratic medical ethic of *primum non nocere* (first, do no harm) could be supplemented by a realizable professional commitment to benefit patients. The theological and practical rationales for LDS cultural skepticism of medicine and aversion to physicians became less compelling as medicine evolved into a profession organized around common standards of practice, developed an ethical code, and offered safer medical treatments that had some prospect of effectiveness.

So significant were these communal and professional transitions that a question faced by the LDS community by the end of the 19th century was whether the emerging scientific medicine would displace faith healing. Ecclesiastical leaders such as George Q. Cannon voiced sentiments that exclusive reliance on medicine with its finite human understanding and fallible human works risked both faith and faith healing. When physicians were immediately summoned as a first resort of care, ill persons can "gain confidence in the doctor and his prescriptions and lose faith in the ordinance. How long would it take," Cannon wondered," if this tendency were allowed to grow among the Latter-day Saints, before faith in the ordinance of the laying on of hands would die out?"[25] Nonetheless, as Stapley observes, "there appears to have been a growing awareness that some of the traditional healing rituals did not have the medical efficacy once thought."[26]

Ironically, as Latter-day Saints followed the broader society in reposing greater trust and confidence in medicine and physicians, healing was stigmatized in medical discourse and literature in large measure because of its religious residue. The scientific medicine of the late 19th and 20th centuries was organized through a method of empirical observation and research that formed the foundations for quantifiable objectives of preventing, treating, and ultimately, curing disease. In this context, the language of healing signified regression to an era of nonscientific superstition and mystery. As articulated by scholar David Barnard, "the metaphor of medical progress [implied] emancipation from the bondage of religious belief and magical thinking," enabling scientific, rationalistic medicine to differentiate itself as a cultural authority from religion and superstition.[27] The quasi-religious and spiritualist background of healing, with its associations of randomness, mystery, and independence from empirical confirmation, did not resonate with an increasingly secularized medicine relying on replicable scientific methodologies. Healing represented an activity carried out with cups and candles, prayers and potions, needles and nerves, and shamans and sacrifices.[28] The emerging self-image of medicine as an applied science ran contrary to the imprecision, absence of a rationalistic method, and enchantment of nature associated with healing, phenomena *against* which the profession of medicine had been organized. The professional and cultural demand for medicine that had a scientific and evidentiary basis offered little conceptual room not only for faith healing practice but for healing within medicine. The art of healing was displaced by the science of medicine.

The associations of healing with charismatic charlatans who perpetuated deceptions on a gullible public necessitated repudiation by professionally organized medicine, but the disavowal of healing as integral to the core
purposes of medicine culminated in scientific reductionism, a professional
focus on diagnosing disease rather than caring for persons, and a medicalization of life experience, especially birthing and dying. Significantly, 21st-
century medical culture has witnessed a professional retrieval of the concept
and language of healing and a renewed emphasis on medicine and nursing
as fundamentally *healing* professions.[29] This professional revival of healing
discourse is reflected in numerous initiatives to reclaim constitutive features
of healing, including personalized relationship, wholeness, witnessing of the
patient's body, caring, therapeutic touch, and attentiveness to the patient's
illness narrative.[30] These initiatives have been supplemented by empirical
inquiry on the influential role of religiosity and spirituality on patterns of
healthy living and a renewed interest in the spirituality of medicine.[31] The
concept of healing as a *defining and unifying* feature of the medical profession is underscored in a collaboration by American and European physicians
to develop a set of professional principles and responsibilities applicable to
the ethical, legal, and regulatory challenges facing 21st-century medicine.
This "Physician's Charter" affirmed, "The medical profession everywhere is
embedded in diverse cultures and national traditions, but its members share
the role of healer, . . ."[32] The reclamation of healing for physician and professional identity has now evolved into the status of a professional universal.

The revival of healing as integral to medical professionalism also reflects
a critique of a central premise of 20th-century scientific medicine, an assumption that patients who are cured are *also* healed.[33] The differentiation of
curing and healing is illustrated by a physician's observation following a diagnosis of prostate cancer: "We can cure your cancer [through surgery]; the
question is, 'What do we leave you with?'" When the purpose and identity
of medicine is defined by curing *disease* rather than making *persons* whole,
those circumstances in which medicine cannot cure, such as the plenitude of
chronic diseases, or the dying process, symbolize professional failure. Curing
in the absence of a commitment to healing manifests an incomplete understanding of the aims of medicine. Physician-author Eric J. Cassell has elo-
quently argued that when the relationship between patient and physician is
constituted as "a healing relationship" rather than a mechanistic relationship
oriented by a "technological fix," it becomes possible for medicine "to heal
the sick, to make whole the cured, to bring the chronically ill back within the

fold, to relieve suffering, and to lift the burdens of illness."[34] This means that healing is a subversive, prophetic ethical value in the context of reductionist and transactional medical practice. The wholeness or integrity of medicine requires not only stories of dramatic cures of disease and technological rescues from death but also narrative quests of patients for new identities that confer wholeness. Physicians as healers are witnesses to the stories of restoration, chaos, quest, and wholeness their patients tell in and through their bodies. Healing in medicine requires healing medicine.

Faith in Medicine: Ritualizing Ethics

The professional organization and scientific basis for medicine initiated in the late 19th century and in full sway in the 20th century transformed the communal understanding of the relationship of faith, medicine, and healing. Consequently, a primary theme of recent ecclesiastical teaching on healing has been to ensure that well-founded confidence in medical treatments does not preclude faith-based blessings of healing, or that the works of medicine do not entirely supplant the faith of believers. The intertwining of faith in God and trust in medical science was reflected in a sermon by ecclesiastical leader Dallin H. Oaks that affirmed the compatibility of different modes of healing effectuated through medical expertise, prayer, and priesthood blessings: "Latter-day Saints believe in applying the best available scientific knowledge and techniques. We use nutrition, exercise, and other practices to preserve health, and we enlist the help of healing practitioners, such as physicians and surgeons, to restore health."[35] While Oaks constructed medicine as a vocation of healing, the primary substantive emphasis of his sermon consisted of pastoral guidance on "healing the sick" through prayer and healing blessings. This is illustrative of an ecclesiastical pattern that can be designated as *ritualizing* ethics.

The pattern of ritualizing ethics means that even in the midst of significant medical advances that may effectuate *cures*, faith remains a necessary component for *healing*. LDS members who are ill may request a special blessing that promises comfort, community presence, divine direction of medical interventions, and divine care through their ordeal, and occasionally may promise healing and restoration of health. By the 1950s, the charismatic and democratized practices of spiritual faith healing had evolved into an institutionalized and rationalized ritual designated as "administering to the sick."[36]

The experience of healing was re-storied through biblical narratives and texts that served as precedent for this ritualized ordinance of healing power.[37] The primary elements of the ritual as it has evolved and is practiced by con- temporary Latter-day Saints include (1) faith on the part of the person who is ill as well as those authorized to provide a blessing; (2) prayers for provi- dential intervention and reassurance to the ill person; (3) an anointing with consecrated oil of the crown of the ill person's head;[38] (4) laying on of hands by church elders to "seal" the anointing, frequently accompanied by words of blessing; and (5) faithful submission of the ill person to the divine will.[39] This ritual expresses profound faith by the believer and the community in providential care and is a powerful symbol of the healing potency of touch. It expresses a communal covenant of solidarity and compassionate presence, a witness that the sick person will not be abandoned. The integration of spir- itual practices and ritualized administration of blessings as a mode of healing manifests a "faith and works" approach to religious life: the faith of the ill person and community is complemented by a call for the ministrations of medicine.

The spiritual practice of prayer and ritualized reassurance is also enacted in the most sacred of LDS worship sites, temples. Even in its earliest form in the 1836 Kirtland, Ohio, temple, the LDS temple was envisioned as a "sacred place of healing," which included healing baptism.[40] A contemporary temple ritual of washing and anointing provides a covenant of health and protection for the sacred character of the body. Worshipers at any LDS temple are also invited to place the name of a person experiencing sickness or affliction on a list of persons for whom prayers will be offered. The temple prayer does not involve any direct intervention as such in contrast to the physical presence required for the ritual of administering to the sick. The person for whom an intercessory prayer is uttered may well be far removed from the temple site of the prayer. The prayer exhibits a collective remembrance that persons who are sick, ill, or experiencing some other adversity are not isolated by their illness or ordeal from the moral memory of the community. The rituals, blessings, intercessory prayers, and ordinances manifest a moral witness to the communal covenantal commitment to share the burdens of others.

These spiritual practices ritualize ethics by providing a ritual resource for patients, their families, LDS health professionals, and priesthood leaders in contexts where moral choices are encountered in medicine, including decisions about seeking medical care, hospitalization, surgical interventions, and cessation of life-prolonging ethical treatments. The moral choice itself may not be removed but the sharp anguish or existential distress of feeling overwhelmed or helpless in the face of disease processes can be alleviated by ritualized reassurance of community comfort, presence, and submissiveness to providential care. The ritualization of ethics is differentiated from a fatalist theology in that medicine is also recognized as enacting divine care and healing processes. The curative powers of medicine and pharmacology and the ethical essence of the healing professions are theologically portrayed as derived from an ultimate source of health and healing: God's love, mercy, and compassion.

The ritualized practices of faith healing do not supplant but instead contextualize scientific medicine and professional expertise: healing is not susceptible to the *technical* expertise of physicians. The spiritual practices and rituals re-story the prospects of restoring health within divine design, faithful commitment, and numerous biblical and contemporary healing narratives. The pattern of ritualizing ethics is intertwined with a profound trust in the integrity and vocational call of the healing professions. LDS ecclesiastical policy on medical, health, and moral issues frequently conjoins appeals to personal or familial moral agency with the necessity of receiving "wise and competent medical advice" from health care professionals.[41] The professional setting is constructed as a forum within which practical medical and moral questions about health care decisions can be more effectively addressed and concrete choices can be more fully informed than is possible through ecclesiastic directives expressing generalized principles.

The general compatibility of faith convictions and trust in scientific medicine reflects a theologically re-storied understanding that the healing potential of medicine is not attributable to scientific knowledge alone but is derived from and inspired by divine knowledge. Recent ecclesiastical teaching has emphasized four sources for healing that augment professional knowledge: divine wisdom, intercessory prayer, blessings of the sick, and the intrinsic healing capabilities of the human body. Spencer W. Kimball, a prophetic leader of the 1970s and 1980s who underwent numerous life-saving medical interventions of his own, asserted the dependence of medical healing on divine sources: "It must be remembered that no physician can heal. He can only provide a satisfactory environment and situation so that the body may use its own God-given power of re-creation to build itself." Kimball also maintained that "numerous healings credited to doctors and hospitals have been the healing of the Lord through the priesthood and by prayer. . . . While the [medical] profession has worked hard to gain the

accumulated knowledge of today, it must be remembered that He who created our bodies has known since the beginning how to remodel, re-create,
and repair them."[42] This theology of healing discerns the presence of divine
providential caring, including the divinely bestowed restorative powers of
the body, in any healing process. This claim is evident in the published response of an ecclesiastical leader to an inquiry about the rationale for medicine given the communal practice of administering to the sick: "The source
of all knowledge comes to man from God for man's benefit, guidance, and
blessing. . . . Medical science is just a link in the whole plan and process of
healing."[43]

This theology of healing implies that even as medicine increasingly
assumes a professional identity of healing, virtues of humility and modesty
are necessary features of the ethos of any *healing* profession. Recent bioethical discussions have similarly emphasized limits on the professional contri-
bution to healing. As developed from ethnographic research, David Schenck
and Larry Churchill contend that a "healer" is distinguished from a competent physician through their recognition that healing occurs in ways beyond their control and expertise; healing is ultimately a "mystery." Healing in
medicine is mediated not through professional expertise but by relationships
that cultivate wholeness and meaning: "Healing . . . always has to do with
connections between people that make us whole and restore us to the deep
sources of meaning for our lives."[44] Healing is about something more ultimate than a physiological condition and invariably opens to narrative
constructions of wholeness, covenantal witness, and solidarity with persons
in the midst of profound vulnerability.

The re-storying of trust in medical expertise through narrative witnesses
to the embodied and divine sources of healing is echoed by current LDS
president (and pioneering heart surgeon) Russell M. Nelson: "Men can do
very little of themselves to heal sick or broken bodies. With an education
they can do a little more; with advanced medical degrees and training, a little
more yet can be done. The real power to heal, however, is a gift from God."[45]
The consistent conceptual claims in LDS theologies of healing are that medicine is a profession oriented to restoring health and healing, professional
efficacy in healing presumes insights into the body's inherent healing capacities and medicine's inherent limits, and the healing potential of medicine
is augmented by divinely bestowed processes of faith, prayer, relationship,
and other spiritual practices. These revealed realities support three ethical
meanings to healing. First, the vocation of the healing professions, including

medicine and nursing, carries out the divine work of healing. Needless to say, the professions often fall short of healing aspirations, which is why wholeness and integrity of the profession are necessary for healing *within* and *through* the profession. Second, healing includes but extends beyond the curative goals of scientific medicine. The profession cultivates its healing potential by commitments to wholeness, attentive listening, relational re-storying, witnessing of the body, and the healing potency of touch. The nonmedical sources of healing and the subversive mystery of all healing processes discloses limits to professional knowledge. Finally, a profession organized through healing relationships and professional integrity is a moral community in whom patients, families, and even ecclesiastical authority can repose trust for bioethical wisdom in conflicts of moral uncertainty. These moral realities of the healing professions underlie respect for the integrity of medicine in LDS moral culture.

Healing as Prophetic Critique

Healing practice has often been considered a subversive threat to more conventional patterns of medicine and caring for the ill. In archaic cultures, the shaman represents a mysterious and mystical way of addressing and eliminating disease from the community. In biblical narratives, Jesus's ministry of healing continually brings him into conflict with religious and political elites. As modern medicine evolved into a scientific practice, professional associations engaged in a "crusade" against healing[46] as a way to both eradicate unreliable methods or quackery and to limit explanations for cures to the empirical, replicable, and scientifically explicable. It is then a significant development for contemporary professional ethical codes to portray that the image of the healer bestows a universal cultural identity to physicians. Insofar as healing is a professional ideal, its absence in practice is grounds for a prophetic critique.

Medical relationships have historically been structured by inegalitarianism and power relationships: the Latin root for doctor, *docere*, refers to "teaching," whereas the root for patient, *pati*, refers to "suffering." The patient-physician relationship was constructed around an active, interventionist, and even heroic role of the physician and a passive role of suffering, undergoing, and disempowerment of a patient.[47] The ethical problem of professional paternalism is that it perpetuates the myth of the physician as the active giver in the relationship who benefits the suffering patient out of philanthropic self-lessness. The paternalistic relationship reflects a deficient moral memory insofar as individual physicians and the profession in general are *first* the recipients of various gifts of society, such as resources for medical education, institutions such as teaching hospitals, a high level of social trust, and the self-governance bestowed on professional associations. Physicians receive benefits from patients who literally offer their bodies for both research and "practice," a pattern initiated by the gifts of whole-body donors for medical education.

The moral reality that persons can become phyns only through first being recipients of various gifts suggests that the more appropriate relational model for physicians in their relationships with patients and the profession in its relationship with society is *covenantal*. The assumption of various professional responsibilities, including truthful disclosure, promise-keeping, respecting confidentiality, promoting patient well-being, compassionate presence to the ill, solidarity and nonabandonment of patients, and advocacy of justice and fairness in access to health care is not a matter of professional discretion but is rooted in a moral indebtedness of physicians to the community. These covenantal commitments constitute medicine's *profession* or witness to society of its defining responsibilities as organized through its distinctive end, an ethic of healing.

The covenantal relationship grounds not only prophetic critique of paternalistic medicine but also of more recent constructions of medicine, including relationship forms embedded in some LDS medical and health policies. Policies that highlight the role of "competent" physicians in medical ethical decision-making bestow on physicians an identity of *technician* whose primary responsibilities are discharged through information disclosure and exercise of professional skill. The social presumption is that any health care professional will be well educated, experienced, and possessed of distinctive knowledge, techniques, and skills so as to merit trust. However, a skilled, competent physician is not necessarily a healer; the science of medicine must be complemented by skills in the art of healing.

Contemporary public discourse that portrays the patient-physician relationship through the language of "consumer" and "provider" claims to correct the historical disempowerment of patients. This model reflects an aspiration to consumer/patient control over health care decisions consistent with the ethical principle of self-determination. However, a consumer-driven medical relationship carries two ethically problematic implications relative to the prophetic criteria of covenant and healing: (1) Health care is construed as a commodity accessible to consumers based on their ability to pay rather than their medical need. This manifests in structural inequities for various vulnerable persons and communities in the health care system as a whole. (2) The consumer/patient–provider/physician relationship becomes a transactional encounter with no more ethical significance than any other form of business or contractual exchange. The consumerist model bestows patient empowerment but at the cost of diminished trust in the relationship and in medicine. The diminishment posed by consumer-provider medicine to the professional identity and ethic of healing are articulated in a powerful moral critique by bioethics scholars Schneiderman, Jecker, and Jonsen: "If the medical profession retreats to the position that it has no internal professional values and merely provides whatever patients, families, or insurers are willing to pay for, it can no longer claim to be a healing profession, that is, a group committed to healing and serving the sick. Instead, medicine becomes a commercial enterprise satisfying the desires of others."[48]

My exposition of the complex relationship between faith and medicine in the LDS community re-stories medicine within a vocational ethic of covenant and healing. This means the LDS community has a stake in the ethical integrity of the medical profession. Insofar as regulatory, commercial, institutional, and professional pressures diminish medicine's ethos of healing, the LDS community has an ethical responsibility to bear witness to the intrinsic value of the healing professions as moral vocations and critique an exclusive professional fixation on technical skills, transactional relationships, and technological care. This prophetic healing critique provides the ecclesiastical and professional context for engaging in the moral morasses of bioethics, beginning with creating new human life.

Notes

1. *Doctrine and Covenants* 35:9; *James* 5:14–16.
2. Courtney S. Campbell, "What More in the Name of God?: Theologies and Theodicies of Faith Healing," *Kennedy Institute of Ethics Journal* 20 (March 2010), 1–25.
3. Dallin H. Oaks, "Healing the Sick," *The Ensign*, May 2010, https://www.lds.org/general-conference/2010/04/healing-the-sick?lang=eng
4. The Church of Jesus Christ of Latter-Day Saints, "Church Policies and Guidelines, Medical and Health Policies: Medical and Health Practices," in *General Handbook: Serving in the Church of Jesus Christ of Latter-day Saints*, Chapter 38.7.7,

February 19, 2020, https://www.churchofjesuschrist.org/study/manual/general-handbook/38-church-policies-and-guidelines?lang=eng#title_number91

5. Courtney S. Campbell, *Bearing Witness: Religious Meanings in Bioethics* (Eugene, OR: Cascade Press, 2019), 127–149.

6. ABIM Foundation, ACP-ASIM Foundation, European Federation of Internal Medicine, "Medical Professionalism in the New Millennium: A Physician's Charter," *Annals of Internal Medicine* 136:3 (February 2002), 243–246.

7. "Healing," https://www.churchofjesuschrist.org/study/history/topics/healing?lang=eng

8. *Doctrine and Covenants* 42:43. The scriptural commendation of herbal treatments has been viewed by some scholars as a condemnation of the mineral-based treatment of "heroic" medicine, such as calomel, a commonly used purgative. Joseph Smith's skepticism about physician competency was influenced by the death of his brother Alvin in 1823 from a physician's administration of calomel. See Robert T. Divett, *Medicine and the Mormons: An Introduction to the History of Latter-day Saint Health Care*, 2nd ed. (Charleston, SC: CreateSpace, 2010), 24–26, 35–37, and Samuel Morris Brown, *In Heaven as It Is on Earth: Joseph Smith and the Early Mormon Conquest of Death* (New York: Oxford University Press, 2012), 22–25.

9. Jonathan A. Stapley, "'Pouring in Oil': The Development of the Modern Mormon Healing Ritual," in *By Our Rites of Worship: Latter-day Saint Views on Ritual in Scripture, History, and Practice*, ed. Daniel L. Belnap (Provo, UT: Religious Studies Center, 2013), 284–286.

10. Stapley, "'Pouring in Oil,'" 287.

11. Cecil O. Samuelson, "Medical Practices," in *Encyclopedia of Mormonism*, ed. Daniel H. Ludlow (New York: Macmillan, 1992), 875, https://contentdm.lib.byu.edu/digital/collection/EoM/id/3912; Lester E. Bush, Jr., *Health and Medicine Among the Latter-day Saints: Science, Sense and Scripture* (New York: Crossroad, 1993), 89–107.

12. Bush, *Health and Medicine Among the Latter-day Saints*, 72–74; Matthew J. Grow, Richard E. Turley, Jr., Steven C. Harper, Scott A. Hales, eds., *Saints: The Story of the Church of Jesus Christ of Latter-day Saints*, v. 1, *The Standard of Truth, 1815–1846* (Salt Lake City: Intellectual Reserve, 2018), 402–403.

13. The Church of Jesus Christ of Latter-day Saints, *Teachings of the Presidents of the Church: Joseph Smith* (Salt Lake City: Intellectual Reserve, 2007), 380–381; Richard Lyman Bushman, *Joseph Smith: Rough Stone Rolling* (New York: Vintage Books, 2005), 383–384.

14. Elaine S. Marshall, "The Power of God to Heal," in *Joseph & Hyrum: Leading as One*, ed. Mark E. Mendenhall et al. (Salt Lake City: Deseret Book, 2010), 175.

15. Marshall, "The Power of God to Heal," 175.

16. Jonathan A. Stapley, Kristine Wright, "The Forms and the Power: The Development of Mormon Ritual Healing to 1847," *Journal of Mormon History* 35 (Summer 2009), 42–87; Stapley, "'Pouring in Oil,'" 285–286; Linda King Newell, "A Gift Given, A Gift Taken: Washing, Anointing, and Blessing the Sick Among Mormon Women," *Sunstone* 6 (1981), 16–25.

17. Jonathan A. Stapley, Kristine Wright, "Female Ritual Healing in Mormonism," *Journal of Mormon History* 37 (Winter 2011), 1–85; Jenne Erigero Alderks,

"Rediscovering the Legacy of Mormon Midwives," *Sunstone*, April 3, 2012, https://www.sunstonemagazine.com/rediscovering-the-legacy-of-mormon-midwives/; Bush, *Health and Medicine Among the Latter-day Saints*, 84–89.

18. Divett, *Medicine and the Mormons*, 118–120.

19. Bush, *Health and Medicine Among the Latter-day Saints*, 91, 94.

20. Steven R. Zimmerman, "The Mormon Health Traditions: An Evolving View of Modern Medicine," *Journal of Religion and Health* 32 (Fall 1993), 189–196.

21. John A. Widtsoe, ed., *Discourses of Brigham Young* (Salt Lake City: Deseret Book Company, 1975), 191–192.

22. Bush, *Health and Medicine Among the Latter-day Saints*, 95.

23. Matthew Wynia, "The Birth of Medical Professionalism: Professionalism and the Role of Professional Associations," in *Healing as Vocation: A Medical Professional Primer*, ed. Kayhan Parsi, Myles N. Sheehan (New York: Rowman and Littlefield, 2006), 23–34.

24. Melanie Thernstrom, *The Pain Chronicles: Cures, Myths, Mysteries, Prayers, Diaries, Brain Scans, Healing, and the Science of Suffering* (New York: Farrar, Straus and Giroux, 2010).

25. Jerreld L. Newquist, ed., *Gospel Truth: Discourses and Writings of President George Q. Cannon*, v. 2 (Salt Lake City: Deseret Book Company, 1974), 188.

26. Stapley, " 'Pouring in Oil,' " 304.

27. David Barnard, "The Physician as Priest, Revisited," *Journal of Religion and Health* 24 (Winter 1985), 274–275.

28. The scientific prejudice against these forms of healing is displayed in Anne Fadiman's remarkable book about the conflict between American medicine and the immigrant Hmong communities of the 1980s. Anne Fadiman, *The Spirit Catches You and You Fall Down* (New York: Farrar, Straus and Giroux, 1997).

29. Some recent illustrative studies are David Schenck, Larry R. Churchill, *Healers: Extraordinary Clinicians at Work* (New York: Oxford University Press, 2012) and Eric J. Cassell, *The Nature of Healing: The Modern Practice of Medicine* (New York: Oxford University Press, 2013).

30. Campbell, *Bearing Witness*, 127–149.

31. Danish Zaidi, "Influences of Religion and Spirituality in Medicine," *AMA Journal of Ethics* 20 (2018), E609–E612, https://journalofethics.ama-assn.org/article/influences-religion-and-spirituality-medicine/2018-07; Aaron Saguil, Karen Phelps, "The Spiritual Assessment," *American Family Physician* 86 (2012), 546–550; Daniel E. Hall, Keith G. Meador, Harold G. Koenig, "Measuring Religiousness in Health Research: Review and Critique," *Journal of Religion and Health* 47 (2008), 134–163.

32. ABIM Foundation, "Medical Professionalism in the New Millennium," 243–246.

33. Eric J. Cassell, *The Nature of Suffering and the Goals of Medicine*, 2nd ed. (New York: Oxford University Press, 2004), 65.

34. Cassell, *The Nature of Suffering and the Goals of Medicine*, 65; cf. Cassell, *The Nature of Healing*, 81–93.

35. Dallin H. Oaks, "Healing the Sick," April 2010, https://www.lds.org/general-conference/2010/04/healing-the-sick?lang=eng

36. The most comprehensive analyses of the historical evolution of administering to the sick are presented by Stapley, "'Pouring in Oil,'" 283–316, and Bush, *Health and Medicine Among the Latter-day Saints*, 69–89.

37. Nephi K. Kezerian, "Sick, Blessing the," in *Encyclopedia of Mormonism*, ed. Daniel H. Ludlow (New York: Macmillan, 1992), 1308–1309, https://contentdm.lib.byu.edu/digital/collection/EoM/id/4194

38. Stapley's historical analysis situates the use of oil as a common part of the ritual, but notes substantial variability in both meaning (19th-century Mormons saw the oil as its own therapeutic agent rather than a symbol) and application (19th- and 20th-century Mormons applied oil to afflicted body parts, or to the entire body, or ingested the oil). See Stapley, "'Pouring in Oil,'" 292–294, 300–301.

39. *Doctrine and Covenants* 46:19–20; Oaks, "Healing the Sick."

40. Stapley, "'Pouring in Oil,'" 287–288, 293–295.

41. The Church of Jesus Christ of Latter-Day Saints, "Church Policies and Guidelines, Medical and Health Policies: Prolonging Life," in *General Handbook: Serving in the Church of Jesus Christ of Latter-day Saints*, Chapter 38.7.9.

42. Spencer W. Kimball, "President Kimball Speaks Out on Administration to the Sick," *New Era*, October 1981, https://www.lds.org/new-era/1981/10/president-kimball-speaks-out-on-administration-to-the-sick?lang=eng

43. Delbert L. Stapley, "Why Are Medical Doctors So Important If the Priesthood Has the Power to Cure Diseases?," *New Era*, March 1971, https://www.lds.org/new-era/1971/03/q-and-a-questions-and-answers/why-are-medical-doctors-so-important-if-the-priesthood-has-the-power-to-cure-diseases?lang=eng

44. Schenck, Churchill, *Healers*, xiii.

45. Melvin K. Gardner, "Elder Russell M. Nelson," *The Ensign*, June 1984, https://www.lds.org/ensign/1984/06/elder-russell-m-nelson-applying-divine-laws?lang=eng

46. Cassell, *The Nature of Healing*, ix, 82–83.

47. James F. Childress, *Who Should Decide?: Paternalism in Health Care* (New York: Oxford University Press, 1982).

48. Lawrence J. Schneiderman, Nancy S. Jecker, Albert R. Jonsen, "Medical Futility: Response to Critiques," *Annals of Internal Medicine* 125:8 (1996), 673.

2

Creating Life

New Conceptions

Beginning the Family

Garrett and Meghan married in their late twenties following the completion of their LDS missionary service. Garrett and Meghan were enthusiastic about beginning their family, but an initial pregnancy ended with a miscarriage after six weeks, a saddening and sobering experience. Following a discussion with their obstetrician-gynecologist, ultrasound imaging unexpectedly revealed that Meghan had a uterine polyp and endometriomas (tissue growing outside the uterus) that diminished the likelihood of a successful pregnancy through natural conception.

Meghan was annoyed and angry with these medical complications, with her body, and even with God, as though a kind of "sick joke" had been played on her. She had received a blessing as a young woman promising that she would be a mother, and she reasoned that she had been a faithful Latter-day Saint her entire life, followed the commandments for worthiness to serve a mission, and had her marriage to Garrett sealed in a temple of The Church of Jesus Christ of Latter-day Saints (LDS), but "then God dealt me this hand." Meghan told Garrett that "she deserved to be a mom," and motherhood for her meant the entire biological and social process, from conception, gestation, birthing, and nursing to nurturing. For his part, Garrett believed there "must be some avenue that would allow us to be parents." The couple discussed their options and resolved to "turn this [infertility] over to God." They received several blessings from congregational leaders and undertook a spiritual practice of service in the LDS temple, acquiring a "sense of peace that things will work out."

Garrett and Meghan were familiar with numerous other young couples who experienced infertility and became parents through assisted methods of reproduction. The couple consulted further with their obstetrician-gynecologist, who indicated the ultrasound results made Meghan a "good candidate" for treatment through a fertility clinic. The physician authorized a referral for the couple

Mormonism, Medicine, and Bioethics. Courtney S. Campbell, Oxford University Press (2021). © Oxford University Press.
DOI: 10.1093/oso/9780197538524.003.0003

to a clinic with high success rates for establishing pregnancies through different methods of assisted reproductive technology. The clinic was located in one of the few U.S. states that mandates insurance coverage for fertility treatments and reproductive technologies, thus sparing the couple of costs that otherwise might have been prohibitive.

Meghan and Garrett initially underwent two cycles of intrauterine insemination (IUI), a procedure that involves placing sperm within a woman's uterus to facilitate fertilization. Although IUI's success rate is as high as 20%, the procedure did not produce a pregnancy. The couple was undeterred and experienced an overwhelming parental impulse: "Whatever it takes, let's do it." They were prepared for the more invasive process and more expensive procedure of in vitro fertilization (IVF), in which fertilization occurs in the laboratory and the fertilized eggs are transferred to the woman's uterus.

The IVF procedure imposed substantial physical burdens for Meghan, including hormonal treatments to hyperstimulate her ovaries and maximize egg retrieval, that added to her emotional burdens. Ultimately, twenty-one eggs were retrieved from Meghan and twelve were fertilized through intracytoplasmic sperm injection (ICSI), a procedure in which a single sperm is injected into the egg. Two of the twelve fertilized eggs were transferred to Meghan to increase the prospects for a singleton pregnancy; the other fertilized eggs were cryopreserved. The arduous ordeal ultimately culminated in an overwhelming result: the birth of a healthy baby girl, Gillian. The following year, a single embryo was unfrozen and transferred to Meghan, and a second girl, Cambria, was born nine months later.

Garrett and Meghan expressed an overriding shared desire to become parents of healthy children. They did not discern any religious or moral distinction between natural and technological methods of conception. Garrett articulated their understanding: "All truth is God's. Medicine and reproductive science have a portion of this truth as inspired by God." The physical and emotional burdens experienced by Meghan made her more aware of supportive moral communities, including religious congregants, couples experiencing infertility, and compassionate medical professionals. "I've not had the experience of being stigmatized, isolated, and silenced. Being open about our experience has brought out compassion and empathy in others, and has made me want to support other couples."

How far might Garrett and Meghan have gone in their quest to become parents of a healthy child? They did not utilize some of the procedures offered by the clinic, such as preimplantation genetic (PGD) screening of the embryos, because the substantial cost of PGD wasn't included in the state mandate for insurance coverage. If conventional IVF procedures had failed, they were willing to consider other options such as egg donation or surrogacy, even though the involvement of third parties would diminish Meghan's biological and gestational mothering experience. The family-creation project is not over for Garrett and Meghan. They have dispositional rights over the remaining nine cryopreserved embryos stored in the fertility clinic.

Reproductive Technologies and the LDS Community

The first child conceived through in vitro fertilization (IVF) was born in the United Kingdom in 1978. IVF and subsequent variations have allowed millions of couples who otherwise might have been childless in their marriage to have the transforming experience of parenting. The U.S. Centers for Disease Control and Prevention reported 284,385 assisted reproductive technology (ART) cycles in 2017, from which 68,908 live births occurred, and due to occasional multiple births of twins or triplets, 78,052 infants were born; nearly 1.7% of all births in the United States occur through procedures such as IVF that involve retrieval and transfer of eggs and embryos.[1]

No clear data exist on use of reproductive technologies among Latter-day Saint populations, but there is also no reason to presume that infertility experience is less frequent among Latter-day Saints than the general population (approximately one in seven couples). What is clear is that after a period of ecclesiastical ambivalence about IUI, IVF, and other medically assisted reproductive procedures, LDS teaching and practice have evolved to a position that recourse to reproductive technologies is a matter for couples to decide without ecclesiastical dictates.[2] Infertility experience is an anguishing but accepted part of LDS familial and moral culture. The official LDS monthly magazine, *The Ensign*, as well as online forums, has included numerous testimonials from married couples, albeit primarily voiced by women, about their ordeals of infertility and the experience of "miracles" of birth through IVF or other reproductive treatments.[3]

A brief summary of the history of LDS ecclesiastical policies on technologically assisted reproduction discloses how ecclesiastical teaching often reflects accommodations to the increasing prevalence of a practice among the medical profession and to broader societal acceptance. As related by LDS historian Lester Bush, Jr., "artificial methods of insemination" were first addressed publicly by the LDS Church through a 1950 editorial in the church-owned newspaper *The Deseret News*. The editorial intimated both principled and pragmatic reasons in asserting that church members should avoid artificial insemination as "the practice is not moral, and frequently would lead to family complications."[4] As medical successes with artificial insemination became common in the late 1950s and 1960s, ecclesiastical teaching clarified that the principal objection was not to an insemination procedure using the sperm of the husband, but rather sperm from a third-party donor, a practice portrayed as potentially creating "problems related to civil law, family harmony, and eternal relationships." Pragmatic concerns over the impact on the family continued to prevail even as the initial principled objection that the "practice is not moral" was abandoned.[5]

The first LDS compilation of guidelines on medical ethics prepared by the Church Commissioner of Health Services in 1974 discarded the concerns about family law and eternal relationships and highlighted only the pragmatic concern about family harmony as the reason "the Church does not approve" of artificial insemination from a third party. Significantly, however, the guidelines incorporated an appeal to moral agency as a decisive consideration: "this [third-party insemination] is a personal matter which must ultimately be left to the determination of the husband and wife with the responsibility of the decision resting solely upon them."[6] The 1974 guidelines also seemed to provide an implicit ecclesiastical endorsement of assisted reproductive methods, even if indirectly, with a statement under the heading of "Sterility Tests": "The Church believes that having children is a blessing and privilege and, that with any abnormal condition, it is appropriate to use medical science to diagnose and restore normal function."[7] The guidelines then not only substantively restricted the scope of ecclesiastical disapproval relative to the two prior decades, with the appeal to agency overriding, but also provided a positive sanction, rooted in parenting and procreative imperatives, for recourse to medicine in circumstances of sterility or infertility. With minor changes in wording, these three features—no statement on artificial insemination (what is publicly rephrased as "IUI") per se, discouragement of third-party IUI, and recognition of the couple's agential responsibility—remain the core of the current policy some four decades later.[8] The pragmatic concerns about the impact of third-party artificial insemination for family harmony were removed from ecclesiastical policy in the mid-1980s as clinical and familial experience revealed the concerns to be more conjecture than empirically based.

A decade of ecclesiastical silence followed the birth of the first baby through IVF in 1978, a decade that witnessed increasing clinical efficacy of IVF methods and use of these procedures by some LDS couples and physicians. Perhaps the statement endorsing a medical approach to sterility and infertility was deemed applicable to this new method of assisted reproduction, but otherwise the silence on IVF specifically is illustrative of ecclesiastical deference to the ethical principles of professional integrity and the responsible exercise of moral agency by couples. The ecclesiastical statement on IVF issued in 1989 reflected the same three features as the policy on IUI: "In vitro fertilization using semen other than that of the husband or an egg other than that of the wife is strongly discouraged. However, this is a personal matter that ultimately must be left to the judgment of the husband and wife."[9] Notwithstanding the subsequent extensive use of IVF and the introduction of numerous reproductive technologies as adjuncts to IVF of the past three decades,[10] the appeal to moral agency has only been supplemented by an implicit privacy appeal in one additional sentence in the current policy: "Responsibility for the decision [third-party IVF] rests solely upon them [husband and wife]."

Current ecclesiastical policies address five procedures in assisted reproductive technology: intrauterine insemination (IUI) and IVF for a married couple, IUI for a single woman, sperm donation, and surrogate motherhood. The first thing of note about these policies is their location in the directives for ecclesiastical administration.[11] It would seem reasonable to expect policies on treatments to alleviate the medical condition of infertility to appear in the directives on "Medical and Health Policies." However, the statements on reproductive technology are found in a different section titled "Policies on Moral Issues." The categorization of a medical procedure as a "moral" issue generally implies that a core ecclesiastical concern is involved, the policy statement has greater weight than simply good advice, and the ecclesiastical policy can become a baseline for assessing public policy. The implications of medical procedures for reproduction for the LDS commitment to the integrity of family life presumably confer to these technologies the status of a "moral" issue.

A second point is that none of the policies are worded in a manner that indicates "support," "permission," or "approval" of the procedure. This may be one reason some LDS couples assume that infertility and IVF are "taboo" topics in LDS culture.[12] Furthermore, as there have been effectively no substantive emendations to these policies in three decades, it is unclear just how extensively the child-bearing portion of LDS membership is aware that there are *any* ecclesiastical statements on reproductive technologies. Garrett and Meghan did not seek ecclesiastical information on IVF but followed the moral culture reflected in decisions of other LDS couples experiencing infertility. The implication of ecclesiastical silence when the gametes of the married couple are used in IUI or IVF and medicine facilitates conception within the marital relationship is that such procedures are permissible provided the procedure otherwise meets medical criteria of safety and effectiveness. The diminished ecclesiastical presence invites both moral agency and moral diversity in familial conversations.

Each of the policies does, however, include an assertion indicating "strong discouragement" or disapproval about a particular aspect of the procedure. IUI or IVF for a married couple receives this assessment only contingently when the procedures include a third-party gamete donor, either donor sperm for IUI or IVF or a donor egg for IVF. Sperm donation and surrogate motherhood are strongly discouraged in themselves. The most restrictive policy concerns IUI for single women, a common practice in reproductive medicine[13] that is explicitly stated as "not approved." Single LDS women can be "subject to Church discipline" if a determination is made that a woman "deliberately" contravenes the restriction. It is unclear to what extent either IUI for a single LDS woman or the accompanying discipline has been enacted, although the issue has been discussed in various blogs and documentaries featuring LDS women who have been gestational surrogates or sought IUI as single women.[14]

The third striking connection between the policies is that no ecclesiastical or ethical rationale is presented for the negative positions of discouragement or disapproval. Historically, not only were the initial ecclesiastical reservations about third-party IUI framed as "problems related to civil law, family harmony, and eternal relationships," but the first statement on surrogate motherhood issued in 1989 discouraged surrogacy on the grounds of "spiritual, emotional, and other difficulties."[15] These are consequentialist rather than principled rationales, and as communal and familial experience displayed the resiliency of civil law and family life, the consequentialist objections could not be sustained. The evident element common to the procedures that are strongly discouraged or not approved is that they involve or imply a relationship with the gametic materials of a person outside of marriage to effectuate fertilization (or gestation in the case of surrogacy). The inference is that the central boundary for ecclesiastical teaching is the sanctity and integrity of marriage and the family.[16] The implicit, unstated principle seems to mirror the explicit endorsement contained in the 1974 statement on sterility tests: recourse to medical procedures is permissible when a physiological abnormality is experienced that prevents fertility and parenting. Reproductive treatments should intend to advance the parenting aspirations of a marriage relationship. However, this rationale cannot support an absolutist prohibition due to the overriding principle of moral agency and to circumstances in which third parties, such as a sibling surrogate, can support marital and parenting goals.[17]

This leads to a fourth important aspect of these policies: Ecclesiastical policy is secondary to respecting the moral agency of the married couple. LDS teaching permits—because it does not discourage—IUI or IVF for married couples using the couple's gametes provided this comports with the moral agency of both husband and wife. The principle of moral agency permits—because it does not forbid—IUI or IVF or surrogacy for married couples using donor gametes or a donor uterus. In circumstances of third-party involvement for purposes of fertilization or gestation, ecclesiastical policy emphasizes the decision is a "personal matter" between the spouses.

The latitude of moral agency also qualifies the otherwise strong discouragement of sperm donation. Sperm donation is the least medically invasive of any of the five procedures, and it is medically required if IUI or IVF with donation of third-party gametes is a matter of moral agency. Why sperm donation is even designated a "moral" issue meriting any ecclesiastical attention is itself of interest. The most plausible reason ecclesiastical policy does not defer to a male's moral agency in the context of a theology of family is that sperm donation can broach "anonymous" fatherhood. Anonymous or nonidentified parenting is laden with not only medical but spiritual values within a religious narrative that emphasizes genealogical lineage and eternal relationships. Notably, a single woman who requests IUI almost certainly desires a biological and social relationship with *her* child in a way that seldom applies to sperm donation. Sperm (and egg) donation also raises most directly the issue of commerce in body tissue and the sanctity of embodied life; "donation" language is frequently a misnomer.

LDS ecclesiastical teaching on assisted reproduction for conception and gestation in cases of medical infertility generally reflects ethical principles of medical professionalism, moral agency, and parental and familial solidarity. Further advances in IVF technology since 1989, such as preimplantation genetic screening of the human embryo (PGD), preconceptual sex

selection, and the disposition of cryopreserved embryos, are more controversial professionally and bioethically and have not been addressed ecclesiastically. Before addressing these bioethical perplexities, I want to identify
some influences in LDS family and moral culture that have made IVF and
other ARTs generally a welcome and even "inspired" medical advancement
because of their healing purpose.

Theologies of Infertility

LDS moral culture is invariably shaped by a pro-family and pro-natalist theology that consists of several features relevant to decisions about utilizing
reproductive technologies.

Procreative Imperatives. LDS interpretations of the Genesis creation narrative portray the first human beings, Adam and Eve, as receiving a covenantal
commandment to "be fruitful, multiply and replenish the earth." Procreation
in this theology of creation is an intrinsic characteristic of human identity and
of marital relationships, and the prospect of having children is a joyful and re-
deeming feature of the human fall from divine presence.[18] Notwithstanding
concerns about worldwide population growth and diminished resources,
the ecclesiastical claim is that fulfilling the procreative imperative in marital life has never been rescinded and "remains in force."[19] Procreative sexuality as a way of living out the parental calling integral to familial life has,
until the last three decades, also informed ecclesiastical perspectives critical
of contraception and family planning.[20] LDS teaching on birth control has
shifted substantially from generalized disapproval in the decade following
the emergence of oral contraceptives, with recommendations for self-control
and allowance for maternal health, to a current position that makes family
planning a matter of moral agency: "The decision as to how many children to
have and when to have them is extremely intimate and private and should be
left between the couple and the Lord."[21] It is perhaps no surprise that the LDS
fertility rate has declined to historic lows in the wake of this ecclesiastical
deference to moral agency on procreation.[22] It would be mistaken nonetheless to understand the procreative imperative as carrying the implication that
sexuality has only a procreative purpose. Procreation is narratively situated
within a theology of human persons as relational, embodied, and generative,
and symbolizes the ultimate meanings ascribed to human aspirations to ful-
fill the purpose of creation.

Re-Storying Infertility. The covenantal character of procreation rooted in
biblical narratives of creation and fall is reinforced by stigmas about female

infertility exemplified in diverse biblical narratives about Sarah, Rebekah, Rachel, Hannah, and Elizabeth. These figures are appropriated in ecclesiastical teaching and in personal witness as teachers of fidelity even in the midst of trials and a resource for experiential solidarity. It is of great meaning for many that these sisters in solidarity eventually gave birth to promised children and prophetic leaders.[23]

The poignancy of the biblical narratives and their embedded stigma has dissipated somewhat in the moral culture with more open acknowledgment of infertility experience, which has generated voluntary support programs, online commentaries, and testimonials. LDS narratives of infertility experience are nevertheless almost invariably articulated by women and often express a sentiment that however much infertility is attributable to frailties inherent in human biology, the inability to conceive or bear children symbolizes an impairment in personal character or even divine disfavor. However, as reproductive impediments are increasingly circumvented through medical technology, moralistic narratives of infertility experience are gradually being supplanted by medicalized "miraculous" narratives. Medical successes, community support, and re-storying infertility as a biological issue rather than a moral failing does not mean that for particular LDS couples the experience of infertility is not deeply distressing and wounding. Infertility generates feelings of resentment, frustration, and anger, and a perception of being ostracized by the community or even abandoned by God. Infertility can become not only a physical ordeal and relational complication but a trial of faith culminating in diminished religious commitments.[24] The re-storying of infertility through medical advances like IVF exhibits how technological innovations become symbols of divine care and inspiration.

Maternal Identities. Although the causes of infertility afflict both females and males in roughly equal proportion, interpretations of women's identity in LDS culture can make infertility an especially anguishing ordeal for LDS females. The indispensable sacred calling of women to be "co-creators" with God in bringing new life into the world means infertility experience can culminate in a devastating perception of incompleteness or a brokenness constructed not as a biological condition beyond one's control but as a circumstance of personal responsibility. This raises profound issues of personal identity, as poignantly expressed by one LDS woman about her infertility: "I felt lost for a long time. I felt I had no purpose. That's the ultimate goal, isn't it, to get married and have a family? I still knew I was a daughter of God, but I hated that I couldn't be a co-creator with Him. I felt broken, like I wasn't a real woman."[25] In this witness, core aspects of personal religious identity—a daughter of God—persist, but this identity is culturally integrated with the bearing of children.

The lamentation narratives of LDS women who experience infertility are permeated with the language of anger, abandonment, failure, and shattering. The anger is directed toward oneself, husbands, community members and their sometimes insensitive comments, mothers who complain of difficulties encountered in raising children, and God. Medical explanations about reproductive biological anomalies notwithstanding, interpreting infertility as something besides a character failure or some form of inexplicable personal responsibility is a difficult ordeal. This identity struggle leads to perceptions of isolation and exclusion, of being "left out" from the experiences of other women or couples. Infertility is a very lonely experience even if a person is not alone.

LDS female ecclesiastical leaders have re-storied the experience of fractured identity and isolation through the salvation narrative of embodied eternal progression and a more expansive interpretation of female and maternal identity as a *spiritual* trust. The biblical narrative designating Eve as "the mother of all living" *prior* to Eve's giving birth to any children offers a theological basis for re-storying motherhood and maternal identity.[26] Motherhood is still constructed as an essential aspect of women's identity, but its essential calling resides in nurturing spiritual and moral needs rather than biological progeny. Motherhood so re-storied is an expansive "calling to love, nurture, and lead those of a younger generation."[27] Similarly, a prominent LDS female writer, Sherri L. Dew, maintains, "Every time we love or lead anyone even one small step along the path [of progress], we are true to our endowment and calling as mothers."[28]

The re-storying of motherhood is connected to a profound revealed reality: the concept of the divine feminine is personified in LDS teaching through a theology of a Heavenly Mother. Although the Heavenly Mother is not worshipped or prayed to, there are consistent references to a Mother in Heaven, and to human beings as children of heavenly "parents," in LDS teachings from Joseph Smith onward. The doctrine of Heavenly Mother is referenced in culturally-meaningful hymns and follows from the moral logic of the salvation narrative that eternal life is comprised of family relationships.[29] The Heavenly Mother deepens LDS meanings of motherhood and maternal identity as creative, caring and nurturing, relational, spiritual presence, and in partnership.

Embodied Selves. The narratives of infertility reflect a question of theodicy, how convictions about the power, knowledge, and benevolent love of God can be harmonized with the mortal experience of frustrated aspirations and fractured identities. The infertility experience seems incoherent with a revealed reality of the soteriological narrative: embodiment is a necessary condition of eternal progression for any person. Birth is a threshold passage by which unembodied spirit beings in the pre-mortal life now acquire embodied experience and wisdom "to progress toward perfection."[30] The physical embodiment of spirit beings fully occurs upon birth into mortal life. The re-storying narrative of spiritual nurture and trust provides a different motivation for procreation and family life than either personal completeness or familial fulfillment, namely, advancing God's work of bestowing eternal life. Earnest reflection on the conviction that the eternal progression of an unborn spirit being may hinge on personal procreative capacity imposes an enormous psychological and spiritual burden. However, as biological parenthood is necessarily subject to embodied mortal contingencies, the salvation narrative likewise warrants the expansive re-storying of parenting, including motherhood and female identity, as a spiritual calling. This primary responsibility in parenting is not fathering or birthing a child but nurturing and educating children in spiritual and ethical paths, a calling that can be enacted in various spiritual, parental, or pedagogical relationships.

Within a communal and cultural setting in which profound matters of salvation, identity, and embodiment are deemed to rest on procreative sexuality, the advent of two generations of reproductive technologies such as IVF situates medicine within a communal narrative of healing deep wounds, restoring wholeness to personal identity, and indirectly advancing the divine plan for human happiness and eternal progression. A resort to IVF in circumstances of medical infertility between a marital union of male and female can be an occasion for expressing communal solidarity. The availability of technologies cannot of course resolve the ecclesiastical question of *who* may assume a parental calling, whether as an anonymous gamete donor or an identified single female. The question to now address is whether any theological and ethical boundaries are crossed by other uses of these new potent powers of medical conceptions.

Preimplantation Genetic Screening and Sex Selection

Garrett and Meghan were offered preimplantation genetic diagnosis (PGD) as part of their infertility treatments but opted not to use this procedure due to cost. IVF is a gateway technology to repro-genetic technologies such as PGD that increase the prospects of having a child with a disease-free genetic composition. A technological advance of the mid-1990s, PGD is a procedure in which cells of an unimplanted fertilized embryo are removed and biopsied to determine whether the embryo has a genetic anomaly. If the cells are free of the genetic-based diseases the parents were concerned about transmitting, the embryo can be transferred to the woman's uterus for gestation to birth (or be cryopreserved for future use); if a genetic anomaly is discovered, the embryo will typically be discarded. PGD has now been utilized to screen embryos for over one hundred different genetic mutations—for such diseases as cystic fibrosis, Huntington's, sickle cell, Duchenne muscular dystrophy, and specific variations of breast and ovarian cancers and Alzheimer's. PGD advances the ethical goal of medicine of disease prevention and permits couples to avoid perpetuating single-gene and sex-linked genetic conditions. However, the prospect of discarding unimplanted embryos with a genetic condition has made PGD ethically objectionable for some religious traditions that hold that a human embryo has full moral standing and rights at conception.

The biomedical rationale for PGD is preventing disease rather than alleviating infertility, although a couple who has already resorted to IVF for infertility may want to utilize PGD to provide assurance that their child will be born free of serious genetically transmitted diseases that could be lifelong in duration. PGD is also an option for couples who have no infertility issues as such but are aware of family or lineage histories of genetic-based disorders and diseases. The limits to utilization of PGD as an adjunct of IVF for infertility treatment are primarily financial: If PGD is not covered by insurance, the procedure costs couples about $5,000 to $15,000, in addition to any noninsurance coverage costs of an IVF cycle. The limits to utilizing PGD principally for disease prevention purposes are the physical and emotional burdens borne by the woman of undergoing conventional IVF hormonal treatments and transvaginal aspiration for egg retrieval as well as any financial burdens assumed by the couple for non-infertility-based IVF procedures. These burdens—physical, emotional, and financial—suggest that PGD will be a rare procedure, far less than the 1.7% of births that occur through IVF, rather than a routine medical practice. PGD can also pose justice issues insofar as the procedure, like IVF, is accessible to some but not all couples who may benefit from it based on financial resources.

No LDS ecclesiastical teaching has formally addressed PGD, a silence that implies that if a couple is considering embryo biopsy for genetic mutations either as part of infertility treatment or for disease prevention, the ethics of moral agency and the integrity of medical professionalism govern decision-making. PGD might also be constructed as a contemporary extension of a historical moral logic in LDS teaching permitting birth control when parents wished to prevent transmitting genetic "impurities" to their children.[31] Ecclesiastical policy affirming "responsibility" of the married couple for decisions about infertility treatments seems applicable to medical circumstances of PGD.

Some LDS couples have found that parental commitments, moral agency, communal support, and medical professionalism provide compelling reasons for recourse to PGD. In one narrative, the Williams family had given birth to a daughter with metachromatic leukodystrophy (MCL), a genetic disorder that diminished the child's ability to move and communicate and led to her death at age ten. The couple confronted intertwined medical, spiritual, and ethical considerations in deliberating over a second child. A couple in such a situation could forgo having any further biological children, seek to adopt, or rely on natural procreation in full knowledge of the genetic risks. As MCL is an autosomal recessive condition, procreation through sexual intercourse holds a 25% chance of a child with MCL, a 50% chance of transmitting the gene to a child who will be a carrier but not experience the condition, and a 25% chance a child will neither be a carrier nor have the disease condition. The couple could also turn to IVF and preimplantation diagnosis to minimize the risks of disease transmission.

The Williams family consulted with faith leaders and fellow congregants and ultimately placed their hopes in the biomedical procedures. It is unclear whether, beyond community solidarity, the couple's decision was informed by any religious considerations distinct from concerns of any parent, religious or nonreligious. Nikki Williams indicated, "We wanted a healthy baby. That's all I cared about." What seems to have been compelling for the couple in their decision to resort to IVF and PGD was a strong moral conviction: "We chose [PGD] because [MCL] is essentially a death sentence."[32]

A second PGD narrative portrayed the ordeal of the Kooymans, who had experienced the death of their son forty-one days following his birth from the rare genetic disease rhizomelic chondrodysplasia punctata (RCDP). Danny Kooyman depicted the loss of his son as "devastating," and the heart-wrenching death motivated the couple to seek out reproductive options from

a fertility clinic to prevent a similar premature death for subsequent children.[33] The IVF procedure produced nine fertilized eggs; the embryo biopsy
revealed that eight had some form of genetic anomaly, including three with
the PEX genetic mutation for RCDP. The single embryo that was transferred
did not produce a pregnancy. The Kooymans' experience likewise focuses
more on motivations for using PGD germane to any parent rather than any
distinctive religious rationale. The fertility clinic director portrayed PGD as
part of a reproductive continuum in which medicine assists biological processes rather than supplanting them: "This technique is similar [to IVF]
where we're helping nature create healthy children," such that the appeal to
"nature" places the ethics of PGD within the preventive ethos of medicine.
However, Mr. Kooyman did allude to a theological narrative for PGD that
reflects the revealed reality on healing in which medicine and reproductive
technologies are divinely inspired vocations: "God is the one who made these
doctors know how to do this." Medicine is a moral vocation that carries out in
its own way the divine work of healing persons in the modern world.

The principles of moral agency and professional calling should not
lead to an uncritical communal endorsement of PGD. The ethos of
preventing disease that warrants PGD requires conceptual clarity as to
what constitutes a "disease," an issue of great significance for current uses
of PGD for sex selection. Communal constructions of disease, disability,
and impairment are embedded in the most ethically controversial use of
PGD, selecting an embryo with a specific genetic anomaly, such as deafness, as a way to cultivate communal identity.[34] In encountering such
parental requests, the ethical integrity of medicine and the use of biomedical technology for healing ends is severely tested. Furthermore, the reality that access to PGD often hinges on financial resources, not medical
need, discloses some of the systemic inequities in access to health care
in general. PGD is also viewed by some critics as a stepping stone to genetic manipulations that will not only diminish genetic-based diseases but
invite the possibility of enhancing familial or socially desired characteristics, a prospect augmented through innovative applications of human
gene editing. A descriptive overview of the ethics of PGD and human gene
editing in the church-owned *Deseret News* foregrounds the prospect of
medical enhancements with a provocative initial paragraph: "Male or female? Brown eyes or green? Intelligent, athletic or both? Most parents
have some idea of who they would like their child to be—and someday
science could let them decide."[35]

A symbolic objection has also been raised against PGD, articulated primarily by advocates of disability rights, about the meanings of a widespread practice of PGD for persons who currently suffer from a disease that could have been prevented had the parents relied on assisted reproductive procedures rather than the genetic lottery of life. This is an especially acute issue in families in which a first child suffers from a disabling or life-threatening condition and the parents use IVF and PGD for subsequent child-bearing. The prospect that PGD can symbolically devalue persons with disabilities is intimated in a comment by Nikki Williams about her first daughter, who died from MCL: "I would never want anyone to feel as though Eliza was not good enough the way she was. She was, and still is, an incredible gift to us."[36] The symbolic critique has been advanced most forcefully by Leon Kass, who served as chair of the President's Commission on Bioethics, in the context of prenatal screening: "How will we come to view and act toward the many disabled persons that will remain among us once we embark on a program to root out genetic abnormality? [Such a person] is liable to be looked upon by the community as one unfit to be alive, as a second-class or even a lower, human type. [He or she] may be seen as a person who need not have been, and who would not have been, if only someone had gotten to them in time."[37] While the critique bears consideration, Kass is engaged in rhetorical overstatement: PGD does not represent a systematic social "program" to eradicate genetic anomalies of the kind Kass fears, insofar as it is a choice for couples who conceive through IVF. PGD is necessarily a rare and hard choice made by conscientious couples after much deliberation rather than a matter of social or government mandate. Still, concerns about fostering a medicalized culture that devalues persons with disabilities should not be dismissed as entirely speculative.[38]

The ethics of discrimination likewise emerge in cultural and bioethical discourse about a further extension of IVF and PGD procedures: preconceptual sex selection. In some circumstances, such as hemophilia and Duchenne muscular dystrophy, genetic screening for the sex of the embryo comports with the medical rationale to prevent sex-linked genetic diseases. There is nothing morally distinctive about medical-based sex selection that is not present in PGD. The much more controversial professional and bioethical question is over nonmedical reasons for sex selection, such as "family balancing." While nonmedical sex selection is prohibited by law in many countries, it is permitted in the United States with variations in state law. An overview of nine Southern California infertility clinics indicated that 20% of couples requesting fertility assistance did not have any infertility complications but rather sought to use IVF and PGD to determine whether the embryos were male or female based on the sex chromosomes. The couples intended to transfer only those embryos of the familially desired sex.[39] The principle of respecting the ethical integrity of medicine is not adequate in these circumstances, as professional societies do not have consensus on sex predetermination for nonmedical purposes of family balancing.

The Ethics Committee of the American Society for Reproductive Medicine in 2015 conferred responsibility for determining provision of nonmedical sex selection to individual fertility clinics.[40] The Ethics Committee maintained that a clinic's decision to provide nonmedical sex selection could be supported by respect for parental reproductive liberty. A parental desire to have a child of a particular sex can express parental love and not reflect gender bias. However, it is morally problematic to require a woman to undergo the physiological burdens required by IVF and embryo transfer when the procedure does not confer *medical* benefits. The committee further observed that nonmedical sex selection can perpetuate "gender essentialism," a concept affirming that "there are certain characteristics inherent in being female and others inherent in being male."[41] Nonmedical sex selection also implicates a social justice issue: For many critics, using scarce medical resources—training and education, research, technological advances, provider specialization—for a nonmedical purpose is contrary to medical stewardship and social justice. This is an especially compelling consideration when many persons in the United States do not have access to basic health services.

There is virtually no discussion or research on the prevalence of nonmedical sex selection in the LDS community. The principles and narratives embedded in the procreative imperative address the parental and soteriological significance of bearing and nurturing children. LDS families certainly express aspirations to have a "balanced" family and experience the diverse nature of parenting that occurs when raising girls and boys, but there are no distinctive religious reasons for using reproductive technologies to achieve this aspiration. However, the social and professional controversy over nonmedical sex selection may connect to LDS ecclesiastical teaching on the concept of gender essentialism. An eternal reality portrayed in the family proclamation is that "Gender is an essential characteristic of individual pre-mortal, mortal, and eternal identity and purpose."[42] This theology of gender identity implies that there are inherent callings and characteristics of females and males, including specific inclinations for marriage and parenting. It is precisely these callings, including the sacred trust of motherhood, that give poignancy to narratives by LDS women about infertility. The theology of gender and the social and professional critique of gender essentialism ultimately move beyond the realm of bioethics discourse and into broader questions about human nature, equality, and inclusion.[43] Communal rethinking on gender identity occurs within a context in which ecclesiastical leaders affirm balancing principles of religious liberty, compassion, sensitivity, and kindness toward persons who identify as transgender, and principles of protection of LGBT communities from discrimination.[44]

Embryonic Stem Cell Research

As illustrated in my initial narrative, one inevitable feature of conventional IVF procedures is the retrieval of more eggs and the fertilization of more embryos than can be safely transferred in any IVF cycle. If PGD is subsequently utilized and the embryos are determined to have a genetic anomaly, they most likely will be discarded. The remaining embryos from either IVF or PGD are cryopreserved for future use. A question that has divided religious communities and policy-makers is the disposition of excess embryos created for purposes of treating infertility when the couple's child-bearing aspirations have been realized.

There are nearly a million human embryos cryopreserved in U.S. fertility clinics, a phenomenon that generates difficult choices for couples and clinics.[45] Couples commonly decide to preserve their embryos for their own future use so that the woman does not have to undergo further burdensome hormonal treatments for hyperovulation and the egg retrieval process. If a couple's family is complete, they may donate their excess embryos to other infertile couples. The disposition option that has stirred the most religious controversy, political debate, and federal law is embryo donation for biomedical research purposes, especially for embryonic stem cell (ESC) research. ESC research disaggregates an embryo created for treating infertility so that researchers can manipulate the stem cells in ways that may eventually advance treatments for lethal or chronic diseases. Embryos no longer needed for infertility treatment, and which might otherwise be discarded, could be donated for scientific research that holds out prospects for treating, preventing, or curing various diseases such as Parkinson's, stroke, diabetes, and Alzheimer's.

The possibilities of embryonic stem cell research emerged in the late 1990s and immediately became situated within a polarized political context. Following his election, U.S. president George W. Bush sought guidance from Pope John Paul II in the summer of 2001 on ESC research. In ensuing controversies over federal funding for such research, however, some commentators claimed that "The LDS Church, not the Vatican, is playing the pivotal role in the struggle over stem cells."[46] More specifically, the positions adopted by congressional representatives who were LDS members (rather than the Church directly) on embryonic stem cell research influenced rethinking on objections to ESC research and to funding research. In the initial years of this controversy, all five LDS U.S. senators supported funding from the federal government for such research. That stance puzzled some pundits because the donation of embryos for research and their subsequent disaggregation seemed contrary to the positions these senators affirmed on the sanctity of human life.

The position on ESC research articulated by the LDS senators was, however, constructed as the "ultimate" pro-life perspective. The critical distinctions were articulated by Senators Orrin Hatch (R-Utah) and Gordon Smith (R-Oregon), who gave congressional testimony that there is an ontological and moral difference between an unimplanted human embryo in a laboratory (infertility or research) setting and an implanted embryo capable of development into a full human being. In a lengthy letter to President Bush supporting federal funding of ESC research, Hatch observed, "To me a frozen embryo is more akin to a frozen unfertilized egg or frozen sperm than to a fetus naturally developing in the body of a mother. . . . I believe that human life begins in the womb, not a Petri dish or a refrigerator."[47] Based on this distinction, Hatch sought to persuade President Bush and other political conservatives that ESC research was morally consistent with sanctity of life and pro-family convictions.[48]

Senator Smith developed further ethical rationales for both ESC research and federal funding: "I believe that life begins in a mother's womb, not in a scientist's laboratory. For me, being pro-life means helping the living as well. So if I err at all on this issue, I choose to err on the side of hope, healing, and health. And I believe the federal government should play a role in research to assure transparency, to assure morality, to assure humanity and to provide the ethical limits and moral boundaries which are important to this issue."[49] Smith's expanded meaning of a pro-life posture avoided the common critique that pro-life views confer more moral protection to unborn human life than post-birth persons. His invocation of the value trinity of "hope, healing and health" implied that conservative objections to ESC research based on the sanctity of the embryo could be overridden by substantive ethical and religious considerations. The position articulated by Hatch and Smith presents affinities with a central principle in Jewish law and bioethics, *pikuach nefesh*, in which saving human life is an overriding imperative requiring abrogation of virtually all other religious commands.[50]

In the midst of the controversy over federal funding, the LDS Church released a statement of "neutrality" on ESC research: "While the First Presidency and the Quorum of the Twelve Apostles have not taken a position at this time on the newly emerging field of stem cell research, it merits cautious scrutiny. The proclaimed potential to provide cures or treatments for many serious diseases needs careful and continuing study by conscientious, qualified investigators. As with any emerging new technology, there are concerns that must be addressed. Scientific and religious viewpoints both demand that strict moral and ethical guidelines be followed."[51] While consistent with ecclesiastical deference to scientific expertise, the statement unfortunately did not identify the "concerns," the requisite "moral and ethical guidelines," or the institutions responsible for oversight. A striking aspect of Smith's position, for example, was his advocacy of the federal government as *moral* arbiter, which reflects a regulatory process with expanded public accountability rather than a privatized research model.[52]

What is clear is that the statement did not reject ESC research as unethical, unprofessional, or contrary to the sanctity of life nor did it assess federal funding as unwise public policy and a form of moral complicity in evil, all critiques advanced by conservative religious denominations. The call for "careful and *continuing* study" indicates indirect support for such research: It would not be possible to assess the safety, benefits, or concerns of stem cell research for disease treatments without ongoing studies. As Lester Bush Jr., has argued, LDS ecclesiastical policies on medical ethics characteristically *follow* the resolution of both scientific and ethical questions about safety and efficacy of innovative research.[53] However, ecclesiastical caution occurs at a considerable ethical and cultural cost of having other stakeholders define the scientific and ethical considerations.

Enduring Issues

The initial reticence and the subsequent embrace of reproductive technologies within LDS ecclesiastical teaching and moral culture turn on several issues, including commitments to the sanctity and integrity of family, the impact such technologies may have on familial relationships, personal and familial moral agency, and the ethical integrity of medicine and fertility specialists in mediating reproductive technologies for purposes of familial healing. Insofar as empirical studies have not disclosed any unique issues with family cohesiveness attributable to reproductive technology as such, it seems that utilization of emerging innovative technologies will fall under the purview of professional integrity and moral agency. The compelling narratives of infertility experience witness to the authority of personal moral experience and the need for inclusion and solidarity of a moral community. The question of the moral status of the human embryo lurks in the background of discussion, practice, and policy regarding preimplantation genetic diagnosis, embryo disposition, and embryonic stem cell research. The posture of ecclesiastical neutrality suggests how commitments to the sanctity of life can be re-storied through the ethical principles of love and covenantal responsibility. The prospect that reproductive technologies can remedy infertility and prevent transmission of genetic diseases illuminates issues in the bioethics of prevention, a subject to which I now turn.

Notes

1. Centers for Disease Control and Prevention, *2017 Assisted Reproductive Technology Fertility Clinic Success Rates Report* (Atlanta: U.S. Department of Health and Human Services), 2019, ftp://ftp.cdc.gov/pub/Publications/art/ART-2017-Clinic-Report-Full.pdf
2. A recent study indicates substantial support among Mormons for IVF as a morally acceptable or morally indifferent practice, ranging between 67 and 85% depending on generation, and 69 and 91% depending on geographic location. Jana Riess, *The Next Mormons: How Millennials Are Changing the LDS Church* (New York: Oxford University Press, 2019), 181–182.
3. Lindsey Redfern, "5 Things Couples Dealing with Infertility in Your Ward Wish You Knew," *LDS Living*, May 28, 2019, http://www.ldsliving.com/5-Things-Couples-Dealing-With-Infertility-In-Your-Ward-Wish-You-Knew/s/80393; Rachel Sheffield, "Finding Peace from Stories of Infertility in the Bible," *The Ensign*, September 2018, https://www.lds.org/ensign/2018/09/finding-peace-

from-stories-of-infertility-in-the-bible?lang=eng; Carolynn R. Spencer, "Learning to Cope with Infertility," *The Ensign*, June 2012, https://www.lds.org/ensign/2012/06/learning-to-cope-with-infertility?lang=eng&_r=1; Melissa Merrill, "Faith and Infertility," *The Ensign*, April 2011, https://www.lds.org/ensign/2011/04/faith-and-infertility?lang=eng; Crystal Nicole Jones, "Our Blessing of Infertility," *The Ensign*, April 2011, https://www.lds.org/ensign/2011/04/faith-and-infertility-expanded/crystal-nicole-jones?lang=eng; Caroline Arveseth, "Five Under Five," *Mormon Women Project*, May 19, 2010, https://www.mormonwomen.com/interview/five-under-five/

4. Lester E. Bush, Jr., *Health and Medicine Among the Latter-day Saints: Science, Sense, and Scripture* (New York: Crossroad, 1993), 170.

5. Bush, *Health and Medicine Among the Latter-day Saints*, 170.

6. "Attitudes of the Church of Jesus Christ of Latter-day Saints Towards Certain Medical Problems," in "Mormon Medical Ethical Guidelines," ed. Lester E. Bush, Jr., *Dialogue* 12 (Fall 1979), 97, 101.

7. "Attitudes of the Church of Jesus Christ of Latter-day Saints Towards Certain Medical Problems," 99.

8. See Appendix A, "Artificial Insemination."

9. The Church of Jesus Christ of Latter-Day Saints, "Church Policies and Guidelines, Policies on Moral Issues: In Vitro Fertilization," in *General Handbook: Serving in the Church of Jesus Christ of Latter-day Saints*, Chapter 38.6.9, February 19, 2020, https://www.churchofjesuschrist.org/study/manual/general-handbook/38-church-policies-and-guidelines?lang=eng#title_number91. See Appendix A, "In Vitro Fertilization."

10. Pamela Mahoney Tsigdonis, "The Big IVF Add-On Racket," *New York Times*, December 13, 2019, A31, https://www.nytimes.com/2019/12/12/opinion/ivf-add-ons.html?smid=nytcore-ios-share

11. The Church of Jesus Christ of Latter-Day Saints, "Church Policies and Guidelines, Policies on Moral Issues," in *General Handbook: Serving in the Church of Jesus Christ of Latter-day Saints*, Chapter 38.6. Minor modifications were made to the ecclesiastical policies in July 2020. See 2020 Appendix.

12. Becca, "I'm a Mormon with Infertility (Infertile in a Family-Focused Faith)," *Love Our Crazy Life*, http://www.loveourcrazylife.com/infertility-mormon-lds/

13. American Society for Reproductive Medicine, Ethics Committee, "Access to Fertility Treatment for Gays, Lesbians, and Unmarried Persons: A Committee Opinion," *Fertility and Sterility* 100:6 (2013), 1524–1527, https://www.asrm.org/globalassets/asrm/asrm-content/news-and publications/ethics-committee-opinions/access_to_fertility_treatment_by_gays_lesbians_and_unmarried_persons-pdfmembers.pdf. See Mary Pflum, "Egg Freezing 'Startups' Have Wall Street Talking—And Traditional Fertility Doctors Worried," *NBC News*, March 4, 2019, https://www.nbcnews.com/health/features/egg-freezing-startups-have-wall-street-talking-traditional-fertility-doctors-n978526; Vanessa Grigoriadis, "IVF Coverage Is the Benefit Everyone Wants," *New York Times*, January 30, 2019, https://www.nytimes.com/2019/01/30/stYle/ivf-coverage.html

14. Independent Lens, "Made in Boise: The Complex World of Surrogacy," http://www.pbs.org/independentlens/films/made-in-boise/; Vanessa Amundson, "My

Position Regarding Surrogacy," on "A Perfectly Imperfect Perfectionist," October 12, 2011, http://myamundson5.blogspot.com/2011/10/my-position-regarding-surrogacy.html?showComment=1330523452486#c4840420031267940597; April Young Bennett, "Five More LDS Church Discipline Policies That Affect Women Unequally," *Exponent II*, February 22, 2014, https://www.the-exponent.com/five-more-lds-church-discipline-policies-that-affect-women-unequally/

15. Bush, *Health and Medicine Among the Latter-day Saints*, 170–172; Courtney S. Campbell, "Sounds of Silence: The Latter-day Saints and Medical Ethics," in *Theological Developments in Bioethics: 1988–1990*, ed. B Andrew Lustig (Dordrecht, The Netherlands: Kluwer Academic Publishers, 1991), 23–40.

16. It is necessary to note here that LDS teachings on reproductive technologies were formulated in the context of legal definitions of marriage as a lawful relationship between a man and a woman. The theology of marriage, family, gender, and sexuality was formally institutionalized in "The Family: A Proclamation to the World," issued in September 1995, as well as by subsequent ecclesiastical commentary (Newsroom, "The Divine Institution of Marriage," https://newsroom.churchofjesuschrist.org/article/the-divine-institution-of-marriage). Following the June 2015 U.S. Supreme Court ruling in *Obergefell v. Hodges* (576 US ___ [2015]) that legalized same-sex marriage, the LDS Church issued a statement acknowledging that "same-sex marriages are now legal in the United States. [However,] The Court's decision does not alter the Lord's doctrine that marriage is a union between a man and a woman ordained by God." Newsroom, "Supreme Court Decision Will Not Alter Doctrine on Marriage," June 26, 2015, https://newsroom.churchofjesuschrist.org/article/supreme-court-decision-will-not-alter-doctrine-on-marriage. Although reproductive technologies advance parenting aspirations for same-sex couples, ecclesiastical policies on reproductive technologies presume heterosexual marriage as the normative context for parenting. The theology of sexuality raises broader issues than can be addressed here. See Riess, *The Next Mormons*, 129–146, and Newsroom, "Church Leaders Counsel Members After Supreme Court Same-Sex Marriage Decision," June 30, 2015, https://newsroom.churchofjesuschrist.org/article/top-church-leaders-counsel-members-after-supreme-court-same-sex-marriage-decision.

17. Mary Madabhushi, "I Feel Like I've Been Here Before," *Mormon Women Project*, November 13, 2017, https://www.mormonwomen.com/2017/11/feel-like-ive/; Joanne Kenen, "Mitt's Son Has Twins Via Surrogate," *Politico*, May 4, 2012, https://www.politico.com/story/2012/05/romneys-son-has-twin-boys-through-surrogate-075939

18. The Pearl of Great Price, *Moses* 5:11–12.

19. The Church of Jesus Christ of Latter-day Saints, "The Family."

20. For a critical analysis of LDS teaching on birth control, see Melissa Proctor, "Bodies, Babies, and Birth Control," *Dialogue: A Journal of Mormon Thought* 36 (2003), 159–175.

21. The Church of Jesus Christ of Latter-Day Saints, "Church Policies and Guidelines, Policies on Moral Issues: Birth Control," in *General Handbook: Serving in the Church of Jesus Christ of Latter-day Saints*, Chapter 38.6.4; cf. Lester E. Bush, Jr., "Birth Control Among the Mormons: Introduction to an Insistent Question," *Dialogue* 10

(1977), 12–44; Jonathan Decker, "Is Birth Control Against the Commandments?," *LDS Living*, November 30, 2018, http://www.ldsliving.com/Ask-a-Latter-day-Saint-Therapist-Is-Birth-Control-Against-the-Commandments/s/89844?utm_source=ldsliving&utm_medium=email; "Birth Control," The Church of Jesus Christ of Latter-day Saints, https://www.lds.org/topics/birth-control?lang=eng; Homer S. Ellsworth, "Birth Control," in *Encyclopedia of Mormonism*, ed. Daniel H. Ludlow (New York: Macmillan, 1992), 116–117 https://contentdm.lib.byu.edu/digital/collection/EoM/id/5515

22. Jana Riess, "The Incredible Shrinking Mormon American Family," Religion News Service: Flunking Sainthood, June 15, 2019, https://religionnews.com/2019/06/15/the-incredible-shrinking-mormon-american-family/; Riess, *The Next Mormons*, 103–105.

23. Sheffield, "Finding Peace from Stories of Infertility in the Bible."

24. The pro-natalist imperative can be experienced as particularly oppressive for a childless woman on Mother's Day. LDS Church websites have recognized this, as have LDS critics. Erin Hallstrom, "When Mother's Day is Hard: A 40-Year-Old Single Women's Thoughts, *LDS Living*, May 10, 2017, https://www.ldsliving.com/When-Mother-s-Day-Is-Hard-A-40-Year-Old-Single-Woman-s-Thoughts/s/85290. Taylor Petrey, "Mormons and the Politics of Mother's Day," *Patheos*, May 12, 2013, https://www.patheos.com/blogs/peculiarpeople/2013/05/mormons-and-the-politics-of-mothers-day/

25. Merrill, "Faith and Infertility."

26. Sherri L. Dew, "Are We Not All Mothers?," *The Ensign*, October 2001, https://www.lds.org/general-conference/2001/10/are-we-not-all-mothers?lang=eng

27. Sheffield, "Finding Peace from Stories of Infertility in the Bible."

28. Dew, "Are We Not All Mothers?"

29. David L. Paulsen, Martin Pulido, "'A Mother There': A Survey of Historical Teachings about Mother in Heaven," *BYU Studies Quarterly* 50 (2011); 70–97; Danielle B. Wagner, "14 Myths and Truths We Know About Our Heavenly Mother," *LDS Living*, May 10, 2018, https://www.ldsliving.com/14-Myths-and-Truths-We-Know-About-Our-Heavenly-Mother/s/88439; "Mother in Heaven," https://www.churchofjesuschrist.org/study/manual/gospel-topics-essays/mother-in-heaven?lang=eng

30. The Church of Jesus Christ of Latter-day Saints, "The Family."

31. The value-laden ecclesiastical language of genetic "impurity" seems to push rationales for both contraception and PGD in the direction of eugenics. For a discussion of historical LDS views on eugenics, see Bush, *Health and Medicine Among the Latter-day Saints*, 167–169.

32. Kelsey Dallas, "Finding God in One of Science's Biggest Debates: Gene Editing," *Deseret News*, January 13, 2016, https://www.deseretnews.com/article/865645387/Finding-God-in-one-of-sciences-biggest-debates-2-Genetic-editing.html. The second child of the Williams family was born free of genetic mutations. See Kelsey Dallas, "The Morality of Playing God with Your Baby's DNA," *Deseret News*, March 28, 2015, https://www.deseretnews.com/article/865666153/The-morality-of-playing-God-with-Your-babys-DNA.html

33. Ed Yeates, "In Vitro Procedure Weeds Out Genetic Defects," *Deseret News*, June 13, 2012, https://www.deseretnews.com/article/865557436/In-vitro-procedure-weeds-out-genetic-defects.html

34. The Hastings Center, "Belonging: On Disability, Technology, and Community," December 3, 2019, https://www.thehastingscenter.org/the-art-of-flourishing-events-series/; Michelle Bayefsky, "Who Should Regulate Pre-Implantation Diagnosis in the United States?," *AMA Journal of Ethics* 20:12 (2018), E1160–E1167, https://journalofethics.ama-assn.org/article/who-should-regulate-preimplantation-genetic-diagnosis-united-states/2018-12

35. Dallas, "The Morality of Playing God with Your Baby's DNA."

36. Dallas, "Finding God in One of Science's Biggest Debates."

37. Leon R. Kass, "Implications of Prenatal Diagnosis for the Quality of, and Right to, Human Life," in *Biomedical Ethics and the Law*, ed. James M. Humber, Robert F. Almeder (New York: Plenum Press, 1976), 313–327.

38. John A. Robertson, *Children of Choice: Freedom and the New Reproductive Technologies* (Princeton, NJ: Princeton University Press, 1994); Kristin Clift, "Glimpses of Eternity: Sampled Mormon Understandings of Disability, Genetic Testing, and Reproductive Choice in New Zealand," M.A. thesis, University of Otago, New Zealand, 2012, https://ourarchive.otago.ac.nz/bitstream/handle/10523/3935/CliftKristin2012MA.pdf?sequence=1&isAllowed=Y

39. Sumathi Reddy, "Fertility Clinics Let You Select Your Baby's Sex," *Wall Street Journal*, August 17, 2015, http://www.wsj.com/article_email/fertility-clinics-let-you-select-your-babys-sex-1439833091-lMYQjAxMTI1MDE1ODkxMjgzWj

40. Ethics Committee of the American Society for Reproductive Medicine, "Use of Reproductive Technology for Sex Selection for Non-Medical Reasons," *Fertility & Sterility* 103 (June 2015), 1418–1422.

41. Ethics Committee, "Use of Reproductive Technology," 1420.

42. The Church of Jesus Christ of Latter-day Saints, "The Family"; Dallin H. Oaks, "Truth and the Plan," https://www.lds.org/general-conference/2018/10/truth-and-the-plan?lang=eng

43. Riess, *The Next Mormons*, 100–103.

44. The February 2020 administrative handbook provided the first formal statement of ecclesiastical policy regarding transgender persons. See The Church of Jesus Christ of Latter-day Saints, "Church Policies and Guidelines, Policies on Moral Issues: Transgender Individuals," in *General Handbook: Serving in the Church of Jesus Christ of Latter-day Saints*, Chapter 38.6.21. For further recent ecclesiastical discussion, see Newsroom, "The Church of Jesus Christ Supports the Federal Fairness for All Act," December 6, 2019, https://newsroom.churchofjesuschrist.org/article/federal-fairness-for-all-support-december-2019; Emma Green, "The Mormon Church Tries to Create a Little More Space for LGBTQ Families," *The Atlantic*, April 7, 2019, https://www.theatlantic.com/family/archive/2019/04/lgbtq-mormons-latter-day-saints-apostasy-child-baptism/586630/; Taylor Petrey, "A Mormon Leader Signals New Openness on Transgender Issues," *Slate*, February 13, 2015, https://slate.com/human-interest/2015/02/

mormons-and-transgender-elder-dallin-h-oaks-says-the-lds-church-is-open-to-rethinking-its-teachings.html; Laurie Goodstein, "Utah Passes Antidiscrimination Bill Backed by Mormon Leaders," *New York Times*, March 12, 2015, https://www.nytimes.com/2015/03/12/us/politics/utah passes antidiscrimination-bill-backed-by-mormon-leaders.html

45. Tamar Lewin, "Industry's Growth Leads to Leftover Embryos, and Painful Choices," *New York Times*, June 17, 2015, https://www.nytimes.com/2015/06/18/us/embryos-egg-donors-difficult-issues.html

46. Drew Clark, "The Mormon Stem Cell Choir," *Slate*, August 3, 2001, https://slate.com/news-and-politics/2001/08/the-mormon-stem-cell-choir.html

47. Lee Davidson, "No LDS Stand on Cell Research," *Deseret News*, July 6, 2001, https://www.deseretnews.com/article/851862/No-LDS-stand-on-cell-research.html

48. Jan Cienski, "Mormons May Be Key in Stem Cell Debate," *World-Wide Religious News*, August 4, 2001, https://wwrn.org/articles/5900/

49. Jim Woolf, "4 of 5 LDS Senators Taking Stand in Support of Stem Cell Research," *The Salt Lake Tribune*, July 19, 2001, http://www.sltrib.com/07192001/nation_w/114784.htm

50. Elliot N. Dorff, "Testimony: National Bioethics Advisory Commission," *Ethical Issues in Human Stem Cell Research: Religious Perspectives*, vol. 3 (Rockville: MD, Government Printing Office, 2000), C3–C5, https://bioethicsarchive.georgetown.edu/nbac/stemcell3.pdf. All commands in the *Torah* can be overridden to save human life except the prohibitions of murder, idolatry, and sexual infidelity.

51. Newsroom, "Embryonic Stem Cell Research," The current ecclesiastical statement simply says the Church has "no position" on the issue, https://newsroom.churchofjesuschrist.org/official-statement/embryonic-stem-cell-research; Davidson, "No LDS Stand on Cell Research."

52. Federal bioethics commissions had previously delineated ethical principles and regulatory guidelines for research on human embryos: National Bioethics Advisory Commission, *Ethical Issues in Human Stem Cell Research*, 1999, https://bioethicsarchive.georgetown.edu/nbac/execsumm.pdf

53. Bush, *Health and Medicine Among the Latter-day Saints*, 201–203.

3

The Wisdom of Prevention

Medicine for an Imperfect World

Kevin and Caitlyn are active members of The Church of Jesus Christ of Latter-day Saints (LDS) and parents of a growing family of six children. They have raised their children with values stressing the importance of church teachings on family unity, preserving good health, and living by the "law of chastity," a commitment of sexual abstinence prior to marriage and sexual fidelity within marriage.

Kevin is a family care doctor with a practice comprising many parents who object to vaccines for their young children. His interactions with vaccine-hesitant parents require multiple conversations over several visits. A similar pattern is evident in Kevin's conversations with families and their "tweens" regarding the HPV vaccine. Scientific studies indicate that the HPV vaccine confers protection against the most prevalent sexually transmitted infection as well as the majority of cases of cervical cancer.[1] Kevin explains that the HPV vaccine is analogous to seat-belt requirements, erring on the side of prevention even if there is minimal risk of actually needing a seat belt in any given instance of vehicular travel. His recommendation frequently receives a negative reaction that reflects aversions to "another shot."[2] He then encourages parents and children to have conversations about the HPV vaccine and consult with him when questions arise.

Within their family setting, Caitlyn attended office visits with her two oldest children when pediatricians recommended the HPV vaccine. While knowing Kevin and Caitlyn had created a family environment emphasizing abstinence from sexual activity prior to marriage, the pediatricians affirmed strong health and disease prevention rationales for the vaccine. Caitlyn initially opposed the HPV vaccine for her children, resistance she attributes to online stories about adverse effects: She commented: "A single story of a negative outcome from a vaccine can have more impact than the recommendations of a thousand physicians." Caitlyn educated herself about the HPV vaccine and, as the older children matured into teenagers, found her views converging with the

Mormonism, Medicine, and Bioethics. Courtney S. Campbell, Oxford University Press (2021). © Oxford University Press.
DOI: 10.1093/oso/9780197538524.003.0004

professional practice of her husband. Their conversations with their oldest chil-
dren about life goals, values, and religious commitments, such as a marriage
covenant in an LDS temple, became an occasion for discussing the HPV vac-
cine. Caitlyn and Kevin agree that their younger children will receive the HPV
vaccine at the recommended age. By providing an environment for open dia-
logue with their children about good health, prevention, and sexual activity,
Caitlyn and Kevin anticipate their children will understand that discussing the
HPV vaccine is a matter of living a healthy life and does not mean parental en-
couragement of sexual activity contrary to the law of chastity. They have com-
municated that these choices are not mutually exclusive: The children can both
receive the vaccine and adhere to Church teachings on virtue, morality, and
chastity.

Caitlyn observes that "in a perfect world, a vaccine like [HPV] would be un-
necessary." Still, she emphasizes in her conversations that even with a child's
current commitment to the covenant of chastity, choices, mistakes, or un-
planned events can leave them at risk of exposure to a sexually transmitted
infection (STI). As the six children mature into adulthood, travel to different
places, and are exposed to different ways of life, they may make different choices
about sexual activity, or may not have full knowledge of the sexual health and
history of their partner. Given these uncertainties, Caitlyn and Kevin believe it
is better for their children to be protected from sexually transmitted diseases so
that they can life the live they choose.

A historical commitment to preventing disease through good health
practices is embedded in LDS teaching and moral culture. One of the more
remarkable illustrations of how community praxis and personal experience
inform scriptural texts and ecclesiastical teaching is embedded in a health
covenant called the "Word of Wisdom."

A Principle with Promise

As with numerous Americans in the early 19th century, early Mormons were
profligate users of tobacco.[3] Church leaders convened regularly in 1833 in
the "School of the Prophets" in Kirtland, Ohio, to discuss spiritual teachings
and receive education in history, grammar, and arithmetic. The male
attendees for these lessons used pipes and chewed tobacco, leaving a smoky
haze and spit from chewing tobacco on the floor when the lessons concluded.
Led by Emma Smith, wife of Joseph Smith, young women would clean the

school room at the end of each session, but the tobacco stains could not be removed. Emma subsequently raised the issue of the uncleanness and odor in the room with her husband, a request that led Joseph Smith to inquire of God about whether tobacco use had divine sanction.[4] This process of practice, questioning, and seeking divine guidance culminated in perhaps the most well-known revelation to Joseph Smith, "a word of wisdom . . . given for a principle with promise . . ." (v. 3).[5] The "Word of Wisdom" remains the primary LDS scriptural text pertaining to healthful living. In a cultural context of the temperance movement, early Mormon leaders such as Hyrum Smith re-storied this revelation within a narrative of God's providential caring: "[God] knows what course to pursue to restore man to his pristine excellency and primitive vigor and health. He has appointed the Word of Wisdom as one of the engines to bring about this thing . . . to restore his body to health and vigor."[6] As scholar Lester Bush Jr., has observed, the revelation was largely consonant with the received medical wisdom of its time period: "Mormonism's health code has never been more in agreement with the views of the medical establishment than it was at the very outset. The recommendations it contains were generally sound medicine at the time they were first set forth."[7]

Although the Word of Wisdom is often formulated in the phrasing used by Bush as a health "code," its four-part structure of invitation, restrictions, permissions, and promises is more illustrative of a covenantal principle. The initial verses invite community members to live according to "the will and order of God" (vv. 1–3). This is followed by a "warning" about "evils and designs which do and will exist in the hearts of conspiring men in the last days" (v. 4) and a set of prohibitions regarding the consumption of alcohol and strong drink, tobacco, and beverages designated as "hot drinks" (vv. 5–9), substances that represented the primary health risk factors of the day. The feature common to these substances was that consumption induced immoderate or excessive stimulation to bodily systems, phenomena considered the primary cause of disease in the 19th century.[8] The prohibitions are then followed by a cluster of permissions that identify foods "ordained for the constitution, nature, and use of man" (v. 10), including wholesome herbs, fruits and vegetables, and grains (vv. 10–17). Another health risk factor in the 19th century, meat consumption, is deemed permissible in seasonal or circumstantial settings, such as winter, cold, famine, and excess hunger (vv. 13, 16). The concluding section identifies several promises or blessings for persons who "remember to keep and do these sayings," including gifts of health

(v. 18), knowledge and wisdom (v. 19), protection in physical exertion that echo biblical promises (v. 20), and deliverance from plague and other health-threatening conditions (v. 21). I first devote attention to the restrictions and permissions that function analogously to a "code" for health and then provide support for understanding the Word of Wisdom as a *health covenant*.[9]

Prohibitions and Permissions. Scholarly research indicates that drinking alcohol, especially whiskey, was a way of life for many Americans in the 19th century, particularly in communities on the American frontier.[10] Social concern about excessive alcohol consumption led to the creation of the American Temperance Society in 1826 and affiliated local organizations that claimed nearly a million members by the mid-1830s. The culture of temperance, which evolved from advocacy of moderate usage to abstinence, resulted in a two-thirds decline in alcohol consumption during the 1840s.[11] This context is perhaps why the revealed wisdom about health is first directed at consuming "wine or strong drink," that is, whiskey and distilled spirits, which is described as "not good, neither meet in the sight of your Father" (vv. 5–7). The revelation did make allowance for "pure wine of the grape of the vine, of your own make," that is, not purchased from others who could not be trusted, as part of the sacramental ritual.

Tobacco, the second substance eliciting warning—"tobacco is . . . not good for man" (v. 8)—had historically been considered to possess medicinal properties, but this was challenged by 19th-century health reformers and some physicians.[12] The prevalent use of tobacco in pipes, cigars, or in chewed form had raised the practical question for the LDS community: When Joseph Smith first presented the Word of Wisdom to a group of twenty-two church elders, twenty were using tobacco.[13] The most common form of contemporary tobacco use, cigarettes, were first distributed during the precise years the Word of Wisdom was promulgated.[14] This evolution of tobacco use has led contemporary LDS commentators to situate tobacco companies and products within a narrative of prescient warning in the Word of Wisdom about emerging evils attributable to conspiring persons,[15] including addiction and its detrimental societal consequences, the role of tobacco in life-threatening cancers and heart ailments, and moral complicity profiting from the poor health practices of vulnerable persons.[16]

A third prohibited practice concerned the consumption of "hot drinks," which were assessed as "not for the body or belly" (v. 9). A much-debated issue in LDS culture concerns the interpretive meaning of "hot drinks." Bush contends that in historical context, the phrasing "hot drinks" referred to coffee and tea, beverages possessed with the stimulating properties against which the medical wisdom of the day warned. "The issue of temperature was itself believed important [by 19th-century thinkers], perhaps even more so than the use of tea or coffee in and of themselves. Hot drinks of any sort were believed potentially quite injurious to the stomach and thereby to health in general."[17] Current ecclesiastical teaching follows this articulation: "The only official interpretation of 'hot drinks' in the Word of Wisdom is the statement made by early Church leaders that the term 'hot drinks' means tea and coffee."[18] The historical rationale concerning the undue stimulating presence of caffeine in coffee and tea was subjected to critical scrutiny when caffeinated soft drinks were marketed in the early 20th century. Community discourse is ongoing about whether the principles behind the Word of Wisdom are violated by other beverages or foods containing substantial amounts of caffeine, such as high-energy drinks. The ethical principles of moral agency and communal self-governance place constraints on legalistic or literalistic interpretations and confer on each person responsibility for specific applications.

The revelation emphasized that herbs, fruits, and grains are ordained by God for "the constitution, nature, and use of man" with "prudence and thanksgiving" (vv. 10–11). Although meat was considered to possess more stimulating, and thus disease-facilitating, properties than plants, fruits, and vegetables,[19] divine authorization for food was extended to "flesh also of beasts and of the fowls of the air, I, the Lord have ordained for the use of man with thanksgiving; nevertheless, they are to be used sparingly; And it is pleasing unto me that they should not be used, only in times of winter, or of cold, or famine" (vv. 12–13). The kind of meat-oriented dietary lifestyle that became a cultural norm in 20th- and 21st-century American life is not sanctioned by this warrant for seasonal usage. The revelation presents a holistic and organic orientation to human nutrition by emphasizing foodstuffs harvested from the earth: "All grain is ordained for the use of man and of beasts, to be the staff of life; . . . All grain is good for the food of man; as also the fruit of the vine, that which yieldeth fruit, whether in the ground or above the ground" (vv. 14, 16).[20]

Community theology has constructed a multifaceted justificatory structure for the Word of Wisdom, a feature continued in present-day ecclesiastical teaching. The commendation of a plant-, herb-, and fruit-based diet, with limited use of meat, deemed adapted to human nature and the human "constitution" conveys that the condemned substances contravene the organic integrity and sanctity of the physical body. This justification is connected with teaching on the embodied sacredness of the person and the human body as a "temple." The consumption of prohibited substances symbolizes defilement and pollution of the self as embodied temple. Human spirituality is thereby in part conditioned by the physiological purity of the body. The embodied justificatory structure encompasses refraining from self-imposed physical harms, expressing reverence for the embodied self, and promoting good health.

The Word of Wisdom initiates and cultivates an ethos of disease prevention relevant for an age when orthodox medicine was an inefficacious, nonscientific practice but of continuing relevance in an era of scientific medicine and industrial production (including genetic modification) of food and potentially addictive pharmacological substances. The consequentialist emphasis on disease prevention and acquiring the blessings promised for adherence is perhaps the more prevalent argument in contemporary teaching. LDS discourse cites scientific studies on the longitudinal health benefits experienced by some church members through adherence, such as increased longevity and substantially diminished rates of death from cancer and cardiovascular disease, and other forms of contemporary scientific vindication.[21]

The community has engaged perennial questions about the bindingness and the scope of this health wisdom.[22] In its original formulation, the Word of Wisdom was promulgated as wise counsel "to be sent greeting; not by commandment or constraint" (v. 2), and there was no insistence on strict conformity. Some early ecclesiastical leaders bound themselves to live according to the teachings, but for first-generation Mormons, a view of substantial tolerance prevailed.[23] Ecclesiastical promotion of greater observance during the 19th century paralleled practices of toleration and support of self-sufficient wine missions and tobacco cultivation.[24] A culture of relative temperance prevailed through the time of Joseph Smith and his successor, Brigham Young—that is, the proscribed substances might be used so long as the person did not violate standards of public decency, such as public drunkenness. However, Young initiated a reform process in the 1860s that "changed the standard from moderate use of tobacco, alcohol, tea and coffee to full abstinence."[25] Commentaries by prophetic successors suggest that relative temperance was a generational necessity lest otherwise observant Mormons be called to a spiritual accounting.[26] It was not until 1933 that adherence to the prohibitions was instituted as an official criterion of worthiness for participation in LDS temple services and for ecclesiastical office. A recent study indicated that adherence is far from universal: "fewer than half of American

Mormons faithfully observe a literal prevailing interpretation of the Word of Wisdom by avoiding each of the substances most commonly understood to be prohibited."[27] The stringency of the permissions, including the consumption of meat, remains a matter of cultural practice and moral agency.

The prophetic "warning" regarding the emergence of modern conspiratorial evils has permitted ecclesiastical policies to extend the basic principle of refraining from harmful substances to include condemning use of illegal drugs while permitting appropriate use of prescription medications under the care of a qualified physician. This broadened scope is clearly significant in a context of deaths from opioid usage at epidemic proportions.[28] Furthermore, a 2019 ecclesiastical statement extended the principle of avoiding "substances that are destructive, habit-forming or addictive" to prohibitions of "vaping or e-cigarettes, green tea, and coffee-based products."[29] There are, however, communal questions about legitimate extensions of the principle of nonharm. Some contemporary Mormons argue that refraining from pollutants of the body supports reservations about the safety of pharmaceuticals, including vaccinations. At the same time, recent legalization of medicinal marijuana has suggested to some commentators that perspective and practice of medicinal marijuana depends on the legal jurisdiction in which LDS members reside.[30]

These communal interpretive disputes reflect an understanding of the Word of Wisdom more as a code of instructions than as a covenant for health and physical well-being. As argued in the introductory chapter, a covenant comprises responsiveness to giftedness as ratified by a promise. A code ethos by contrast tends to reduce moral responsibilities to legalistic minimalisms or the specific performance of actions stipulated by explicit wording. This legalistic application is reflected in communal concerns to define the *minimal* boundaries for observance and, paradoxically, a *maximalist* approach that everything health-related be aligned with Word of Wisdom precepts. A code construction focuses only on what is required with respect to visible behaviors, not on what is just and ethical. Consequently, a health code interpretation diminishes the moral value of the health teachings to an information transaction and interprets a gift of love and wisdom as an authoritative contract. A covenant by contrast focuses on the internal and evolving nature of relationships between covenantal partners and so "cut[s] deeper into personal identity."[31] A covenant manifests endurance over time that allows for the covenantal relationship and its meaning to evolve in unspecifiable ways. A covenantal context for health wisdom necessarily moves beyond transactional exchanges of information to transforming personal identities.

The Word of Wisdom teachings are not a comprehensive guide to good health. As Bush observes, in the mid-19th century, the quest for good health, longevity, and reductions of infant and maternal mortality would have advanced more significantly by communal promotion of sanitation, safe water, personal hygiene, and better nutrition, measures that did not take hold publicly until the 20th century.[32] Despite ongoing debates, it is clear that the health covenant affirms physical health as an important part of religious commitment. The theological understanding of the human soul as an embodied person imbued with sanctity and purity means it is not possible to uncouple spiritual well-being from physical health. The principles underlying the prohibitions, permissions, and promises of the health ethic elevate a communal ethos of health promotion and disease prevention in the moral culture recently exemplified in discourse on vaccination ethics.

Vaccine Resistance

The health covenant and medical immunizations share a common objective of preventing disease, but the preventive methods are not only different but, for some persons, morally different. The Word of Wisdom emerged in an era of nonscientific medicine, and its restrictions require persons to refrain from ingesting into the body substances that can be physically toxic, addictive, and even lethal. Vaccinations are a central accomplishment of scientific-based modern medicine and public health and confer immunity through direct injections into the body of chemically altered benign strains of a disease or virus such that antibodies are produced in the event of disease exposure. Vaccines appropriate medical and pharmacological expertise even while infringing on the organic integrity of the body. However, the inherent risks of immunization raise for some faith communities issues of trust in the medical profession, the integrity of pharmaceutical manufacturers, biomedical research on vaccine safety, and government mandates.[33]

Religious resistance to vaccines is embedded within historical narratives that have shaped both influential legal cases permitting mandatory immunizations and current statutes permitting parents to claim exemptions from mandatory vaccines for their children on religious grounds. In the early 20th century, a virulent smallpox epidemic occurred in the northeastern United States, leading the city of Cambridge, Massachusetts, to pass an ordinance requiring vaccination of all adults against smallpox. Lutheran pastor Henning Jacobson objected to the requirement, claiming he had experienced "great and extreme suffering" from a smallpox vaccination in his homeland of Sweden and that the mandate violated his liberty. The U.S. Supreme Court affirmed in a 7-2 ruling on Jacobson's objection a precedent-setting public health principle that "the state may be justified in restricting individual liberty . . . under the pressure of great dangers" to public safety from contagious disease. Since smallpox was an ongoing epidemic, the Cambridge ordinance satisfied the high standard of preventing "great danger" to the public.[34] When a mandatory vaccination program has a reasonable chance of preventing or alleviating an epidemic, the public interest in eradicating contagion has primacy over the ethical principle of personal liberty and freedom from bodily invasive medical interventions.[35]

A situation parallel to the Massachusetts epidemic was transpiring in Utah in the same time period. During 1899 and 1900, Utah reported four thousand cases of smallpox and twenty-six deaths. The Utah Board of Health submitted a mandate that children lacking a smallpox vaccination could not attend public school. This requirement prompted resistance from some LDS leaders predisposed to distrust government due to the legacy of governmental interference with LDS religious liberty. Apostle Brigham Young Jr. asserted that smallpox vaccinations signified pollution and defilement of the body: "Gentile doctors [are] trying to force Babylon into the people and some of them are willing to disease the blood of our children if they can do so, and they think they are doing God's service."[36] Charles M. Penrose, editor of the church-owned *Deseret News*, asserted the same principle contested in *Jacobson*, calling mandatory vaccinations an "encroachment" on personal liberty. The policy and public health outcomes of the Utah controversy were tragically dissimilar to the legal resolution of Pastor Jacobson's objection. Despite ecclesiastical recommendations supporting mandatory smallpox vaccination, the Utah legislature passed a law in 1901 eliminating compulsory vaccination requirements. Children were thus exposed to smallpox for a substantial period of time; during the period between 1900 and 1925, some 15,000 persons in Utah died from smallpox.[37]

These community conflicts over vaccination occurred in the era of transition from faith-healing rituals to the embrace of scientific-oriented medicine.[38] The narrative of suspicion and distrust of medical interventions

in bodily wholeness, of immunization as an insidious form of paternalistic governmental control, or of vaccines as a money-making enterprise
for pharmaceutical companies has not entirely ceased within LDS culture.
A small antivaccine subculture interprets vaccines through the "warning"
instructions of the Word of Wisdom about conspiratorial evils in the contemporary world. Utah has followed national trends of increasing numbers of parents claiming exemptions—religious and personal belief—from
vaccinations for their young children.[39]

In a compelling personal narrative, a prospective missionary crystallized a
"vivid" internal debate as well as fears, both realized and actual, that underlie
vaccine hesitancy or vaccine refusal generally. The prospective missionary
did not receive any immunizations as a child stemming initially from a severe adverse reaction to a vaccine shot experienced by her older brother. The
missionary's mother subsequently declined vaccinations for her daughter
attributable to the absence of evident risk of contracting a disease without
an immunization and her review of studies indicating that vaccines could
contribute to weakened immune systems. Distrust of medicine also was a
consideration as the mother wondered how certain physicians could be that
immunizations would not cause cancer. A final influence was embodied organic integrity: "the less you infuse into your body, the more healthy you'll
be." Ultimately, the missionary's disposition to service overrode familial
and personal fears and she received the immunizations necessary for a missionary call outside the United States.[40]

This narrative of vaccine hesitancy reflects experience, some empirical
evidence, fears, and views about bodily integrity that could be affirmed by
any person requesting a nonmedical religious or personal exemption from
mandated vaccines. A national study disclosed that the most prevalent rationale behind parental refusals of vaccines for children is that vaccines are
not necessary.[41] This perception is a function of the *success* of immunization
programs over the past half-century, which have made contagious childhood
diseases a receding memory. This generational shift is displayed in an un-
scientific poll conducted through an online newsletter about LDS culture.[42]
Ninety-six percent of the approximately three thousand respondents indicated they received childhood vaccinations, but only 67% supported their
children receiving vaccinations. This vaccine reticence highlighted concerns
about ingredients (20%), allergic reactions or general safety concerns (19%),
and connections between vaccines and autism or mental health (7%).[43]
A smaller ratio of responses (4%) expressed an aversion to consuming

pharmaceuticals that implicitly appeals to the organic integrity and healing capacity of the body.

The Prevention Ethic: Ecclesiastical Teaching

It is striking that the subject of immunization does not appear in current ecclesiastical policies on medical, health, or moral issues. Vaccines were mentioned in the 1974 compilation of LDS positions on medical ethics in a category that included narcotics and blood: "The Church regards the use of these substances, as prescribed under medical supervision for the treatment or prevention of disease, as wholly a medical question."[44] The designation of vaccines as a "wholly medical" issue rather than an ecclesiastical or ethical consideration indicates immunization is a matter between physicians and the moral agency of parents and patients and may have reflected a sentiment that vaccination was a settled question in LDS culture. The posture of ecclesiastical deference to medicine soon shifted. The First Presidency issued three statements over the next decade advocating immunization and communal participation in vaccine education. A 1976 statement encouraged Mormons to "carefully consider the benefits and risks of [the swine flu] vaccination to the health of themselves and their families." Ecclesiastical recommendations to receive "competent medical advice" for personal health was supplemented with encouragement of persons who are "technically qualified and . . . feel so inclined" to participate in community immunization initiatives.[45]

The most extensive ecclesiastical endorsement of childhood vaccination was issued in a 1978 statement that combined appeals to empirical information, personal experience and empathy, and ethical imperatives to prevent disease and premature death.[46] It affirmed a twofold call to church members to (1) "protect [your] own children through immunization," and (2) provide community service to "eradicate ignorance and apathy" contributing to decreasing rates of childhood immunization. The statement cited health reports that some 40% of children fourteen years and younger in the United States had not been adequately immunized against preventable diseases such as polio, measles, rubella, diphtheria, pertussis, mumps, and tetanus. This culture of inadequate immunization thereby imposed involuntary risks of "preventable lifelong physical or mental impairment" on thousands of children. The statement appealed to parenting experience and the moral virtue of empathy, citing the anguish of parents whose children had died prematurely from diseases for which vaccinations were available. The painful lessons from such deaths converged with sentiments of ecclesiastical leadership in a call for popular mobilization against "these deadly enemies." Ethical values of preventing disease, suffering, and premature death were presupposed in the ecclesiastical claim that immunization is "a small price to pay for protection against these destroying diseases."

In 1985, a "reminder" about immunization appeared in the official church magazine[47] that is notable for two features. It identified reasons for parental laxity in having their children immunized: contagious diseases are "uncommon," making vaccines unnecessary, and vaccines evoke "fear" about adverse side effects. These sentiments resonate directly with the rationales of vaccine-hesitant parents in the 21st century. The statement also intimates an evolution in thinking about vaccination. The 1974 statement depicting vaccination as entirely a medical question deserving careful consideration implied that vaccination is morally *discretionary*. The 1985 "reminder" portrays vaccination as morally *required*: "parents have an obligation to protect their families through immunization."

Despite current public controversy over vaccinations and specific outbreaks of measles,[48] in the three decades since these ecclesiastical statements were issued national rates of immunization have increased such that the Centers for Disease Control and Prevention goals for immunization by 2020 are now within reach.[49] The Healthy People 2020 public health initiative set out national goals of 90% vaccination coverage for children between nineteen and thirty-five months for the DPT vaccine for diphtheria, pertussis, and tetanus; the MMR vaccine for measles, mumps, and rubella; the varicella vaccine against chicken pox: and the polio vaccine: as well as for three recent vaccines: Hepatitis B, Pneumococcal conjugate vaccine (PCV), and Haemophilius influenzae type B (HiB). These national rates of vaccination should not be mistaken for sufficient herd immunity of specific communities. Some communities in which vaccination rates of kindergarten children have regressed to approximately 40% are at increased risks of disease outbreaks or serious threats of preventable lethal illness for children.[50]

As anticipated in the 1978 ecclesiastical statement, immunization has a pressing global dimension. In 2012, LDS Humanitarian Services announced that immunizations would be included with other sponsored humanitarian initiatives such as clean water, food production, and vision care. In the decade between 2003 and the 2012 announcement, the LDS Church contributed $16 million to a global initiative for vaccines for measles and rubella and donated $1.5 million in 2012 to increase access to vaccines for pneumonia and diarrhea in five African countries. Ecclesiastical encouragement of engaged community service resulted in approximately sixty thousand volunteers participating in immunization initiatives in thirty-five countries.[51] The LDS Church displays an increasingly practical commitment to vaccinations to prevent disease and premature death internationally,[52] even though antivaccination subcultures persist where infectious childhood disease has largely been eradicated.

An outlier in the vaccination compliance narrative of particular relevance for the LDS community is the Gardasil vaccination for immunity from HPV (human papillamovirus) during early adolescence. HPV is the most common sexually transmitted disease in the United States, with 14 million new cases diagnosed annually. More than forty-three thousand persons were diagnosed with an HPV-associated cancer in 2015. HPV is also a primary cause of cervical cancer in women: some twelve thousand women receive a cervical cancer diagnosis annually, and HPV is responsible for four thousand deaths a year. The Healthy People 2020 program set a goal of 80% coverage for adolescent females and males, but as of 2017, only 53% of adolescent females and 44% of adolescent males were "up to date" with the recommended HPV vaccination schedule.[53] Given the prevalence of HPV in the United States, increasing immunization rates through the HPV vaccine arguably has greater immediacy than vaccinations for diseases that have largely been eradicated. Once the HPV vaccine received approval for both adolescent girls and boys in 2011, physicians and health care leaders argued that "given the fact that we have a safe and effective vaccine, there's little reason why parents and providers aren't vaccinating every single child."[54] Yet, as illustrated in the initial narrative, the HPV vaccine can present an ethical challenge to covenantal commitments of LDS families and practitioners.

Medical and Moral Immunization

The initial LDS reception to the HPV vaccine was decidedly mixed. Shortly after the Food and Drug Administration approved Gardasil as a vaccine for HPV in 2008, the *Deseret News* portrayed a California pediatrician, a former LDS congregational leader, as "adamant about young girls receiving the series . . . of vaccinations that prevent cervical cancer." In the course of a clinical visit with an LDS teen, the pediatrician commented, "I don't care if a girl is

entering a convent, each girl should have this vaccination. This is a vaccination that actually saves lives and prevents cancer." Other LDS physicians
compared the HPV vaccine to immunizations for polio or measles and drew
out policy analogies to argue that HPV vaccination should become a routine
part of preteen health care for all age-appropriate girls (at the time, the HPV
vaccine was approved for use only in females).[55] Other LDS forums were
noticeably less receptive. A story in the Brigham Young University campus
newspaper encouraging HPV vaccination related the comments of three female students who believed the vaccine recommendation was a good idea
but personally inapplicable because of their commitment to church teachings
on chastity.[56] Also, a bill before the Utah legislature that requested $1 mil-
lion in funds for vaccination received only $25,000, and those funds were for
educational programs to inform women that abstinence before and fidelity
after marriage are "the surest prevention of sexually transmitted diseases including the human papillomavirus." [57]

An HPV vaccine mandate for all adolescent children can present an ethical conflict for LDS parents in light of ecclesiastical teachings on sexual morality. The clearest moral standard affirmed continuously in contemporary
teaching is the "law of chastity," which signifies a covenantal commitment
of sexual abstinence prior to marriage and complete sexual fidelity within
marriage (defined as a relationship between a man and a woman). As a moral
"law," the standard of chastity is an ecclesiastical nonnegotiable value, a central covenant for participation in sacrament and temple rituals, missionary
service, and callings of service and ministry. An ecclesiastical and cultural
hesitancy persists regarding social or medical practices that have as their primary intention to prevent sexually transmitted diseases but may secondarily
imply the possibility of STD-free sexual relations prior to the covenantal
commitment of marriage.

LDS students, parents, and practitioners can find themselves in compromising ground when considering whether HPV is a "cancer vaccine" as
claimed by physicians or a "promiscuity vaccine" as argued by some evangelical religions.[58] While HPV vaccination rates among adolescents have
been steadily increasing,[59] vaccination rates in Utah lag behind national
trends: Vaccination coverage in 2017 among all Utah teens ages thirteen to
fifteen was 37.4%, placing Utah forty-seventh among U.S. states.[60] These
lower vaccination rates may partly be attributable to attitudes in the moral
culture about premarital sexual activity. While Mormon students may believe the vaccine is not applicable to their sexual activity, Mormon parents

may believe moral immunization is sufficient and ask of pediatricians, "Why [would] I want to vaccinate my child against a sexually transmitted disease?"[61]

The vaccine prevalence rate among young women ages eighteen to twenty-six in Utah has been the subject of one scientific study that explored the religious influence on attitudes toward HPV in three aspects: (1) *awareness*: participants' awareness of both HPV and an HPV vaccine; (2) *knowledge*: participants' level of understanding about how HPV is transmitted and the relationship of HPV and cervical cancer; and (3) *receipt*: a physician recommendation for an HPV vaccine and receipt of at least one dose of HPV vaccine.[62] The researchers noted that prior studies on religious belief and HPV vaccination had displayed a bifurcated attitude toward the vaccine. Various religious communities have argued that "religious norms regulating sexual activity of unmarried women render the HPV vaccine unnecessary." However, these norms hold a tenuous status among adherents such that "various religious groups have voiced concern that the HPV vaccine will promote sexual disinhibition among teens."[63] The researchers selected Utah to provide more understanding about religious concerns on both the necessity of the HPV vaccine since "nearly 80% of [Utah] residents practice an organized religion, with a high density of individuals who identify as members of The Church of Jesus Christ of Latter Day Saints."[64] The researchers speculated that "the religious climate in Utah may underlie low vaccination rates, and could also affect HPV vaccine-related awareness and knowledge."[65]

The study displayed a high level of HPV literacy and vaccination rates among all 326 participants. However, in the two principal cognitive areas of HPV literacy, research participants who practiced an organized religion had somewhat lower levels of awareness, knowledge, and receipt: Awareness of HPV (89.9% awareness) and of the HPV vaccine (86.4%), knowledge of HPV transmission (72.3% knowledge) and of the relationship between HPV and cervical cancer (74.8%), and receipt of a physician recommendation for the vaccine (85.1%) ranged between 7 and 12% lower than the awareness and knowledge levels of those without an organized religious practice. The study showed a more significant disparity on enacting the recommendation and whether the participants had initiated the HPV series of vaccine doses. Almost 49% of the religiously affiliated participants had received one vaccine dose in contrast to 72.5% for persons who did not have an organized religion practice.[66]

These findings led the researchers to conclude that young women with an organized religious practice "were significantly less likely" to possess HPV literacy and less likely to follow a recommendation for a vaccine with a subsequent vaccination: "the religious participants in this study were less likely to exhibit a basic awareness of the virus, the vaccine, or the mechanism of transmission, revealing that their decisions not to vaccinate may not necessarily have been informed. [These] findings may expose a failure of providers to inform their patients, as well as a failure of patients to educate themselves."[67] Apart from the issue of sexual activity, a clear ethical issue is raised if there are reasons to suspect a decision has not met informed consent criteria. As a remedy, the researchers advocated "educating religious young women in Utah about HPV and the HPV vaccine," a proposed collaboration between "religious young women, their health care providers, [and] church leaders" as well as policy makers and public health officials.[68]

A rationalistic assumption is embedded in proposals that education will necessarily increase vaccination rates, an assumption that is not supported by the reasons parents give for refusing vaccines for their children. However, the most striking recommendation is the call for involvement in HPV literacy programs by "church leaders." This extends the conflicts experienced by parents and by LDS health care professionals to the ecclesiastical level. It also raises the issue as to whether vaccinations are entirely a "medical question," per the 1974 compilation on LDS medical ethics policies, or instead broach core religious and moral values. The medicalized HPV narrative portrays vaccination as addressing a disease reality that afflicts a good portion of the sexually active population. HPV prevalence could be prevented by incorporating the HPV vaccine into the routine schedule of vaccinations for adolescent children. Moreover, as a "cancer vaccine," the HPV vaccine is deemed an essential medical intervention for cancer prevention. The narrative of religious moralism by contrast differentiates HPV from, for example, the MMR vaccine on the basis of moral immunization. Parents can protect their children and individuals can protect themselves through sexual literacy, disease screening (pap smears for cervical cancer screening), and abstinence. The moralistic narrative contends that though HPV vaccination is a *practical* necessity for many persons, it is not *medically* necessary for HPV prevention.

Despite the potential for conflicts of familial, religious, and professional values, it is important to not overlook substantial common ground. The medicalized HPV narrative appeals to precisely the same values underlying the general LDS posture regarding childhood immunizations: a shared commitment to *preventing disease* and avoiding premature death. Furthermore, whether the question concerns medical interventions or participating in sexual activity, the bioethical principle of respect for personal self-determination and the LDS commitment to moral agency each emphasize *informed choices*. There is also common ground regarding the integrity of the *medical profession*. The intention of the medical profession to administer the HPV vaccine as a preventive measure against a common sexually transmitted disease and a lethal cancer must be honored. The HPV vaccine may not be medically necessary given alternative behaviors to avoid sexually transmitted infections (STIs), but the practical necessity for the vaccine is of paramount consideration due to the realities of sexual relationships prior to marriage.

A fourth shared interest is that compelling medical reasons exist for the HPV vaccine *independent of religious norms* regarding sexual activity. Despite religious critiques of public health promotion of safer sex practices, HPV literacy and immunization do not entail premarital or nonmarital sexual activity. A Canadian study reported "strong evidence that HPV vaccination does not have any significant effect on clinical indicators of sexual behavior among adolescent girls."[69] There are many influences on adolescent and teen sexual activity, including biology and hormones, socialization, curiosity, and a sex-saturated media environment. It is problematic to claim that a vaccine intended to prevent disease is a tipping point for initiating sexual activity. As reflected in the educational approach of Kevin and Caitlyn, it is entirely possible for a young person to receive the vaccine and remain sexually abstinent. Resistance to the HPV vaccine based on concerns about promoting sexual permissiveness likely masks deeper reservations.

Personal and familial experience and the principles of moral agency, family integrity, disease prevention, and the covenant of chastity are weighed differently by LDS families and lead to communal moral diversity regarding the HPV vaccine. It is possible to discern seven different patterns in the LDS community:

- *Life Conversation:* As illustrated by the initial narrative, some families discuss the HPV vaccine in conversations with children about life values and goals. The health aspects of the vaccine are secondary to parental discussion of correct moral principles and behaviors that constitute a way of life authentic to their child's values. This conversational approach is consistent with allowing the preadolescent child as a maturing moral person to ultimately make the decision about the vaccine.

- *Medical Health.* Other families take a medically oriented approach and discuss with their children the importance of protecting themselves from HPV and from potentially lethal cancers by receiving the HPV vaccine. The shared commitment to good health and the precedent of prior vaccinations provides context for this medicalized approach and also is consistent with leaving the decision to receive the vaccine to the child.

- *Continuum.* A third pattern displayed in other families is to consider the HPV vaccine part of the immunization continuum from infancy, the medical equivalent for adolescents of the MMR vaccine for infants. Parents may not find it necessary to have a fuller discussion of HPV or of the rationale and values behind the vaccine. In this continuum approach, parents may make the decision for the child, as with vaccines of childhood, or in collaboration with their child.

- *Reversibility.* Some families find that the relationship with the pediatrician can influence parental and preadolescent choices, especially when the relationship is mediated by common values about health, disease prevention, and teen sexuality. In trusting relationships, families have asked the pediatrician, "What would you do if you were in my position?" The parental appeal to professional empathy by requesting the professional to imaginatively reverse places with the parent has invariably led to a recommendation to have the vaccine administered, and acceptance of the recommendation by parents and the child.

- *Paternalism.* By contrast, some LDS families have found their pediatrician to be "really pushy" and "placing pressure" that their child receive the vaccine. These conversations are perceived by parents to flow not from shared values and trust but from professional paternalism. In these relationships, families have made provisional decisions to encourage their children to become HPV literate and come to their own decision about whether to receive the vaccine.

- *Education.* Some parents acknowledge being influenced by stories of adverse reactions to the HPV vaccine and have decided against having the vaccine for their child. Since this resistance is an empirical judgment, not a matter of ethical or religious principle, sensitive education from empathic pediatricians and parental familiarity with the nature and frequency of risks could culminate in a different decision.

- *Moral Immunization.* Other families have determined as a matter of principle to decline the vaccination and rely on the moral immunization provided by family teaching about sexuality and the covenant of chastity.

This diversity of family patterns is not surprising given the ecclesiastical insistence on the principles of moral agency and professional integrity. Resistance to the HPV vaccine reflects a common initial ambivalence to innovative medical interventions, which is heightened on a subject so charged as teen sexual activity. However, ongoing education and the routinization of the HPV vaccine as part of preadolescent health care are likely to diminish religious or moral hesitations and make the primary issues, as with other vaccinations, matters of safety and efficacy. Familial education about the HPV vaccine should not construct a "false" choice: the vaccine is not incompatible with a religious-based commitment to sexual abstinence, and empirical evidence indicates that receiving the vaccine does not have a threshold influence on decisions about sexual activity by teens.

My typology of familial patterns leaves open the role of church leaders in education about the HPV vaccine as recommended by researchers. Ecclesiastical leaders will likely continue to affirm core principles about family, sexuality, and moral agency, and highlight immunization benefits in general, leaving conversations specific to the HPV vaccine to be matters of agency, self-governance, and professionalism. The moral wisdom of preventing disease through observing the Word of Wisdom or through vaccinations reflects fundamental covenantal commitments within LDS culture. Through covenantal relationships, the family emerges as the primary context for shaping moral life, an emphasis reinforced by ethical decisions regarding the genetic legacy of family history.

Notes

1. Melanie Drolet, Elodie Benard, Norma Perez, Marc Brisson, "HPV Vaccination Programmes: Updated Systematic Review and Data Analysis," *The Lancet*, June 26, 2019, https://doi.org/10.1016/S0140-6736(19)30298-3
2. Bruce Y. Lee, "Should Kids Be Required to Get the HPV Vaccine?," *Forbes*, February 13, 2018, https://www.forbes.com/sites/brucelee/2018/02/13/hpv-vaccine-should-kids-be-required-to-get-the-vaccine/#26b8f914a6df

3. Steven I. Hadju, Manjunath S. Vadmal, "A Note from History: The Use of Tobacco," *Annals of Clinical & Laboratory Science* 40 (2010), 178–181; Paul H. Peterson, Ronald W. Walker, "Brigham Young's Word of Wisdom Legacy," *BYU Studies Quarterly* 42 (2003), 29–64.

4. Matthew McBride, James Goldberg, eds., *Revelations in Context: The Stories Behind the Sections of the Doctrine and Covenants* (Salt Lake City: Intellectual Reserve, 2016), 183–191; Matthew J. Grow, Richard E. Turley, Jr., Steven C. Harper, Scott A. Hales, *Saints: The Story of the Church of Jesus Christ of Latter-day Saints,* vol. 1, *The Standard of Truth, 1815–1846* (Salt Lake City: Intellectual Reserve, 2018), 167–168.

5. This and the following quotations followed by verses come from *Doctrine and Covenants* 89.

6. Elaine S. Marshall, "The Power of God to Heal: The Shared Gifts of Joseph and Hyrum," in *Joseph and Hyrum—Leading as One,* ed. Mark E. Mendenhall et al. (Salt Lake City: Deseret Book, 2010), 176.

7. Lester E. Bush, Jr., "The Word of Wisdom in Early Nineteenth-Century Perspective," *Dialogue* 14 (1981), 58.

8. Bush, "The Word of Wisdom in Early Nineteenth-Century Perspective," 49.

9. For discussion of the limits of ethics "codes," see William F. May, *The Physician's Covenant: Images of the Healer in Medical Ethics* (Louisville, KY: Westminster John Knox Press, 2001), 112–154; and Carol Mason Spicer, "Nature and Role of Codes and Other Ethics Directives," *Encyclopedia of Bioethics,* 3rd ed., vol. 5, ed. Stephen G. Post (New York: Macmillan Reference USA, 2004), 2621–2629.

10. Alcohol consumption in this era has been estimated at seven gallons annually for persons above the age of fifteen. William J. Rorabaugh, "Alcohol in America," *OAH Magazine of History* 6:2 (1991), 17–19; William J. Rorabaugh, *The Alcoholic Republic: An American Tradition* (New York: Oxford University Press, 1979).

11. Lester E. Bush, Jr., *Health and Medicine Among the Latter-day Saints: Science, Sense, and Scripture* (New York: Crossroad, 1993), 49.

12. Hadju, Hadmal, "A Note from History"; Eric Burns, *The Smoke of the Gods: A Social History of Tobacco* (Philadelphia: Temple University Press), 2007.

13. Peterson, Walker, "Brigham Young's Word of Wisdom Legacy," 31.

14. Steven C. Harper, *Setting the Record Straight: The Word of Wisdom* (Orem, UT: Millennium Press, 2007), 29–31.

15. Harper, *Setting the Record Straight,* 39–43.

16. Scott A. Johnson, *The Word of Wisdom: Discovering the LDS Code of Health* (Springville, UT: Cedar Fort Press, 2013), 33, 49.

17. Bush, *Health and Medicine Among the Latter-day Saints,* 50.

18. The Church of Jesus Christ of Latter-Day Saints, "Church Policies and Guidelines, Medical and Health Policies: Word of Wisdom," in *General Handbook: Serving in the Church of Jesus Christ of Latter-day Saints,* Chapter 38.7.12, February 19, 2020, https://www.churchofjesuschrist.org/study/manual/general-handbook/38-church-policies-and-guidelines?lang=eng#title_number91

19. Bush, "The Word of Wisdom in Early Nineteenth-Century Perspective," 53–54.

20. Rachel Hunt Steenblik, "A Mormon Ethic of Food," *Dialogue: A Journal of Mormon Thought* 48 (2015), 59–74.

21. William T. Stephenson, "Cancer, Nutrition, and the Word of Wisdom," *The Ensign*, July 2008, https://www.lds.org/ensign/2008/07/cancer-nutrition-and-the-word-of-wisdom-one-doctors-observations?lang=eng; Joseph Lynn Lyon, "Word of Wisdom," in *Encyclopedia of Mormonism*, ed. Daniel H. Ludlow (New York: Macmillan, 1992), 991–992, https://contentdm.lib.byu.edu/digital/collection/EoM/id/4353; Heather May, "What Science Says About Mormonism's Health Code," *Salt Lake Tribune*, October 6, 2012, http://archive.sltrib.com/article.php?id=54897327&itype=CMSID. For scholarly examination, see James E. Enstrom, Lester Brenslow, "Lifestyle and Reduced Mortality Among Active California Mormons, 1980–2004," *Preventive Medicine* 46 (2008), 133–136; Ray A. Merrill, "Tobacco Smoking and Cancer in Utah," *Dialogue: A Journal of Mormon Thought* 35 (2002), 73–77; Ray A. Merrill, Hala N. Mandanat, Joseph L. Lyon, "Active Religion and Health in Utah," *Dialogue: A Journal of Mormon Thought* 35 (2002), 78–90; James E. Enstrom, "Health Practices and Cancer Mortality Among Active California Mormons," *Journal of the National Cancer Institute* 81 (1989), 1809–1810; Mormon Newsroom, "Health Practices," https://newsroom.churchofjesuschrist.org/article/health-practices.

22. John E. Ferguson, III, Benjamin R. Knoll, Jana Riess, "The Word of Wisdom in Contemporary American Mormonism: Perceptions and Practice," *Dialogue* 51 (2018), 39–77.

23. Harper, *Setting the Record Straight*, 45–64.

24. Bush, *Health and Medicine Among the Latter-day Saints*, 52–60; Peterson, Walker, "Brigham Young's Word of Wisdom Legacy," 38–39, 44.

25. Peterson, Walker, "Brigham Young's Word of Wisdom Legacy," 29, 42–43.

26. Joseph F. Smith, LDS Church president from 1901 to 1918, commented that a stringent interpretation of the Word of Wisdom in the 19th century "would have brought every man addicted to these noxious things under condemnation; so the Lord was merciful and gave them a chance to overcome, before he brought them under law" (Joseph F. Smith, *Conference Report* 1913, as quoted in Harper, *Setting the Record Straight*, 64).

27. Ferguson III, Knoll, Riess, "The Word of Wisdom in Contemporary American Mormonism," 49.

28. National Institute on Drug Abuse, "Overdose Death Rates," August 2018, https://www.drugabuse.gov/related-topics/trends-statistics/overdose-death-rates

29. Newsroom, "Statement on the Word of Wisdom," August 15, 2019, https://newsroom.churchofjesuschrist.org/article/statement-word-of-wisdom-august-2019; "Vaping, Coffee, Tea, and Marijuana," *New Era*, August 2019, https://www.churchofjesuschrist.org/study/new-era/2019/08/vaping-coffee-tea-and-marijuana?lang=eng

30. J. Kael Weston, "The Mormon Church vs. Pot," *New York Times*, November 6, 2018, https://www.nytimes.com/2018/11/06/opinion/mormon-church-medical-marijuana-utah-referendum.html. See Chapter 8 for further discussion of LDS policy on medical marijuana.

31. May, *The Physician's Covenant*, 127.

32. Bush, "The Word of Wisdom in Early Nineteenth-Century Perspective," 59.
33. These religious concerns are illustrated generally by the resistance of the Jewish Hasidic community to the measles vaccine. Tyler Pager, "'Monkey, Rat and Pig DNA': How Misinformation Is Driving the Measles Outbreak Among Ultra-Orthodox Jews," *New York Times,* April 10, 2019, https://www.nytimes.com/2019/04/09/nyregion/jews-measles-vaccination.html
34. *Jacobson v. Massachusetts,* 197 U.S. 11 (1905).
35. "Toward a Twenty-First Century *Jacobson v. Massachusetts*," *Harvard Law Review* 121 (2008), 1820–1842.
36. Geoffrey S. Nelson, "Mormons and Compulsory Vaccination," *MormonPress*, March 30, 2015, http://www.mormonpress.com/mormon_vaccination.
37. Ardis E. Parshall, "Even with Proven Smallpox Vaccines, 19th-Century Utahns Balked," *Salt Lake Tribune,* December 2, 2011, http://archive.sltrib.com/article.php?id=53038477&itype=CMSID
38. Robert T. Divett, *Medicine and the Mormons,* 2nd ed. (Charleston, SC: CreateSpace, 2010), 156–158; Bush, *Health and Medicine Among the Latter-day Saints,* 63.
39. Jacqueline K. Olive, Peter J. Hotez, Ashish Damania, Melissa S. Nolan, "The State of the Anti-Vaccine Movement in the United States: A Focused Examination of Nonmedical Exemptions in States and Counties," *PLOS Medicine*, June 12, 2018, https://doi.org/10.1371/journal.pmed.1002578; Daphne Chen, "With More Parents Choosing Not to Vaccinate, Utah on Brink of Losing 'Herd Immunity,'" February 9, 2016, KSL.com, https://www.ksl.com/?sid=38451248&nid=148&fm=home_page&s_cid=toppick2
40. Katie Barlow, "My Vaccination Decision," *LDS Living*, http://www.ldsliving.com/-LDSL-Blog-My-Vaccination-Decision/s/65450
41. Catherine Hough-Telford, "Vaccine Delays, Refusals, and Patient Dismisses: A Survey of Pediatricians," *Pediatrics* 138 (September 2016), http://pediatrics.aappublications.org/content/pediatrics/138/3/e20162127.full.pdf; Chephra McKee, Kristen Bohannon, "Exploring the Reasons Behind Parental Refusals of Vaccines," *Journal of Pediatric Pharmacology and Therapeutics* 21 (March–April 2016), 104–109, https://www.ncbi.nlm.nih.gov/pmc/articles/PMC4869767/
42. "Vaccines: Poll," *LDS Living,* http://www.ldsliving.com/-Poll-Vaccines/s/65438
43. The purported link between the MMR vaccine and autism has been repudiated and retracted due to scientific falsification of data in the initial study but the perception still lingers in public consciousness. Jonathan D. Quick, Heidi Larson, "The Vaccine-Autism Myth Started 20 Years Ago. Here's Why It Still Endures Today," *Time*, February 28, 2018, http://time.com/5175704/andrew-wakefield-vaccine-autism/
44. "Attitudes of the Church of Jesus Christ of Latter-day Saints Towards Certain Medical Problems," in "Mormon Medical Ethical Guidelines," ed. Lester E. Bush, Jr., *Dialogue* 12 (Fall 1979), 100.
45. "Attitudes of the Church of Jesus Christ of Latter-day Saints Towards Certain Medical Problems," 104.
46. "Immunize Children, Leaders Urge," *Liahona*, July 1978, https://www.lds.org/liahona/1978/07/immunize-children-leaders-urge?lang=eng. See Appendix A, "Vaccinations."

47. "Immunization—A Reminder," *The Ensign,* July 1985, https://www.lds.org/ensign/1985/07/random-sampler/immunizations-a-reminder?lang=eng

48. PBS, "Frontline: The Vaccine War," April 27, 2010, https://www.pbs.org/wgbh/frontline/film/vaccines/; Centers for Disease Control and Prevention, "Measles Cases and Outbreaks: 2019," https://www.cdc.gov/measles/cases-outbreaks.html

49. Centers for Disease Control and Protection, "Healthy People 2020 Mid-Course Review: Immunization and Infectious Diseases," January 20, 2017, https://www.cdc.gov/nchs/data/hpdata2020/HP2020MCR-C23-IID.pdf

50. Jonathan Lambert, "Measles Cases Mount in Pacific Northwest Outbreak," *NPR: Public Health,* February 8, 2019, https://www.npr.org/sections/health-shots/2019/02/08/692665531/measles-cases-mount-in-pacific-northwest-outbreak

51. "Church Makes Immunizations an Official Initiative, Provides Social Mobilization," The Church of Jesus Christ of Latter-day Saints, June 13, 2012, https://www.lds.org/church/news/church-makes-immunizations-an-official-initiative-provides-social-mobilization?lang=eng.

52. Newsroom, "Spreading the Word of Immunization," https://www.churchofjesuschrist.org/media/image/information-statistics-graphics-lds-church-aa8cc67?lang=eng

53. Centers for Disease Control and Prevention, "National, Regional, State, and Selected Local Area Vaccination Coverage Among Adolescents Aged 13–17 Years—United States, 2017," *Morbidity and Mortality Weekly Report* 67:33 (2018), 909–917, https://www.cdc.gov/mmwr/volumes/67/wr/mm6733a1.htm#T1_down; Centers for Disease Control and Protection, "Healthy People 2020 Mid-Course Review: Immunization and Infectious Diseases," https://www.cdc.gov/nchs/data/hpdata2020/HP2020MCR-C23-IID.pdf

54. Laurie McGinley, "HPV-Related Cancers Are Rising. So Are Vaccine Rates, Just Not Fast Enough," *Washington Post,* August 23, 2018, https://www.washingtonpost.com/news/to-your-health/wp/2018/08/23/hpv-related-cancer-rates-are-rising-so-are-vaccine-rates-just-not-fast-enough/?utm_term=.26a38bef2943

55. Sherry Young, "HPV Vaccinations Are a Choice to Ponder Seriously," *Deseret News,* July 7, 2008, https://www.deseretnews.com/article/700240635/HPV-vaccinations-are-a-choice-to-ponder-seriously.html

56. Emily Webster, "Health Department Recommends HPV Vaccine," *Daily Universe,* January 23, 2008, https://universe.byu.edu/2008/01/23/health-department-recommends-hpv-vaccine/

57. April Young Bennett, "'Irrespective of Cause': A Story about Jon Huntsman, Sr. and HPV Vaccine," *The Exponent,* February 3, 2018, https://www.the-exponent.com/irrespective-of-cause-a-story-about-jon-huntsman-sr-and-hpv-vaccine/

58. Jennifer Graham, "Can Your Child Wait Until College to Get the HPV Vaccine?," *Deseret News,* April 28, 2017, https://www.deseretnews.com/article/865678797/Can-your-child-wait-until-college-to-get-the-HPV-vaccine.html; Nancy Gibbs, "Defusing the War over the 'Promiscuity' Vaccine, *Time,* June 21, 2006, http://content.time.com/time/nation/article/0,8599,1206813,00.html

59. Centers for Disease Control and Prevention, "More US Adolescents Up to Date on HPV Vaccination," August 23, 2018, https://www.cdc.gov/media/releases/2018/p0823-HPV-vaccination.html

60. Ben Lockhart, "Report: More Utah Youth Getting HPV Vaccine for Cancer Prevention," *Deseret News*, September 22, 2018, https://www.deseretnews.com/article/900033147/report-more-utah-youth-getting-hpv-vaccine-for-cancer-prevention.html

61. Daphne Chen, "What You Don't Know about the HPV Vaccine May Hurt You," *Deseret News*, January 15, 2016, https://www.deseretnews.com/article/865645597/What-you-dont-know-about-the-HPV-vaccine-may-hurt-you.html

62. Julie Bodson, et al., "Religion and HPV Vaccine-Related Awareness, Knowledge, and Receipt Among Insured Women Aged 18–26 in Utah," *PLOS One* 12 (2017), 1–11, https://www.ncbi.nlm.nih.gov/pmc/articles/PMC5571930/

63. Bodson, "Religion and HPV Vaccine-Related Awareness," 2.

64. Bodson, "Religion and HPV Vaccine-Related Awareness," 3.

65. Bodson, "Religion and HPV Vaccine-Related Awareness," 6.

66. Bodson, "Religion and HPV Vaccine-Related Awareness," 6.

67. Bodson, "Religion and HPV vaccine-related awareness," 8.

68. Bodson, "Religion and HPV Vaccine-Related Awareness," 8. Cf. Alicia L. Best et al., "Examining the Influence of Religious and Spiritual Beliefs on HPV Vaccine Uptake Among College Women," *Journal of Religion and Health* 58 (2019), 2196–2207.

69. Leah M. Smith et al., "Effect of Human Papillomavirus (HPV) Vaccine on Clinical Indicators of Sexual Behavior Among Adolescent Girls: The Ontario Grade 8 HPV Vaccine Cohort Study," *Canadian Medical Association Journal* 187 (February 3, 2015), E74–E81, http://www.cmaj.ca/content/187/2/E74; Robert A. Bednarczyk et al., "Sexual Activity–Related Outcomes After Human Papillomavirus Vaccination of 11- to 12-Year-Olds," *Pediatrics* 130 (November 2012), http://pediatrics.aappublications.org/content/130/5/798

4

Back to the Future

Genes, Disease, and Future Generations

The Genealogy of Genetic Disease

"How can a religious culture contribute to science?"[1] The following narrative illustrates how the moral culture of The Church of Jesus Christ of Latter-day Saints (LDS) has facilitated biomedical research on genetic-based diseases.[2]

Gregg Johnson is a sixty-one-year-old married man with two sons. Johnson's mother, Sandra, died of colon cancer at age forty-seven. A similar fate befell both his grandmother, who died at age forty-two, and his great-grandmother. In the decade following his mother's death, meticulous research in laboratories, in genealogical records, and in field studies led to the discovery of a mutation in a tumor-suppressing gene, APC, associated with colon cancer in numerous families. The common ancestors for these families were Lyman and Aurelia Hinman, who converted to the LDS Church in 1840. Gregg Johnson is a great-great-great-great-grandson of Lyman and Aurelia Hinman and likewise inherited the APC mutation.

This genetic legacy meant that Johnson and his extended family required proactive measures to avoid an early death from colon cancer. Since his mid-twenties, Johnson has undergone annual colonoscopies that detected and removed nearly 150 precancerous polyps. These medical surveillance measures have allowed all of Johnson's extended family to live into their sixties, well past the early deaths of their familial predecessors. One of Johnson's two sons also inherited the APC mutation. The prospect of transmitting the APC mutation to yet another generation of Johnson progeny was a consideration in reproductive decision-making. Johnson also advances research on the APC mutation by participating in a clinical trial that uses two drugs to prevent polyp growth in the small intestine. Deborah Neklason, director of the Utah Genome Project, commented, "We have prevented almost all the [potential] cancers in [the Johnson] family. This is an incredible story of the impact of genetic testing."[3]

Mormonism, Medicine, and Bioethics. Courtney S. Campbell, Oxford University Press (2021). © Oxford University Press.
DOI: 10.1093/oso/9780197538524.003.0005

LDS ecclesiastical policies have never addressed bioethical issues in genetic testing, treatment, or developments in human germ-line gene editing
that may eradicate some genetically transmitted diseases. However, the LDS
practice of researching family history and genealogical lineage has generated
significant health-related implications. The research that benefited Gregg
Johnson and his extended family was possible only because the LDS Church
agreed in the 1970s to contribute the family records "of all pioneers who came
[west] on the Mormon trail"—records that trace back eight generations—to
the Utah Population Database. The database, now comprising the life and
health records of 7.3 million persons, combines genealogical records with
cancer registry data. This information has assisted biomedical researchers
in identifying genetic factors in more than thirty diseases, including colon
cancer, breast cancer, cardiac arrhythmia, melanoma, asthma, and diabetes.[4]

The cultural emphasis on family heritage was critical to research identifying the BRCA mutations for hereditary breast and ovarian cancer. A genetic
test to determine whether a woman had inherited BRCA genetic mutations
was first made available in 1996 through Myriad Genetics. Myriad's founder,
Mark Skolnick, had assumed a research position at the University of Utah
because the large LDS communities were "a dream population for an epide-
miologist trying to track down genetic diseases."[5] Skolnick sought to "use the
power of the Mormon family history and pedigrees to help map and clone
genes," and the genetic test for inherited breast cancer relied on a study of
LDS families with multiple cases of breast cancer.[6] Myriad Genetics attributes
its success in discovering genetic mutations for these cancers and forms of
heart disease to the genealogical commitments and cooperative moral culture of Mormonism: "If the original pioneer had a mutation responsible
for cancer that spread through the family, that can be traced fairly easy. . . .
You couldn't do the same thing in many places in the world."[7] Consequently,
"more human disease genes have been discovered in Utah than in any other
place in the world."[8]

These medical breakthroughs are intertwined with the LDS salvation narrative: research on ancestry is central to theological teaching on salvation for
the deceased and the binding together of families in covenantal relationships
of eternal life. Biomedical acquisition of information about health and disease patterns within families is one medical consequence of family history
and genealogical lineage research.[9] The medical knowledge gleaned from
studies of genetically inherited disease can impact decisions about reproduction, lifestyle, and behavior change to prevent disease as well as lead to

medical diagnoses and potential treatments. Remarkably, a form of religious practice concerned with the spiritual status of the *deceased* possesses significant benefits for the physical health of the *living*. The health futures of persons and their posterity are anticipated and sometimes altered through information about their predecessors.

This chapter examines the ethical dimensions of genetic testing and interventions in the human genome. The ethical questions raised by various forms of genetic testing for diseases carry over into innovative developments of human gene editing technologies. Gene editing also offers the prospect of enhancing human physiological and cognitive capabilities. The enticement of enhancements is a feature of contemporary technology appropriated by LDS theologies of transhumanist perfectionism. My analysis concludes with a framework of normative ethical principles for personal, professional, and policy decisions on these controversial questions.

Ethical Conundrums of Genetic Testing

The ethics of genetic intervention display the intertwining of the revealed reality and the restored morality. A central practice in LDS moral culture, extensive commitments to family genealogical and ancestral history, is directed by a theological narrative in which genealogical research is carried out not as a "hobby" but as vital to *salvation*. The theological narrative gives a *why* for the *what* of family history research that serves not only soteriological but also biomedical interests. These practices display the propensity in LDS discourse to construct an ultimate harmony between religious teaching and scientific truth. Nonetheless, medical appropriation of genealogical knowledge used to advance scientific knowledge and health ends raises significant ethical questions for and within the LDS community.[10] Such issues include but are not limited to genetic screening of embryos for disease-bearing traits (as with preimplantation genetic diagnosis), genetic testing of asymptomatic persons whose family history suggests they have a genetic predisposition to a particular disease in the future, genetic discrimination against both persons with genetic-based disabilities and against otherwise healthy persons, and genetic selection in the human embryo of desired social traits.

While genetic testing is not without risks of foreknowledge about a genetic predisposition becoming a health or disease destiny, as illustrated by the Johnson family history, genetic testing offers several benefits for persons,

including understanding their possible health futures such that they undertake lifestyle changes to prevent a disease, prevent genetic transmission of
a disease to offspring through informed reproductive decision making, and
prepare for the onset of certain diseases. *Carrier* testing designates medical
procedures addressing whether a person carries a specific genetic mutation
that may be expressed in a disease in themselves or in their children. A dominant disorder, such as the APC genetic mutation or Huntington's disease,
occurs when a person has inherited an abnormal chromosome from one
of their parents. A recessive disorder, such as cystic fibrosis, occurs when a
person inherits an abnormal chromosome from both parents. Genetic testing
then offers prospective parents the possibility of preventing genetic transmission of disease to offspring. *Predictive* genetic testing determines whether
a person has a genetic mutation that confers risks to themselves of a middle-
age or late-onset disorder. Predictive genetic testing is most commonly
recommended in otherwise healthy and asymptomatic persons who are
aware of a family history of a particular disease, such as cancer or Alzheimer's.
Genetic testing practices and medical interventions in the human genome
such as gene transfer research are ethically warranted by theological, social,
and professional commitments to advance biomedical knowledge and by the
principles of respect for agency and self-determination. The availability of a
medical technology such as a genetic test sometimes carries its own utilization imperatives, but these should not supplant ethical responsibilities.

Carrier Testing: Disease and Disability

*McKay and Sue were beginning their family, and in the course of their initial
pregnancy, Sue was invited to participate in research assessing the prevalence of
the genetic mutation for cystic fibrosis (CFTR) in a regional population.[11] Sue's
consent to the research authorized genetic analysis of a blood sample and disclosure of any health-related results. Much to her surprise, the testing revealed she
was a carrier of the CFTR mutation. As cystic fibrosis is a recessive condition,
the couple's physician recommended that McKay be tested for the mutation to
determine any risk of CFTR transmission to the developing child. The genetic
test revealed McKay also had the CFTR mutation, an overwhelming disclosure
as the couple contemplated that they may have inadvertently transmitted a disease to their offspring. Their genetic counselor indicated that as both McKay
and Sue were CFTR carriers, their child had a 25% chance of contracting cystic*

fibrosis, a 50% chance of being a carrier of the mutation but not having the disease, and a 25% chance of having neither the mutation nor the disease.

The counselor recommended genetic testing of the gestating fetus through amniocentesis. The results of this test stunned the couple: Sue was carrying twin girls, both of whom tested positive for cystic fibrosis. McKay and Sue faced an arduous decision about whether to continue the pregnancy, knowing their daughters would experience a lifelong disease requiring frequent medical interventions to clear their lungs of mucus, or to terminate the pregnancy. Ultimately, the experience of parenting twins was the overriding factor in their decision to continue the pregnancy; McKay and Sue reasoned their daughters would experience a unique bonding through their shared condition.[12]

Several ethical issues are embedded in the genetic legacy of this family. In certain respects, biology qualified their moral agency. They inherited an unchosen genetic risk profile that constrained their reproductive choices. One ethical issue with broad-reaching implications then concerns the ethics of genetic testing of each parent *prior* to procreation to determine their carrier status or their predispositions to an early-onset disease that might impair their capacity for caring for their children.[13] A second conundrum concerns procreative choices when parents already have knowledge of the possibility of transmitting a genetic disease to offspring. Assuming a couple is committed to the parenting experience, they could engage in so-called genetic roulette, procreating naturally in the hopes they and their child beat the genetic probabilities, or utilize in vitro fertilization and preimplantation diagnosis to screen out embryos with a genetic mutation. The IVF alternative could be extended further to use of egg or sperm donors to avoid transmitting a recessive disease like cystic fibrosis. A couple could also forgo the biology of parenting and adopt a child. A third ethical challenge is manifested in the moral choice McKay and Sue faced directly, whether to terminate a pregnancy of a fetus with a known genetic condition, a procedure known as *selective abortion.*[14] The prevalence of this procedure can hinge on the nature of the disease. Studies indicate that pregnant women with a definitive prenatal diagnosis of Down syndrome choose selective abortion in approximately 70% of cases.[15] Circumstances of selective abortion present moral conflicts between parental autonomy and the equality of persons who may be born with disabilities.[16] This presents a fourth ethical conundrum: communal and societal constructions of disability.

LDS ecclesiastical policy has indirectly addressed the issue of selective abortion: one of three exceptions to the policy opposition to abortion is the determination by a competent physician that "the fetus has severe defects that will not allow the baby to survive beyond birth."[17] The circumstances in which a fetus has a genetic condition that is incompatible with life, such as the neural tube defect known as anencephaly, are extremely rare. The ecclesiastical exception for pregnancy termination based on inability to survive beyond birth does not encompass circumstances when prenatal diagnosis reveals a condition of Down syndrome or cystic fibrosis. These conditions impact quality of life for both the parents and the child but not survival. Various educational and medical interventions have extended the life span of a child born with Down syndrome over the past three decades from twenty-five to sixty, and from twenty to thirty-seven for a child born with cystic fibrosis.

There is nonetheless a conceptual and procedural anomaly in the ecclesiastical position. No ecclesiastical policy has been articulated on amniocentesis, or other prenatal diagnostic procedures, such as ultrasonography. Amniocentesis is recommended in prenatal care if the pregnant woman has experienced a child born with a chromosomal condition or neural tube defect from a previous pregnancy, has a family history of a specific genetic condition, the woman or her partner is a known carrier of a genetic condition, or the woman is above the age of thirty-five.[18] Since amniocentesis involves an invasive procedure to retrieve amniotic fluid from the uterus and slightly increases risks of miscarriage, Lester Bush Jr. argues that amniocentesis is "a procedure for which there is virtually no justification if abortion is not considered an acceptable alternative."[19] It might then seem reasonable to expect some expression of ecclesiastical concern about prenatal diagnostic testing. An analysis of amniocentesis by one LDS physician contended that important prenatal technological advances such as amniocentesis and ultrasonography "have been paralleled by an increasingly common disregard for the sanctity of unborn human life." The author did not recommend that LDS couples refuse such testing, however, since prenatal diagnostic information can help parents "better prepare themselves for the physical and financial challenges which await them in caring for a handicapped child."[20] The restrictive indications for a genetic-based exception to the ecclesiastical abortion prohibition are not readily reconciled with the increasingly customary use of prenatal genetic diagnostic procedures such as amniocentesis about which there is ecclesiastical silence. This silence implies that carrier testing for reproductive choices is appropriately directed by professional values and parental agency.[21]

The practices of prenatal diagnosis, selective abortion, and eventual gene editing to eliminate disease-bearing genetic traits hold significant implications for an implicit theology of disability in LDS moral culture.[22] The experience of disability is commonly re-storied through the salvation narrative and the salvific necessity of embodiment, as reflected in the observations of current prophetic leader Russell M. Nelson: "For reasons usually unknown, some people are born with physical limitations. . . . Nevertheless, the gift of a physical body is priceless. . . . A perfect body is not required to achieve a divine destiny. In fact, some of the sweetest spirits are housed in frail frames. Great spiritual strength is often developed by those with physical challenges precisely because they are challenged. Such individuals are entitled to all the blessings that God has in store for His faithful and obedient children."[23] The salvific re-storying of disability means that persons born with significant disabilities, physical or mental, are commonly conferred a special spiritual status and considered to be among the most perfected of God's spirit children. A person whose disabilities render them incapable of moral and religious accountability, whether through lack of ability to make choices between right and wrong or physical incapacity to enact those choices, is theologically re-storied as the spiritual equivalent of a little child who receives eternal life through the sacrificial atonement of Jesus Christ. A disability in mortal life is not an impediment to the promise of eternal progression in and through a perfected body in post-mortal life.

The re-storying of disability also means appropriating and emulating the healing narratives of Jesus, who throughout his ministry healed persons with various disabilities and stigmatizing diseases of the era, including blindness and leprosy; various physical incapacities; and demonic possession. These healing stories provide a touchstone for the community's re-storying of its own moral calling of inclusion and welcoming of persons with disabilities: "Church members are encouraged to follow the Savior's example of offering hope, understanding, and love to those who have disabilities."[24] Although these ideals are not always enacted, and the re-storying at times is seen as evasive in meeting the ongoing practical needs of persons with disabilities,[25] research indicates that the covenantal commitment to care for the vulnerable is especially enacted in the context of families of children born with disabilities, for whom practical caregiving and theological re-storying opens to a "spiritual experience."[26] Latter-day Saint families facing such circumstances find that their "religious beliefs provided both a personal and family philosophical context for dealing with daily events of life and a viewpoint for the larger meanings of life" and that such convictions can reframe "devastation" into a "blessing."[27] The re-storying of disability can facilitate discovering new resources of faith and meaning and enacting cove-nantal responsibilities even in the midst of personal, parental, and caregiving ordeals.

Predictive Genetic Testing: Knowledge and Power

In concert with knowledge of family genetic history, predictive genetic testing provides knowledge of disease predispositions so that lifestyles can be modified to minimize risk or to better manage symptoms over time. The availability of predictive testing is illustrative of a maxim embedded in the progressive ideology of biomedicine: "knowledge is power."[28] The empowering assumption is that more knowledge of nature, including sci-entific knowledge of human biological nature, provides greater control and mastery of the contingencies of life, including minimizing risks to health and avoiding unintended transmission of genetic mutations. Knowledge of one's genetic health status and possible health futures expands the scope of moral agency and responsibility by providing information important to preventive care. When a genetic-based disease can be managed through medication and behaviors, it is reasonable for persons aware of a family history of disease sus-ceptibility to undergo genetic testing to determine their own risk level and assume greater responsibility for their own health.

The ethics of carrier and predictive testing are also shaped by distinctive implications for other family members. Genetic diseases are commonly *family* diseases, and genetic knowledge can be *shared* knowledge. When family lineage discloses a history of debilitating chronic disease or fatal diseases, the repercussions of an individual choice to opt for or to decline ge-netic testing are contextually shaped by familial relationships. An individual's decision is protected by rights of privacy and autonomy, and testing results are confidentially protected medical information. However, insofar as a ge-netic legacy is shared, responsibilities to be tested may emerge accompanied by a responsibility to disclose the results to relevant family members.[29]

The familial context of genetic testing was illustrated in one research study examining interactions between forty-nine sibling pairs who had undergone testing for the BRCA 1 and BRCA 2 genes that confer increased risk of hereditary breast and ovarian cancer.[30] Ninety percent of the study participants were Latter-day Saints. The study explored differences in psy-chosocial interactions between the siblings depending on whether their genetic testing results were common (either negative: negative or posi-tive: positive) or mixed (negative: positive). Support behaviors and friendly interactions in response to a sibling health problem were less common when the sibling pair had a mixed result rather than either one of the shared results. The researchers speculated that some study results may have been influenced by how, relative to other study populations, "Mormons spend more time with extended family members and as such, may have more involved relationships with their adult siblings."[31]

The ethical implications of the ideology of empowering knowledge shift substantially for conditions that lack reliable preventive interventions or ef-fective treatments. Disclosure of information about a person's predispositions to a genetic disease in the absence of treatments, therapies, or behavioral modifications can be disempowering and even paralyzing. This concern is especially relevant in predictive genetic testing to assess predispositions to lethal and chronic diseases such as Huntington's and Alzheimer's. One study reported that though nearly three-quarters of at-risk persons for Huntington's indicated prospectively that they would seek genetic testing to determine their health status and future should a predictive HD test become available, only 15 to 20% of these persons actually had the test subsequent to its availability.[32] While genetic testing is a responsible exercise of agency when a person can make constructive life changes with the information ac-quired from testing for preventable or treatable disorders, it is less beneficial and may violate the professional ethic of "do no harm" to recommend a ge-netic test when "there is currently no treatment . . . , [nor] anything an af-fected person can do to prevent the inevitable onset of symptoms."[33]

A person who refuses a recommendation for genetic testing, grounded in "a right not to know," must integrate their choice with a broader conception of personal responsibility for one's own health.[34] The LDS construction of moral agency confers moral responsibility on persons for those matters in life over which they can exert some control. While we have no control over the genetic lottery of life, including genetic-based disease mutations, agency can be exercised in response to this lottery, including taking preventive, diag-nostic, or treatment measures to minimize risk and the impact of the disease personally and within a family. Forgoing recommended genetic testing in the context of a family history of inherited mutations imposes a personal respon-sibility for awareness of a genetic legacy and enacting a lifestyle emphasizing nutrition, exercise, self-health practices, and preventive care with these genetic risks in mind. The question of personal moral responsibility for health will surely increase as more mutations are discovered and treatments are developed. However, as we do not choose our genes, unless society is to treat genetic probabilities by a different standard than how it treats many other health and disease conditions over which persons can have causal responsibility (e.g., smoking prevalence in lung cancer), it is arbitrary and unfair to insist that a person has an ethical obligation to be tested to assess their personal genetic risk.

Human Germ-Line Gene Editing

While it would be unwise to edit a family history in such a way as to become oblivious to potential health risks, innovative genetic technologies promise to permit editing of heritable genetic mutations. Human germ-line gene editing takes both the science and the ethics of genetic manipulation across thresholds established by embryo, carrier, and predictive testing. The prospects, hope, and hype about gene editing are significant: "No scientific discovery of the past century holds more promise—or raises more troubling ethical questions."[35]

A research university convened its Innovative Research Advisory Panel (IRAP) to make ethical recommendations regarding a groundbreaking research procedure to edit human embryos to correct genetic mutations in a specific type of heart ailment, familial hypertrophic cardiomyopathy (FHC). FHC restricts the heart's capacity to pump blood, creating chest pain and abnormal heart rhythms, and is a leading cause of sudden death in young athletes. The gene editing procedure to correct the mutation would occur in a laboratory setting and would not involve transferring the embryos for gestation. The procedure was considered "innovative" as it represented the first research proposal for a germ-line or reproductive cell genetic editing intervention.

The IRAP heard compelling testimony from persons afflicted with FHC who had unknowingly passed the genetic mutation to their children, necessitating behavioral restrictions on strenuous exercise and athletic activities. As one parent exclaimed, "This condition needs to end with me!" The committee agreed the research proposal had scientific merit and that crossing the conceptual and technical boundary from somatic to germ-line genetic modification could be justified by ethical principles of generating knowledge, disease prevention, and alleviating suffering. Some committee members expressed concern that germ-line editing for disease would inevitably open a door to genetic enhancements (such as improved cognition and memory). Ultimately, the IRAP recommended that the proposal proceed because it was limited to an assessment of the efficacy of gene editing in laboratory research. Two years later, the research team reported it had successfully corrected the FHC mutation in fifty-eight embryos, demonstrating the efficacy of germ-line gene editing and refining the procedure to improve safety.[36]

Prior gene transfer research and clinical treatments dating to the early 1990s addressed somatic (bodily) cell genetic modifications to ameliorate a disease condition in a living person. This research meant that any successful interventions would end with the person upon their death. Germ-line interventions have historically been considered intrinsically fraught with scientific uncertainty and ethical questions because the modifications are transmitted to future generations. With the advent of gene editing, federal advisory panels deliberating on the ethics and policy issues have considered faith-based perspectives as part of broader public and professional discussions.[37] One ethical question builds off of limited scientific understanding of genetic interactions. Certain genes that cause debilitating or lethal diseases may also confer immunity from other diseases, such as the immunity from malaria conferred by sickle cell anemia. The progressivist scientific understanding of knowledge then must confront limited scientific understanding, rooted ultimately in human finitude, a pervasive theme in religious bioethical discourse about germ-line gene editing. Lutheran theologian Ted Peters presents one speculative scenario: "Suppose we edit the human genome to get rid of the gene for Huntington's disease. . . . The problem is we don't know whether that particular gene works in concert with other genes to produce something positive. By eliminating Huntington's, we might have screwed something else up."[38] However, given what we *do know* about genetically inherited diseases such as FHC, it would be a serious moral mistake to let what we *do not know* about genetic interactions inhibit gene editing.

Although some religious communities have expressed concern that human germ-line gene editing (and genetic manipulation generally) reflects human arrogance in the absence of knowledge and wisdom (or "playing God"), religious acceptance of genetic modifications for disease treatment or prevention have already been ethically established. Core religious commitments of compassion, relief of unnecessary suffering, and obligations to future generations, as with the Johnson family narrative, support genetic interventions

in principle. These imperatives suggest it would be callous to forgo opportunities to acquire new genetic knowledge that could address the profound
human burdens of genetic-based diseases, especially when germ-line gene
editing heralds the possibility of avoiding transmitting genetic mutations to
progeny. The overriding professional, medical, and religious imperative to
relieve preventable suffering entails an ethical responsibility to proceed with
gene editing research and eventually with clinical applications.

A further bioethical consideration, similar to that encountered in the
ethics of preimplantation genetic diagnosis, is that germ-line gene editing
may seem contrary to the principle of human equality. Some arguments
claim gene editing symbolizes a form of social intolerance toward persons
currently living with the disease that gene editing seeks to eradicate and so
undermines the sensibility that a more moral society is committed to inclusion and respect for human diversity.[39] A justice-based disability critique of
germ-line editing appeals to social commitments of equality and solidarity
with vulnerable persons and communities. Cultural identities have been
formed around conditions like deafness and dwarfism that manifest the variability of cultural interpretations of disease. Moreover, LDS culture has cultivated its own constructions of disability in the context of salvation, healing,
and covenantal narratives. The prospect of germ-line editing to eliminate
serious hereditary diseases and conditions is not simply a technological imperative but a narrative embedded in the theological calling of medicine as
a healing profession. However, the potential for exclusionary or discriminatory attitudes to underlie widespread applications of these technologies
means that the ethical and policy feasibility of germ-line gene editing must
confer paramount importance to the voices of persons who live with the
diseases.[40]

The most contentious issue in the IRAP deliberations recounted above
concerned whether germ-line gene editing will open the door to enhancement editing. A recent study revealed that Americans express more war-
iness than support for enhancement capabilities, including gene editing to
diminish a baby's risk of disease, brain chip implants to increase cognitive
abilities, or use of synthetic blood to increase physical abilities.[41] A majority
of Americans who identify as "highly religious" believed these enhancement
scenarios "crossed a line we should not cross." In relevant respects, however, the enhancement threshold has *already* been crossed through vaccines
to enhance the immune system and provide protection against various
diseases. The issue is whether there is a defensible differentiation between

immunization enhancements and other genetic enhancements with which the public, especially the religious public, is concerned. Somatic cell gene editing seems no morally different than altering bodily chemistry through vaccinations, while the intention behind germ-line gene editing is to modify the genetic inheritance of future generations. The question is whether resistance from religious communities or the broader society to germ-line gene editing is primarily a function of objections to *enhancements* or to modifying *genetic* inheritance. The former objection seems difficult to sustain given the strong religious (and LDS) support for vaccines, while the latter objection must counter religious argumentation about promoting health, preventing disease, and relieving suffering.[42]

A related question concerns the fluidity of cultural conceptions of "health," "disease," and "disability," constructs susceptible to reinterpretation by technological advances. Even if professional regulation and public policy constrain germ-line gene editing to disease prevention,[43] expanded or individualized conceptions of "disease" can open a path to enhancement. Using the example of short stature that led to increased use of human growth hormone, bioethicist Marci Darnovsky asks, "How would we draw the distinction between a medical and enhancement purpose for germ-line modification?"[44] Darnovsky contends that such a distinction is critical since treatments for disease fall within the vocation of medicine while enhancement endeavors (vaccination aside) generally do not. Since parents already have alternatives to avoid transmitting a genetic disease to offspring, including embryo screening in preimplantation diagnosis or use of third-party eggs or sperm, some policymakers advocate a moratorium on clinical applications of gene editing. However, appealing to the conceptual fluidity of disease cannot provide grounds for an indefinite moratorium on germ-line gene editing given the enhancement practices society already tolerates.

For reasons I address in the concluding section, it is unlikely that LDS ecclesiastical policy will address human germ-line gene editing, at least in the formative stages of the discussion when defining scientific, policy, and ethical issues are framed. LDS scientists, researchers, and physicians may engage in these deliberations through professional associations, and patients may become participants in research studies on the presumption of research integrity. However, the prospect of genetic and other forms of enhancement has entered LDS intellectual culture through the discourse of Mormon transhumanism.

Enhancement and Perfectionism

In critiquing contemporary "Promethean" aspirations to master human contingency through biomedical technologies, philosopher Michael Sandel formulates the core issue in enhancement ethics: "We usually admire parents who seek the best for their children, who spare no effort to help them achieve happiness and success. Some parents confer advantages on their children by enrolling them in expensive schools, hiring private tutors, sending them to tennis camp, providing them with piano lessons, ballet lessons, swimming lessons, SAT-prep courses, and so on. If it is permissible and even admirable for parents to help their children in these ways, why isn't it equally admirable for parents to use whatever genetic technologies may emerge (provided they are safe) to enhance their child's intelligence, musical ability, or athletic skill?"[45] This core question intersects with the contemporary philosophy of transhumanism.[46] Transhumanist philosophy is a quintessential expression of the progressive maxim that knowledge is power and that emerging technologies in genetics, robotics, information and computers, nanotechnology, cryogenics, and artificial intelligence possess potency to infinitely expand the realms of human abilities and capacities and transform human nature.[47] The core themes of philosophical transhumanism include the intrinsic progressive nature of technology; expanding human intelligence; technologically eradicating causes of pain and suffering, including poverty, disease, and mortality; accessible information about lineage and posterity; overcoming both the vicissitudes of aging and aging itself; and a quest for immortality indebted to transformative technological advances.[48] Mormon transhumanism appropriates this expansive vision of an ever-evolving human being into the narrative of eternal progress and the aspirations for godhood in LDS scripture and ecclesiastical teaching.[49] Mormon transhumanism constructs deification as a technological project and ethical responsibility: "we should ethically use our resources including religion, science, and technology to improve ourselves and our world until we become Gods ourselves."[50]

The Mormon Transhumanist Association (MTA), an organization self-defined as "the world's largest advocacy network for ethical use of technology and religion to extend human abilities,"[51] reflects a convergence of philosophical transhumanism with LDS teachings. The MTA recognizes two foundational documents, the secular "Transhumanist Declaration" and a "Mormon Transhumanist Affirmation." The declaration affirms the scope of the transhumanist project: "We envision the possibility of broadening human potential by overcoming aging, cognitive shortcomings, involuntary suffering, and our confinement to planet Earth."[52] The MTA Affirmation roots transhumanism within faithful emulation of Jesus Christ, a religious quest facilitated by progressive science and enhancement technologies: "Mormon teachings about humans becoming gods connect with transhumanist ideas about using technology to enhance the human condition."[53]

MTA founder Lincoln Cannon identifies four premises that support a claim that "Mormonism actually mandates transhumanism."[54] These include a religious ethical claim that Mormons receive divine commands to use "ordained means" to participate in the work of God, an interpretation that science and technology are divinely ordained methods, a theological claim that God's "work" involves aiding other persons to attain godhood, and a conceptual claim that an essential feature of the divine is "a glorified immortal body."[55] These premises culminate in a "religious mandate": "God commands [Mormons] to use science and technology to help each other attain a glorified immortal body."[56] Honoring the mandate to obtain this glorified body means that cryonics or uploading the contents of mind to a digital substrate can be "the most faithful choice for Mormons."[57]

Two propositions in the MTA affirmation are illustrative of the significant conceptual, religious, and ethical problems in this eschatological vision. The fourth proposition asserts, "We believe that scientific knowledge and technological power are among the means ordained of God to enable [spiritual and physical] exaltation, including realization of diverse prophetic visions of transfiguration, immortality, resurrection, renewal of this world, and the discovery and creation of worlds without end."[58] This statement situates basic themes in the LDS salvation narrative—prophetic vision, resurrection, eternal life, and ongoing creative processes—as human achievements facilitated by science and technology. This is not a totalizing scientism, for the claim implies other divinely authorized means can realize "exaltation" (a Mormon metaphor for godhood). However, the proposition articulates an engineering model for concepts of embodiment, immortality, and resurrection and thereby places science and technology in a salvific role. Transcending the human condition is mediated through scientific progress and technological advancement rather than faith and divine grace.

The fifth proposition stipulates, "We feel a duty to use science and technology according to wisdom and inspiration, to identify and prepare for risks and responsibilities associated with future advances, and to persuade others to do likewise."[59] The language of "duty," "risk," and "responsibility" reflects ethical concepts, but the proposition inverts the standard ethical criterion of "ought implies can" to "can implies ought." The affirmation thereby expresses the essence of the technological imperative: If technology is available that could confer some kind of benefit by enhancing human capacities, there is an ethical imperative to use the technology. The epistemic methods that inform this duty consist of "wisdom," including insights developed within other religious or philosophical worldviews, and "inspiration," the cultural signifier for divine epistemic guidance. What the proposition entirely neglects is the specification of any ethical criteria to determine how this duty can be responsibly enacted, what the risks and responsibilities are, and how these are to be balanced relative to anticipated "advances." The advocacy of ethical uses of technology necessarily presupposes unethical applications, but no standards are presented to indicate how persons, researchers, or citizens are to make ethical judgments about a particular technological application. Cannon's expositions offer no insights on this ethical agnosticism. He interprets the moral point of several biblical narratives to be that human actions uninformed by ethical values "beyond egoism" can culminate in cultural destruction but are likewise agnostic about communally shared ethical criteria to determine unethical uses of technology.[60]

Mormon transhumanism pushes ultimately in the direction of a teleological ethic: any use of technology that contributes to humanity's inevitable "future advances" is morally justifiable. Transhumanist ends justify the scientific, technological, and religious means. The position does not permit any intrinsic restriction on a specific technology. However, technology is not value-neutral but is imbued with and reflects human values. The development of any technology requires ascertaining the purposes or ends of the technology and identifying who controls the technology. Though the transhumanist teleology of extending human capacities reflects the ethical maxim that ends justify the means, not all ends are justifiable, nor do legitimate ends justify any and all means. The primary ethical difficulty with Mormon transhumanism is that ethical responsibility is instrumentalized to serve aesthetic, religious, and scientific values.

Biomedical technologies premised on advocacy of expanding human capabilities risk dehumanizing characteristics. Mormon transhumanism frames the goals of science and biotechnologies to be relieving the human condition *of* the human condition, including our finitude, fallibility, aging, experience of pain and suffering, and mortality. In a critical assessment, Adam Miller contends that some enhancements envisioned in Mormon transhumanism, such as a digitalized repository for immortal consciousness, may so alter human experience as to "mask the fundamentally temporal, material, and agential character of life."[61] It is an existential and an ethical problem when the fundamental characteristics of life experience are modified beyond recognition. LDS conceptualizations of the sacred integrity of the body and the pedagogical nature of suffering generate ethical resistance to transhumanist aspirations.

Transhumanist philosophy, including Mormon transhumanism, ultimately presents an eschatological vision that selectively appropriates LDS theological teachings on immortal life, the salvific acquisition of knowledge about self and technology, and ongoing eternal progression. However, the faith tradition broadly considers immortality, knowledge, and eternal progress as gifts of divine grace and human covenantal responsiveness rather than isolated human works of technological achievement. This requires *living with* the finitude, frailty, and fallibility of the human condition rather than seeking a mechanical remedy that transitions humans to a post-human reality. The envisioned eschatological ends risk displacing nature, denying our intrinsic embodied self, compromising professional integrity, violating moral agency, and perpetuating injustices. What citizens, researchers, professionals, and saints must avoid is ceding the shape of the future to imperatives of technological innovation in the absence of clear ethical principles.

Back to the Future: Criteria for LDS Genetic Ethics

The bioethical controversies illustrated in genetic testing, gene editing, and genetic enhancements reflect a substantive temporal continuity of the self, as a person may find their personal health futures shaped by their ancestral and genealogical lineage and encounter meaningful moral questions about the genetic legacy they wish to bestow to their posterity. I propose here several ethical criteria for responsible use of genetic technologies that incorporate relevant LDS communal values, including knowledge, professionalism, moral agency, reversibility, responsibility to future generations, and relational commitment. The accessibility of personal genetic information can be situated within ecclesiastical efforts to advance population genetic knowledge as manifested through the integration of family history and genealogical records with public health and cancer registries. These practices reflect a formative commitment to *acquiring knowledge* as divinely mandated and as a basic human good. Scientific and biomedical research that contains the possibility of developing preventive, treatment, curative, or healing medical interventions for genetic-based disease instantiates an implicit partnership with the divine.

The moral autonomy of biomedical research and clinical application of genetic interventions honor the *integrity of the medical profession*. This principle presumes that research and applications are directed and constrained by ethical safeguards of safety, efficacy, and respect. Genetic interventions must be justified by the broader purposes and goals of medicine, including healing, preventing disease, and developing effective treatments. The corollary of communal trust in medical professionalism is *ecclesiastical deference* to scientific and biomedical expertise. This does not mean that the realm of genetic interventions is entirely a professional question devoid of religious or ethical content. It does imply that biomedical commitments to knowledge, healing, and respectful consultation should be sufficient to address many moral issues in genetic testing or editing within the patient-physician or participant-researcher relationship. When scientific finitude and uncertainty about safety, risks, or effectiveness of an innovative genetic intervention prevails, the moral burden falls to advocates of the intervention to provide demonstrable evidence of its safety. The cautious approach of the scientific and biomedical communities on human gene editing entails a moral preference for *reversible* technological interventions. A core moral issue posed by germ-line gene editing, especially in enhancement applications, is their irreversible impact on the genomes of subsequent generations.

This emphasis on professional integrity and autonomy means that personal and communal *moral agency* is essential for responsible genetic interventions. The exercise of moral agency in participating or dissenting from genetic studies or genetic tests requires a sufficient level of genetic literacy such that a person can comprehend the benefits and risks for themselves, their extended family, and their prospective progeny, and risk-minimization alternatives to genetic testing. Genetic literacy comprehension is required of the community and research participants regarding the impact of genetic interventions now for both present and future generations. Current generations have a responsibility to future generations to ensure the protection of a "right to an open future."[62] This "open" future consists in a moral landscape comprising a diverse array of choices at least equivalent to what current generations possess. The randomness of the genetic and social lotteries that bequeath an involuntary legacy of health and relationship means no person can lay claim to an entirely open future. However, respecting moral agency and securing a similar breadth of moral choice for progeny is an essential feature of current ethical choices. Securing equivalent autonomy and agency for progeny and future generations requires extending the empathic imagination to persons we may never meet but whose lives we will impact by our decisions now.

Broad professional and societal acceptance of diagnostic, carrier, and predictive testing accompanies somatic cell genetic interventions. The ecclesiastical conferral of moral authority to personal agency regarding medical and health practices that do not present ethically or legally controversial issues warrants utilizing genetic testing procedures, even if circumstantial controversies arise that require negotiation within the patient-physician or participant-researcher relationship. The genetic nuance of moral agency involves appreciating that genetic information is not individualistic but bears outward to the familial and relational. Genetic testing shares with the cultural commitment to genealogical research a witness that each person is enmeshed in a web of *identity-forming relationships* that transcend personal choice and bestow a powerful reminder of responsibilities to lineage, ancestry, and posterity.

The ethical criteria of knowledge, professionalism, moral agency, reversibility, responsibility to future generations, and relationship imply that generating an ethic of genetic testing, editing, or enhancements will be more experientially informed and communally practiced than theologically driven. LDS professionals in genetic science, medicine, or genetic counseling may play a prominent role in shaping the moral landscape of genetic ethics. The emerging possibilities of human germ-line genetic editing to prevent disease will likely be organized according to principles of medical professionalism rather than ecclesiastical teaching. Somatic gene editing to address a predisposing genetic condition experienced by an individual possesses no substantial moral difference than other preventive or treatment measures while germ-line gene editing to minimize the possibilities of intergenerational transmission of genetic disease bears on the LDS ethics of disease prevention. The 2001 ecclesiastical statement on embryo research might well be repurposed to address ethical or policy controversies over somatic or germ-line editing for purposes of disease prevention: the "newly emerging field of [human gene editing] merits cautious scrutiny. The proclaimed potential to provide cures or treatments for many serious diseases needs careful and continuing study by conscientious, qualified investigators. As with any emerging new technology, there are concerns that must be addressed. Scientific and religious viewpoints both demand that strict moral and ethical guidelines be followed."[63]

The patterns of ecclesiastical caution identified by Lester Bush respecting ecclesiastical policy formulation on medical ethics in general seem particularly applicable to human gene-editing technologies: (1) Even when there are significant ethical or theological interests at stake, ecclesiastical policy tends to be deferred "when the issues are extraordinarily complex and when important scientific questions remain unanswered"; (2) developed policies "generally do not appear until relatively late in the public discussion"; and (3) policies invariably evolve in "the direction of greater conformity to the medical and social consensus on the subject."[64] The issue of human germline editing to alleviate disease meets the condition of "extraordinary complexity" insofar as there remain scientific questions about safety and efficacy that require independent scrutiny of preliminary research studies. The "relatively late" timing of a policy statement entails that, as with other innovative medical technologies, personal agency and professional engagement will be a living laboratory for generating LDS genetic ethics.

The ethical difficulties with arriving "late" to an ethical controversy include the risk of supporting a previously established medical and social consensus and diminishing the capacity for an effective moral prophetic critique. Ecclesiastical deference may generate a reactionary rather than proactive ethics and function as a priestly blessing for already established practices. A prophetic ethical tradition should not be content with the answers to difficult bioethical questions developed by others but should be engaged in the initial framing of the questions. The ethical trade-off to ecclesiastical caution is that questions or values of religious importance may not be articulated in the formative stages of the professional or research discussion. The consequence is that the religious community may be on the receiving end of a medical and moral fait accompli. This dialectic between professional autonomy, moral agency, and ecclesiastical prophetic witness is illustrated in LDS teaching regarding the ethics of organ donation.

Notes

1. Genetics Science Learning Center, "Genetics in Utah," https://learn.genetics.utah.edu/content/science/utah/
2. This story has been compiled from two sources: Jennifer Dobner, "Genetics and U," *Continuum*, Spring 2014, https://continuum.utah.edu/features/genetics-and-u/; Sarah Zhang, "What Mormon Family Trees Tell Us About Cancer," *The Atlantic*, June

23, 2017, https://www.theatlantic.com/science/archive/2017/06/mormon-genetic-testing/530781/

3. Dobner, "Genetics and U"; cf. "The Genetics of Genealogy," *The Liahona,* August 1979, https://www.lds.org/liahona/1979/08/discovery/the-genetics-of-genealogy?lang=eng&clang=ara

4. Genetics Science Learning Center, "Genetics in Utah"; Kirk Johnson, "By Accident, Utah Is Proving an Ideal Genealogical Laboratory," *New York Times,* July 31, 2004, https://www.nytimes.com/2004/07/31/us/by-accident-utah-is-proving-an-ideal-genetic-laboratory.html. The prospects of determining risks of genetic-based diseases through family genealogical records have been continually cited on the "Newsroom" public information website of the LDS Church. Newsroom, "Church Family History Records Lead to Groundbreaking Genetic Research," July 18, 2008, https://newsroom.churchofjesuschrist.org/article/church-family-history-records-lead-to-groundbreaking-genetic-research; Newsroom, "Major Medical Findings Aided by Church History Records," https://newsroom.churchofjesuschrist.org/ldsnewsroom/eng/news-releases-stories/major-medical-findings-aided-by-church-family-history-records (accessed August 3, 2020).

5. Thomas H. Maugh, II, "The Man Who Makes the Pieces of the Puzzle Fit," *Los Angeles Times,* March 20, 1994, http://articles.latimes.com/1994-09-20/news/ls-40953_1_breast-cancer-gene

6. DNA Learning Center, "Mark Skolnick," https://www.dnalc.org/view/15718-Mark-Skolnick.html; Sherry I. Brandt-Rauf et al., "Ashkenazi Jews and Breast Cancer: The Consequences of Linking Ethnic Identity to Genetic Disease," *American Journal of Public Health* 96:11 (November 2006), 1979–1988, https://www.ncbi.nlm.nih.gov/pmc/articles/PMC1751808/

7. Bill Hockett, spokesman for Myriad Genetics, Inc., as cited in Kristen Moulton, "Large Families Make Utah Fertile Ground for Genetics Researchers," *Los Angeles Times,* October 25, 1998, http://articles.latimes.com/1998/oct/25/local/me-36126; Natalie Anger, "Fierce Competition Marked Fervent Race for Cancer Gene," *New York Times,* September 20, 1994, https://www.nytimes.com/1994/09/20/science/fierce-competition-marked-fervid-race-for-cancer-gene.html

8. Ray Gesteland, University of Utah, as quoted in "Genetics in Utah."

9. A prominent LDS medical researcher, James O. Mason, reported that LDS prophetic leader Gordon B. Hinckley commented, "[God] has more than one purpose in mind for family history." Newsroom, "Mormon Physicians Pioneer Research in Genetics," https://newsroom.churchofjesuschrist.org/ldsnewsroom/eng/news-releases-stories/mormon-physicians-pioneer-research-in-genetics (accessed August 3, 2020).

10. Devyn Smith, "The Human Genome Project, Modern Biology, and Mormonism: A Viable Marriage?," *Dialogue: A Journal of Mormon Thought* 35 (2002), 61–71.

11. Cystic fibrosis is a chronic disease characterized by breathing difficulties due to persistent lung infections from mucus accumulations.

12. This narrative is presented in a documentary, Oregon Public Broadcasting, *A Question of Genes: Inherited Risks,* https://repository.library.georgetown.edu/handle/10822/549317

13. The most common example of premarital genetic screening of prospective marriage partners occurs among the Ashkenazi Jewish population due to a significantly increased risk of transmitting the gene for Tay Sachs disease, a fatal illness in young children preceded by significant mental deterioration. Komal Bajaj, Susan J. Gross, "Carrier Screening: Past, Present, and Future," *Journal of Clinical Medicine* 3:3 (September 2014), 1033–1042.

14. Amy Julia Becker, "The Social Construction of Selective Abortion," *The Atlantic*, January 22, 2013, https://www.theatlantic.com/sexes/archive/2013/01/the-social-construction-of-selective-abortion/267386/

15. Jaime L. Natoli et al., "Pre-Natal diagnosis of Down Syndrome: A Systematic Review of Termination Rates (1995–2011)," *Prenatal Diagnosis*, March 14, 2012, https://obgyn.onlinelibrary.wiley.com/doi/full/10.1002/pd.2910

16. Chris Bodenner, "When Does Abortion Become Eugenics?," *The Atlantic*, May 24, 2016, https://www.theatlantic.com/notes/all/2016/05/when-does-an-abortion-become-eugenics/483659/; Mark Lawrence Shrad, "Does Down Syndrome Justify Abortion?," *New York Times*, September 4, 2015; https://www.nytimes.com/2015/09/04/opinion/does-down-syndrome-justify-abortion.html

17. The Church of Jesus Christ of Latter-Day Saints, "Church Policies and Guidelines, Policies on Moral Issues: Abortion," in *General Handbook: Serving in the Church of Jesus Christ of Latter-day Saints*, Chapter 38.6.1, February 19, 2020, https://www.churchofjesuschrist.org/study/manual/general-handbook/38-church-policies-and-guidelines?lang=eng#title_number91

18. The Mayo Clinic, "Amniocentesis," https://www.mayoclinic.org/tests-procedures/amniocentesis/about/pac-20392914

19. "Attitudes of the Church of Jesus Christ of Latter-day Saints Towards Certain Medical Problems," in "Mormon Medical Ethical Guidelines," ed. Lester E. Bush, Jr., *Dialogue* 12 (Fall 1979), 100. Somewhat ironically, LDS physician J. Edwin Seegmiller is credited for helping develop amniocentesis procedures. See Ashley Evanson, "Mormons and Science," *LDS Living*, September 15, 2008, http://www.ldsliving.com/Mormons-and-Science/s/4696

20. Stephen L. Clark, "A Doctor Looks at Amniocentesis," *The Ensign*, April 1985, https://www.lds.org/study/ensign/1985/04/random-sampler/a-doctor-looks-at-amniocentesis?lang=eng

21. Relying on admittedly incomplete knowledge about the frequency of amniocentesis in the LDS community, Bush asserts, "It is not as rare as one might suppose for LDS women discovering significant fetal abnormalities to have these pregnancies terminated" (Lester E. Bush, Jr., *Health and Medicine Among the Latter-day Saints: Science, Sense, and Scripture* [New York: Crossroad, 1993], 166); cf. Rayna Rapp, "Refusing Prenatal Diagnosis: The Meanings of Bioscience in a Multicultural World," *Science, Technology, and Human Values* 23:1 (Winter 1998), 45–70; Jennifer Graham, "Behind the Numbers: How to Make Sense of Utah's Abortion Rate for Married Women," *Deseret News*, June 14, 2017, https://www.deseretnews.com/article/865681905/Behind-the-numbers-How-to-make-sense-of-Utahs-abortion-rate-for-married-women.html

22. See Anne Leahy, "Mormonism Dysembodied: Placing LDS Theology in Conversation with Disability," *element* 5:1 (2009), 29–42; Kristin Clift, "Glimpses of Eternity: Sampled Mormon Understandings of Disability, Genetic Testing, and Reproductive Choice in New Zealand," M.A. thesis, University of Otago, New Zealand, 2012; https://ourarchive.otago.ac.nz/bitstream/handle/10523/3935/CliftKristin2012MA.pdf?sequence=1&isAllowed=Y

23. Russell M. Nelson, "We Are Children of God," *The Ensign*, November 1998, https://www.churchofjesuschrist.org/study/general-conference/1998/10/we-are-children-of-god?lang=eng

24. The Church of Jesus Christ of Latter-Day Saints, "Members with Disabilities," in *Handbook 2: Administering the Church*, Section 21.1.26, https://www.lds.org/handbook/handbook-2-administering-the-church?lang=eng. The LDS Church set up an extensive "Disabilities Resources" website in 2009: https://www.churchofjesuschrist.org/life/disability

25. Jana Riess, "How Welcoming Are Mormons to People with Disabilities?," *Religion News Service,* October 17, 2016, https://religionnews.com/2016/10/17/how-welcoming-are-mormons-to-people-with-disabilities-at-new-hartford-temple-some-room-for-improvement/; Clift, "Glimpses of Eternity," 47–56.

26. Elaine S. Marshall et al., "'This Is a Spiritual Experience': Perspectives of Latter-day Saint Families Living with a Child with Disabilities," *Qualitative Health Research* 13:1 (2003), 57–76.

27. Marshall, "'This Is a Spiritual Experience,'" 70, 67.

28. This maxim is commonly attributed to philosopher of science Francis Bacon and to political philosopher Thomas Hobbes.

29. Benjamin E. Berkman, Sara Chandros Hull, "The 'Right Not to Know' in the Genomic Era: Time to Break from Tradition?," *American Journal of Bioethics* 14 (March 2014), 28–31; Shawneequa Callier, Rachel Simpson, "Genetic Diseases and the Duty to Disclose," *AMA Journal of Ethics*, August 2012, https://journalofethics.ama-assn.org/article/genetic-diseases-and-duty-disclose/2012-08

30. Heidi A. Hamann et al., "Interpersonal Responses Among Sibling Dyads Tested for BRCA1/BRCA 2 Genetic Mutations," *Health Psychology* 27 (2008), 100–108.

31. Hamann, "Interpersonal Responses," 106.

32. S. Creighton, E. W. Almqvist, D. MacGregor, et al., "Predictive, Pre-Natal and Diagnostic Genetic Testing for Huntington's Disease: The Experience in Canada from 1987 to 2000," *Clinical Genetics* 63 (2003), 462–475; Bettina Meiser, "Psychological Impact of Genetic Testing for Huntington's Disease: An Update of the Literature," *Journal of Neurology, Neurosurgery, and Psychiatry* 69 (2000), 574–578.

33. Karen Norrgard, "Ethics of Genetic Testing: Medical Insurance and Genetic Discrimination," *Nature Education* 1 (2008), 90.

34. David B. Resnik, "Genetics and Personal Responsibility for Health," *New Genetics and Society* 33:2 (June 30, 2014), 113–125.

35. Michael Specter, "How the DNA Revolution Is Changing Us," *National Geographic*, August 2016, https://www.nationalgeographic.com/magazine/2016/08/dna-crispr-gene-editing-science-ethics/

36. Heidi Ledford, "CRISPR Fixes Embryo Error," *Nature* 548 (August 3, 2017), 13–14. This narrative is a condensed version of my experience on this ethics panel in 2015–2016.

37. Committee on Human Gene Editing: Scientific, Medical, and Ethical Considerations, *Human Genome Editing: Science, Ethics, and Governance* (Washington, D.C.: National Academies Press, 2017), https://www.nap.edu/catalog/24623/human-genome-editing-science-ethics-and-governance

38. Kelsey Dallas, "Finding God in One of Science's Biggest Debates—Genetic Editing," *Deseret News*, January 13, 2016, https://www.deseretnews.com/article/865645387/Finding-God-in-one-of-sciences-biggest-debates-2-Genetic-editing.html

39. Erika Check Hayden, "Should You Edit Your Children's Genes?," *Nature* 530 (February 23, 2016), https://www.nature.com/news/should-you-edit-your-children-s-genes-1.19432

40. Emily Beitiks, "5 Reasons Why We Need People with Disabilities in the CRISPR Debates," Center for Genetics and Society, September 7, 2016, https://www.geneticsandsociety.org/biopolitical-times/5-reasons-why-we-need-people-disabilities-crispr-debates

41. Cary Funk et al., "U.S. Public Wary of Biomedical Technologies to 'Enhance' Human Abilities," Pew Research Center, July 26, 2016, http://www.pewinternet.org/2016/07/26/u-s-public-wary-of-biomedical-technologies-to-enhance-human-abilities/; Cary Funk et al., "U.S. Public Opinion on the Future Use of Gene Editing," Pew Research Center, July 26, 2016, http://www.pewinternet.org/2016/07/26/u-s-public-opinion-on-the-future-use-of-gene-editing/

42. David Masci, "Human Enhancement: The Scientific and Ethical Dimensions of Striving for Perfection," Pew Research Center, July 26, 2016, http://www.pewinternet.org/essay/human-enhancement-the-scientific-and-ethical-dimensions-of-striving-for-perfection/

43. Committee on Human Gene Editing: Scientific, Medical, and Ethical Considerations, *Human Genome Editing*, 29–60, 181–194.

44. Marcy Darnovsky, "Do Not Open the Door to Editing Genes in Future Humans," *National Geographic*, August 2016, https://www.nationalgeographic.com/magazine/2016/08/human-gene-editing-pro-con-opinions/

45. Michael Sandel, *The Case Against Perfection* (Cambridge, MA: The Belknap Press of Harvard University Press, 2007), 51. Cf. John Harris, *Enhancing Evolution: The Ethical Case for Making Better People* (Princeton, NJ: Princeton University Press (2007), 1–5.

46. Allen Porter, "Bioethics and Transhumanism," *Journal of Medicine and Philosophy* 42 (June 2017), 237–260.

47. Connor Simpson, "Google Wants to Cheat Death," *The Atlantic*, September 18, 2013, https://www.theatlantic.com/technology/archive/2013/09/google-wants-cheat-death/310943/

48. C. Christopher Hook, "Transhumanism and Posthumanism," in *Bioethics*, 4th ed., vol. 6, ed. Bruce Jennings (New York: Macmillan Reference, 2014), 3096–3102; Dawn Chan, "The Immortality Upgrade," *The New Yorker*, April 20, 2016, https://www.newyorker.com/tech/annals-of-technology/mormon-transhumanism-and-the-immortality-upgrade

49. Richard Lyman Bushman, *Joseph Smith: Rough Stone Rolling* (New York: Vintage Books, 2007), 533–537; Samuel Morris Brown, *In Heaven as It Is on Earth: Joseph Smith and the Early Mormon Conquest of Death* (New York: Oxford University Press, 2012), 248–278.

50. Mormon Transhumanist Association, "What Is Mormon Transhumanism?," https://transfigurism.org/primer-mormon-transhumanism

51. Lincoln Cannon, "Mormonism Mandates Transhumanism," in *Religion and Human Enhancement: Death, Values, and Modernity,* ed. Tracy J. Trothen, Calvin Mercer (New York: Palgrave Macmillan, 2017), 61.

52. Humanity+, "Transhumanist Declaration," https://humanityplus.org/philosophy/transhumanist-declaration/. Cf. Henry Greely et al., "Towards Responsible Use of Cognitively Enhancing Drugs by the Healthy," *Nature* 456 (December 11, 2008), 702–705.

53. Mormon Transhumanist Association, "The Basics of Mormon Transhumanism," https://transfigurism.org/primers/1

54. Lincoln Cannon, "What Is Mormon Transhumanism?," *Theology and Science* 13 (2015), 213–214.

55. Cannon, "What is Mormon Transhumanism?," 203.

56. Cannon, "What is Mormon Transhumanism?," 214.

57. Cannon, "Mormonism Mandates Transhumanism," 57, 60. Cannon understands "resurrection" to refer to transference of mind into a "non-biological substrate" or a "spiritual" body.

58. Mormon Transhumanist Association, "Mormon Transhumanist Affirmation," https://transfigurism.org/mormon-transhumanist-affirmation

59. Mormon Transhumanist Association, "Mormon Transhumanist Affirmation."

60. Cannon, "Mormonism Mandates Transhumanism," 54; Cannon, "What Is Mormon Transhumanism?," 205.

61. Adam S. Miller, "Suffering, Agency, and Redemption," in *Transhumanism and the Body: The World Religions Speak,* ed. Calvin Mercer, Derek F. Maher (New York: Palgrave Macmillan, 2014), 132, 134.

62. Dena S. Davis, *Genetic Dilemmas: Reproductive Technology, Parental Choices, and Children's Futures* (New York: Oxford University Press, 2010).

63. I have inserted "human gene editing" in place of "stem cell research" in this statement. Cf. Lee Davidson, "No LDS Stand on Cell Research," *Deseret News,* July 6, 2001, https://www.deseretnews.com/article/851862/No-LDS-stand-on-cell-research.html

64. Bush, *Health and Medicine Among the Latter-day Saints,* 202.

5

Gifts of Life

Organ and Tissue Donation

Something Wonderful

Cynthia Bowers learned that a coworker had two daughters experiencing kidney failure and in need of a kidney transplant. A colleague had become a living organ donor and provided a kidney to one of the daughters. Ms. Bowers had experienced the loss of her own daughter from cancer two years previously and expressed an empathic connection with her coworker's family: "I didn't have a chance to save my own daughter. How could I let someone lose a daughter when I knew what it was like to lose a daughter?" Ms. Bowers's family supported her desire to be a living organ donor and her daughter Laura also offered to donate a kidney. Laura's kidney was used for the transplant procedure for the coworker's daughter as her organ was a closer match. Laura Bowers subsequently met the recipient and observed, "It just gives you such a thrill to know that you can save someone's life." Cynthia Bowers subsequently donated a kidney to a person she did not know and reflected, "I have been able to do something wonderful with my life."[1]

Overwhelmed by the unanticipated death of her son, Conrad, Lucille Jensen observed, "It's just an impossible time to make any kind of decision. It's a tragedy when things like that happen and usually without a lot of time for you to think about it. I can't even describe the anxiety and the fear and the dread and the loss. . . . Your body and your mind can't think right." In the midst of this existential ordeal, donating her son's organs became a source of consolation and blessing: "[donation] gives you . . . a co-experience with joy as you go through your loss." Jensen maintains contact with three recipients of her son's organs, including a woman who received her son's heart and has since married and adopted a child.[2]

Mormons—members of The Church of Jesus Christ of Latter-day Saints (LDS)—increasingly share their experiences as organ donors or as proxy decision-makers for donating organs from a deceased relative. Compelling

Mormonism, Medicine, and Bioethics. Courtney S. Campbell, Oxford University Press (2021). © Oxford University Press.
DOI: 10.1093/oso/9780197538524.003.0006

organ donation narratives have been related by ecclesiastical leaders in worldwide conferences to illustrate profound relational ties and as metaphors for spiritual transformation.[3] As LDS ecclesiastical teaching initially discouraged organ donation, these contemporary narratives manifest a remarkable shift in attitude and practice toward organ donation and transplantation within LDS moral culture. This chapter situates the ethics of organ and tissue donation within an LDS communal ethos of love of neighbor, altruism, and offering "gifts of life." I begin with a broad-based bioethical analysis of the ethics of organ transplantation and then turn to an exposition of ethical themes in Christian and LDS teaching that warrant organ donation as a morally and spiritually valuable action and as a matter for individual agency rather than state or ecclesiastical mandate.

Priorities, Procurement, and Procedures

Organ transplantation is an exemplary illustration of the so-called miracles of modern medicine, allowing persons who are literally at the brink of death to undergo a life-saving procedure that restores them to good health and a desirable quality of life for many years. The entire donation and transplantation process seems suffused with ethical virtues. Whether authorized by living donors or by proxy decision-makers on behalf of the deceased, organ donation expresses the virtues of love, altruism, and caring for strangers. Donation is a potent symbol of moral community as it offers a rare opportunity for persons living in large, impersonal, and bureaucratic societies to benefit strangers whom they may never meet.[4] The process of retrieving an organ from a donor and transplanting the organ in a recipient represents sophisticated application of medical expertise to provide inestimable physiological and emotional benefits for donors, families, and recipients. Organ transplantation thereby exhibits a unique convergence of moral and medical virtues.

This does not mean that organ transplantation is immune from ethical scrutiny. The increasing prevalence of successful transplants creates a cluster of ethical and policy questions. Three paramount issues emerge in bioethics discussion: (1) the *priority* of transplantation relative to other health care needs in a society with limited resources, (2) policies for increasing *organ procurement* in the context of a chronic scarcity of transplantable organs, and (3) the need for fair *procedures* in distributing scarce transplantable organs.

I briefly comment on the justice issues of priority and procedures, and then provide a more extensive analysis of the ethics of organ procurement.

Priority. Organ transplantation is a life-saving procedure with exorbitant costs. From medical preoperation preparation of the recipient to hospital costs to post-operative care, the cost of procedures for the three most common transplants range from about $415,000 for kidney transplants to $815,000 for liver transplants, to $1.4 million for heart transplants.[5] The aggregate number of kidney, liver, and heart transplants performed on an annual basis generates a cost of approximately $16 billion. A significant commitment of resources—financial, institutional, professional—is invested to reinforce a cultural myth that no amount of money is too high to save a human life. However, the presumptive good of transplantation has been bioethically challenged on the basis of distributing limited resources for basic health care to a broader population. It can seem morally problematic to provide an expensive life-saving benefit to the approximately twenty-five thousand transplant recipients of kidneys, livers, and hearts when, as of 2017, some 28.5 million Americans lacked health insurance coverage.[6] In one policy experiment, the state of Oregon redirected federal Medicaid funds historically dedicated for approximately thirty-five organ transplants to provide access to basic health care for an estimated fifteen hundred persons, primarily pregnant women and children from low-income families. This policy had a distinct utilitarian rationale to provide health care benefits for a large medically underserved population while restricting access to life-saving transplants. The controversy that ensued from this policy culminated in a new state health care program that included a rationing scheme in which preventive medical interventions such as annual exams, maternity care, and vaccinations received higher priority than transplants, even though most transplants received state funding.[7] It is ultimately difficult to justify rationing transplants in the absence of both universal access to basic health care services and policy discourse about the relative priority of different forms of medical treatments, including emergency, curative, preventive, rehabilitative, chronic, and palliative medicine. It cannot be claimed that organ transplantation uniquely contributes to health care stratification when inequalities of access are embedded in health care structures.

Equitable Procedures. The fair allocation of scarce organs is a contemporary manifestation of an original bioethical question about just distribution of limited health care resources: "Who shall live when not all can live?"[8] The distribution procedure initiated by the U.S. National Transplant Act of 1984 placed medical criteria, including need for a transplant and prospect of ben-efit, and justice as guiding criteria for organ allocation to increase organ transplants and express fairness, equity, and respect for the dignity of donors and recipients.[9] The ethical integrity of the allocation process is important for the feasibility of the entire transplantation enterprise; perceptions of un-fair distribution of organs erode public trust and undermine procurement efforts. The allocation issues are conditioned by the inadequate supply of transplantable organs to meet demand, a phenomenon that underlies ethical inquiry of procurement processes.

Organ Procurement. As of early 2020, 112,717 persons were registered on the national organ transplant waiting lists, with over 102,900 persons in need of a kidney transplant. Nearly two-thirds of the prospective recipients have been on a waiting list for one to five years, with the average wait time for persons needing kidneys (who are on dialysis procedures in the interim) at 679 days or close to two years.[10] The number of transplants performed in 2018 reached an all-time high of 39,718, but remains below the rate of per-sons needing a transplant. This scarcity creates a tragic outcome: more than 7,000 persons die annually while on the transplant waiting list.[11] The chronic scarcity in transplantable organs is attributable to several factors, including the successes of medicine in making transplants safe and effective; societal increases of chronic diseases, especially end-stage kidney and liver failure; and policy reliance on individual choice to donate an organ. Since passage of the organizing statutes for organ transplantation—the 1968 Uniform Anatomical Gift Act and the 1984 National Transplant Act—numerous proposals have been advanced to increase donation rates. A recurrent ques-tion is whether current laws should be preserved, modified, or substantially changed to address the scarcity of transplantable organs.

Alternatives for organ procurement are defined by a cluster of ethical values[12] that correspond with LDS ethical principles. The value of societal beneficence entails maximizing donated organs so that the greatest number of persons can benefit, a moral commitment that corresponds with LDS principles of justice and stewardship. The value of respecting donor choice, reflecting the principle of moral agency, means procurement reliance on in-formed decisions to "opt in" to donation by persons who make an explicit choice or respecting the decisions of family members about donation in the absence of a decision by a deceased relative. A corollary of the agential prin-ciple is respecting the decisions of persons to decline to become donors.[13] The value of generosity, corresponding with LDS principles of love, hospitality, and covenant, confers moral priority on procurement policies that reflect a donative or gift intent, a giving of part of oneself to others without expectation of tangible self-benefit. The policy commitment to honor the integrity and dignity of bodily life, which corresponds with communal values of embodiment, entails refraining from procurement policies that treat the human body as a commercial resource.

The values of societal beneficence, respect for choice, generosity, and bodily integrity ethically support the prevalent organ procurement framework in the United States of *explicit choice or opt-in*. The framework's principal ethical inadequacy is its failure to generate a sufficient number of transplantable organs for all persons who could benefit from them. Three ethically defensible procurement alternatives to remedy this deficiency have been proposed. A procurement framework present in many European and some Asian countries known as *presumed consent* is essentially an opt-out rather than an opt-in method. The legal and ethical presumption is that, unless persons explicitly register their dissent or refusal with an organ retrieval agency, they are presumed to have authorized organ retrieval upon death. Empirical studies show that many countries with presumed consent frameworks have higher organ retrieval rates than those achieved under explicit consent.[14] Procedural safeguards ethically differentiate presumed consent from organ conscription: Persons can dissent, and in practice, retrieval agencies request family authorization even though under no legally requirement to do so. The ethical concern with this framework is whether presuming consent cultivates the value of generosity insofar as it allows organs to be retrieved from persons who are apathetic or indifferent about donation but didn't make an explicit choice to dissent.

A second alternative to explicit consent enacted in pilot studies in some U.S. jurisdictions and in other nations is designated as *rewarded gifting*. In this framework, persons who are organ donors or who authorize others to donate their organs receive some form of limited compensation, such as payments for funeral expenses, tax credits, or educational tuition vouchers for children. A variation of rewarded gifting has been adopted in Israel through a 2010 law that confers higher priority on patients in need of a transplant who have a personal or familial history of donation. During the first year following enactment, the number of Israelis registering as donors doubled and "the total number of candidates waiting for a transplant fell for the first time ever."[15] The U.S. experience of rewarded gifting, however, has had very mixed results regarding whether this framework contains incentives necessary to increase organ retrieval rates. The feasibility of rewarded gifting ultimately hinges on assumptions about the reasons many persons currently do *not* authorize organ donation. It is ethically questionable whether compensation or a "reward" dilutes the virtuous character of organ donation as an act of giving and frames it more as a commercial transaction.

A third alternative, an *organ market*, has been the subject of vigorous bioethical debate but has yet to be fully incorporated in any health care system. The philosophical advocates of an organ market contend it will increase the supply of transplantable organs (a claim again reflecting assumptions about incentives, moral or financial, necessary for organ retrieval) and affirm respect for individual self-determination by allowing but not mandating personal commerce in the body.[16] The self-determination argument presents more compelling rationale insofar as the case for increasing the supply of transplantable organs through a market is currently entirely speculative and requires empirical research. Critics contend, moreover, that an organ market compromises some core values of the donation process. An organ market would not necessarily have to supplant the ethos of generosity, but establishing parallel systems of donation and commerce would make a clear symbolic claim that generosity and donation are not intrinsic features of organ retrieval. An organ market may fail to honor the dignity of bodily life by permitting commerce in bodily organs; societal permission of bodily commerce for replenishable bodily tissues such as reproductive tissue, plasma, bone marrow, and breast milk does not provide a comparable analogue for commerce in nonregenerative organs. An organ market inevitably confronts issues of fairness and equity in access and socioeconomic stratification in the distribution of organs. Insofar as trust and confidence in the equitable allocation of organs is necessary for popular support for the overall transplantation endeavor, the prospect of stratified distribution could diminished general participation.

The explicit donation framework best instantiates many of the core values of organ donation but for the fact that so many preventable deaths occur because altruism and generosity do not provide a sufficient supply of organs. A presumed consent approach is an ethically acceptable method as empirical studies show increases in organ retrieval rates in various countries without substantial moral compromise. However, in a society such as the United States, in which a polarizing cultural chasm opened over mandated health insurance coverage, presuming consent is not politically feasible. Although there is a substantial Latter-day Saint presence in countries with presumed consent procedures, LDS teaching, when permitting organ donation, has portrayed the practice as a discretionary gift of life and has not addressed the structural issues of organ procurement.

A Selfless Gift

LDS understandings of organ donation have paralleled advances in transplantation medicine increasing the safety (to donors and recipients) and the efficacy of transplantation. Early transplant procedures, beginning with kidney transplants in 1954 and heart transplants in 1968, were rarely successful in prolonging life for more than a few weeks and were invariably accompanied with substantial complications, pain, suffering, and eventual death. LDS resistance to organ transplantation during this experimental era was, however, based more on theological reasons than medical risks. Lester Bush, Jr., contends that organ transplantation was "one of the last subjects of disagreement between the [LDS] church and modern medicine."[17] During the 1960s, when transplantation presented substantial medical risks to donors and recipients, LDS members who participated in temple rituals were advised "to not will their organs for medical use."[18] No explicit rationale was offered for this recommendation although such resistance may have reflected a common religious concern that organ removal postmortem fails to respect the integrity of the body or otherwise constitutes desecration. A second reservation voiced by diverse religious communities, including LDS moral culture, during this era concerned the impact of transplantation on bodily resurrection.[19] These concerns persist among some individuals with religious convictions even as widespread religious denominational support for donation as a practice of charity has supplanted skepticism. A recent study disclosed that roughly 26% of the reasons offered by persons who declined to be organ donors had a religious valence, including belief in miracles, general religious convictions, and bodily integrity after death.[20]

The next phase of organ transplantation was heralded by the passage of the Uniform Anatomical Gift Act (UAGA) in 1968 that made donation a matter of individual choice. This individualistic focus also permeated the 1974 LDS ecclesiastical statement that accommodated transplantation within the ethics of moral agency: "The question of whether one should will his bodily organs to be used as transplants or for research after death must be answered from deep within the conscience of the individual involved. Those who seek counsel from the Church on this subject are encouraged to review the advantages and disadvantages of doing so, to implore the Lord for inspiration and guidance, and then to take the course of action which would give them a feeling of peace and comfort."[21] The transplantation ethic was the sole instance in the ecclesiastical guidelines that invoked the language of personal "conscience." The appropriation of the legalistic language of "will" intimated that the decision about donation was analogous to the provisions for transferring one's legal estate at death. This did not provide a conceptual justification of organ transplantation based on appeals to "gifts" or "donations." The process of individual decision-making about "willing" organs for transplant or research adhered to the conventional LDS pattern for acquiring spiritual and moral insight: education to achieve transplant literacy, study and mindful pondering, prayer for divine guidance, and a decision manifesting emotional serenity. However, reliance on personal inspiration did not explicitly account for the views of family members.

A noticeable shift in the moral tone of LDS teaching on body and organ donation occurred in the mid-1980s,[22] which not coincidentally was the decade in which organ transplantation became both a safer, more efficacious intervention due to immunosuppressive medications that made it less likely the recipient's body would reject a transplanted organ and a more frequent medical procedure subsequent to passage of brain death statutes.[23] This moral shift was exemplified in an essay by Cecil O. Samuelson, dean of the medical school at the University of Utah (and, later, president of the church-owned Brigham Young University) that addressed a published inquiry about whether it was "wrong" to consider organ donation.[24] Samuelson's very first sentence situates transplantation not in the risk-benefit framework of the 1974 policy but within the narrative of medical progress: "Organ transplantation is one of the true medical wonders of our age." While acknowledging that transplantation presented "economic, ethical, moral, [and] religious questions," Samuelson rejected historical communal concerns over the postmortem impact of donation for bodily integrity on pragmatic biological grounds: "Organ transplantation does not affect one's resurrection, since the organ would soon have returned to the basic elements of the earth following death anyway." Samuelson shifted from speculative critiques to supportive relational rationales, citing the "tremendous blessings" families experience as proxy donors, including consolation when mourning for a deceased relative, as well as life prolongation for recipients. The commentary gave explicit attention to "the selfless love that is evident in this gift of life and health" such that donation reflected core communal values rather than a "willed" individual property transfer. Furthermore, Samuelson invoked definitive authorization by situating organ donation within biblical narratives of miraculous healings. The appeals to "wonder," "blessing," "love," "gift," and healing provide distinctive religious motivations for donation.

Recent ecclesiastical policies on organ donation emphasize these religious and moral rationales. The current statement begins with a prefatory comment highlighting the altruistic intention embedded in organ donations: "The donation of organs and tissues is a selfless act that often results in great benefit to individuals with medical conditions."[25] An ethic of gift-based altruism thus contextualizes the ethics of personal agency. Organ donation is not only a praiseworthy act but expresses the virtuous, selfless, and charitable character of a moral *agent*. The policy also offers the first ecclesiastical acknowledgment of the medical value of tissue donation. The frequency of tissue transplants from blood, bone marrow, corneas, skin, heart valves, veins, ligaments, and bones far exceeds the number of organ transplants performed in the United States.[26] Notably, some societal procurement policies considered for organ transplants have been utilized for tissue procurement. State statutes permit presumed consent for corneal donation, and a recent court decision legalized financial compensation for bone marrow donations in the form of scholarships, housing payments, or gifts to charity up to three thousand dollars, a version of rewarded gifting.[27]

LDS ethics on organ and tissue donation has thus evolved from substantive theological reservations to procedural respect for individual choice to encouragement insofar as donation is framed as a "selfless" gift to benefit others, be it a family member or strangers. This moral evolution parallels advancements in the sophistication, success, and safety of transplantation medicine that have increased the seven-year survival rate for all common transplants above 75%.[28] The characterization of donation as selfless giving implies that donation is a praiseworthy or supererogatory act: a moral agent goes above and beyond what is ethically required and literally embodies altruism in their religious and moral aspirations. The moral culture of benefitting strangers when there is minimal prospect of relationship, let alone reciprocity, is precisely why organ and tissue donation are ecclesiastically commended as selfless: the relational context where commitments to benefit others are part of a personal moral landscape, such as everyday family interactions, is missing in organ donation to strangers. This interpretation of organ donation as an altruistic alternative that is neither legally presumed nor ethically mandated extends the ethics of covenantal care, kinship, and special relationships into a moral community with strangers.

The corollary of both the covenantal and agency principles, however, is that organ and tissue donation are not ethical obligations. There are circumstances in which individuals do have a moral responsibility to come to the assistance of strangers. The ethical rule of rescue, developed in part from the paradigmatic ethic of love displayed in the biblical parable of the Good Samaritan, delineates specific conditions whereby a moral agent has an ethical responsibility to provide assistance to others with whom they have no special relationship: (1) *Vulnerability*: The stranger requires assistance to prevent serious harm, injury, or death; (2) *Effectiveness*: The actions of a moral agent likely will prevent the harm or minimize the risk to the stranger; (3) *Necessity*: The moral agent is the only person, or at least the person best positioned, to provide assistance to the stranger; (4) *Risk-Assumption*: The risks assumed by a moral agent through their assistance fall within a reasonable range; and (5) *Proportionate Benefit*: The benefits to the stranger of receiving the assistance, such as being rescued from death by a donated organ, are proportionally greater than the risks or costs incurred by the moral agent.

The moral logic of the rule of rescue generates a positive ethical duty to benefit strangers in specific circumstances, such as when a child is at risk of drowning in a swimming pool and bystanders can readily rescue them. These criteria do not, however, generate a robust personal ethical responsibility for organ donation. Tens of thousands of vulnerable strangers are in need of a transplant procedure to prevent their death, an organ donation would likely be an effective life-saving intervention, and the benefits are substantial for a potential recipient. However, a case for donation as a specific morally required responsibility is complicated by the conditions of necessity and risk assumption. A postmortem donation of various organs, such as kidneys, liver, heart, lungs, and pancreas, could save the lives of eight persons,[29] but a deceased individual is not the only person whose organs could provide this life-saving intervention. The symbolism of sharing the bodily self encounters an equally powerful constraint of respecting the body as a symbol of the person.

The circumstantially dependent nature of risk in donation is highlighted by meaningful differences in risk to deceased organ donors, living organ donors, and blood donors. There is no meaningful physiological risk to a person who authorizes the retrieval of their organs after they are deceased, though an individual donation will not impact the mortal needs of

the thousands of people on the transplant waiting list. A living organ donation, while similarly morally commendable, cannot be morally required
since it involves risks of general complications from anesthesia as well as
postsurgical recovery that exceed the threshold of reasonable minimal risk.
A person cannot have a required moral responsibility to undergo an invasive
medical intervention with no medical benefit, and some prospect of harm,
to themselves. This moral calculus is altered from minimal to reasonable risk
when the person in need of an organ transplant is a relative with whom a
moral agent has a special relationship and is the best organ match. The donation of blood to strangers, while still not rising to the level of personal ethical
obligation, is nonetheless organizationally coordinated within the religious
community as a matter of promoting civic welfare precisely because of the
minimal risks to donors and the life-saving benefits for recipients.

The pattern of ecclesiastically supported *organized* donation of blood displayed in annual blood drives in LDS congregations[30] is not paralleled by
communal discourse regarding organ donation. The LDS Church has not participated in the annual "donor Sabbath," a nationwide educational initiative
on organ donation in conjunction with the days of worship of Muslim, Jewish,
and Christian religions to encourage "opportunit[ies] for faith communities
to share their views and join in the conversation."[31] However, a compelling
argument can be constructed for something resembling a moral rule of rescue
for religious communities that relate narratives like the Good Samaritan parable, stress sacramental symbols of bodily sharing, and enact principles and
covenants of love, hospitality, and solidarity with the vulnerable as part of
their formative moral fabric.

Organized Donation: The Sacramental Motivation

In a pioneering essay on religious attitudes toward transplantation, bioethicist William F. May examined the religious warrants for organ donation.[32]
May formulated his position that donation within constraints was a permissible middle path in contrast to religious understandings that either opposed donation in principle or supported organ retrieval irrespective of the
person's choice. These contrasts between religious positions reflect different
theologies of embodiment on whether the body is real, good, and intrinsic
to personal identity. In May's view, these characteristics of embodiment

justify and constrain retrieval of bodily organs based on the principle of ex-plicit consent: "the view of life in the body in the Western religious tradition authorizes medical intervention, but, at the same time, shies away from all interventions ... that do not enjoy the full consent of all parties in question."[33]

Given the inescapability of embodied life in Western religiosity, May maintained that organ recovery invariably invoked powerful symbols and sentiments ranging from revulsion to generosity. Such symbols and their ritual instantiation within faith communities can motivate practices of *or-ganized* donation. The potency of religious symbolism led May to differen-tiate alternatives for organ procurement in terms of ethical acceptability and preferability: "In matters so fundamental as the donation of human organs, ... *giving and receiving* are better than routine *taking and getting* [i.e., pre-sumed consent] and certainly to be preferred to *buying and selling* [i.e., organ market]."[34] The moral resources and symbols of religious communities ex-tend beyond simply commending organ and tissue donation and generate positive reasons for an organized communal donative culture. Rejecting a tepid, legalistic permission for donation, May develops a morally potent analogy by drawing on a central Christian symbol and liturgical ritual, the sacramental or communion ritual: "Christ shares, under the form of bread and wine, his body and blood with his disciples. He invites and bids his followers to share in this life of service. What more fitting and direct sign of this sacrifice than the believer's sharing of a portion of his body and blood?"[35] A Christian, called to an ethic of emulation of Christ, embodies donative love through offering gifts of the body, whether it be blood, tissue, or organ dona-tion, in selfless service to others.

Insofar as LDS moral culture enacts communal organization for *blood* do-nation while discourse on *organ* donation is noticeable by its absence, May's appeal to embodiment and symbolism should hold special resonance for LDS communities. The meaningful ethic of service in LDS moral culture[36] is embedded within a commonly invoked scriptural warrant: "when you are in the service of your fellow beings, you are only in the service of your God."[37] However, in practice the culture of service is characterized by insularity to-ward callings and stewardship within the religious community. As May suggests, moral power resides in reversing this theological rationale: rather than situate care for the neighbor or stranger as an occasion for service in one's path to God, the exemplary sacrifice of Jesus becomes a model for caring and service of others, including the self-giving of one's body and blood to save life.

The same scriptural sermon invoked for the ethic of service presents an analogy comparable to the religious grounding for organ and tissue donation that May finds in the sacramental symbol. As delineated previously in the principles of hospitality to the stranger and covenantal care, the prophetic teacher Benjamin reminds community members perplexed about communal responsibilities for the poor: "Are we not all beggars? Do we not all depend upon the same Being, even God, for all the substance which we have, for both food and raiment? . . . Now, if God, who has created you, on whom you are dependent for your lives and for all that ye have and are, doth grant unto you whatsoever you ask that is right, . . . then, how you ought to impart of the substance that ye have." one to another."[38] The experiential dependency of each person and the community on gifts beyond our control for life, sustenance, and nurture is vital to cultivating empathy with the vulnerable who are in need of bodily sustenance. This interdependency creates and sustains a moral culture organized around gifts and donations. An ethos of love, stewardship, and care for the vulnerable stranger will express responsive generosity to the divine gifts of care and sustenance each person receives from God. A culture of service grounded in responsive generosity rather than atomized individual choices provides a stronger religious warrant for an organized pattern of donation of bodily tissues and organs than currently prevails in ecclesiastical teaching and communal practices.

A Donative Culture

The LDS community has an ethical responsibility to cultivate a donative moral culture regarding organ and tissue donation within which persons and families can freely exercise their moral agency. The principle of respect for personal self-determination and the commitment of explicit or proxy consent appropriately guide public policy on organ donation and transplantation. However, communal narratives of self-giving to the stranger or beggar and a community characterized by a covenantal gift ethos and donative culture provide a moral motivation for organ donation that is absent in the ecclesiastical ethic of individual choice. This will not resolve the broader organ scarcity crisis. Perhaps the ethical best society can achieve is laden with a tragic dimension of preventable deaths of prospective recipients. A donative culture is most compatible with the underlying ethical principles and values of organ donation and also of stories of dying that the community can appropriate into its ongoing collective life.

Notes

1. Sarah Jane Weaver, "Mother, Daughter Reach Out to Strangers," *Church News*, February 2, 2006, https://www.thechurchnews.com/archive/2006-02-04/mother-daughter-reach-out-to-strangers-28305

2. Allison Pond, "Myths About Religion and Organ Donation Cause Hesitation," *Deseret News*, August 5, 2001, https://www.deseretnews.com/article/700168701/Myths-about-religion-and-organ-donation-cause-hesitation.html

3. Dale G. Renlund, "Family History and Temple Work: Sealing and Healing," *The Ensign,* May 2018, https://www.lds.org/general-conference/2018/04/family-history-and-temple-work-sealing-and-healing?lang=eng; Dale G. Renlund, "Preserving the Heart's Mighty Change," *The Ensign,* November 2011, https://www.lds.org/general-conference/2009/10/preserving-the-hearts-mighty-change?lang=eng

4. Thomas H. Murray, "Gifts of the Body and the Needs of Strangers," *Hastings Center Report* 17:2 (April 1987), 30–38.

5. Millman Research Report, "2017 U.S. Organ and Tissue Transplant Cost Estimates and Discussion," August 2017, http://us.milliman.com/uploadedFiles/insight/2017/2017-Transplant-Report.pdf; United Network for Organ Sharing, "Transplant Costs," https://transplantliving.org/financing-a-transplant/transplant-costs/

6. United States Census Bureau, "Health Insurance Coverage in the United States: 2018," November 8, 2019, https://www.census.gov/library/publications/2019/demo/p60-267.html

7. Daniel Fox, H. M. Leichter, "Rationing Care in Oregon: The New Accountability," *Health Affairs* 10:2 (Summer 1991), 7–27.

8. Shana Alexander, "They Decide Who Lives, Who Dies," *Life* 53 (1962), 102–25.

9. Organ Procurement and Transplantation Network, "Ethical Principles in the Allocation of Human Organs," June 2015, https://optn.transplant.hrsa.gov/resources/ethics/ethical-principles-in-the-allocation-of-human-organs/

10. Organ Procurement and Transplantation Network, "National Data," https://optn.transplant.hrsa.gov/data/view-data-reports/national-data/

11. United Network for Organ Sharing, "Transplant Trends," https://unos.org/data/transplant-trends/

12. President's Council on Bioethics, "Organ Transplantation: Ethical Dilemmas and Policy Choices," https://bioethicsarchive.georgetown.edu/pcbe/background/org_transplant.html, accessed August 3, 2020.

13. Tiffanie Wen, "Why Don't More People Want to Donate Their Organs?," *The Atlantic*, November 10, 2014, https://www.theatlantic.com/health/archive/2014/11/why-dont-people-want-to-donate-their-organs/382297/

14. R. Matesanz et al., "How Spain Reached 40 Deceased Organ Donors per Million Population," January 9, 2017, https://onlinelibrary.wiley.com/doi/full/10.1111/ajt 14104

15. Wen, "Why Don't More People Want to Donate Their Organs?"

16. Stephen Wilkinson, "The Sale of Human Organs," *Stanford Encyclopedia of Philosophy*, 2015, https://plato.stanford.edu/entries/organs-sale/

17. Lester E. Bush, Jr., *Health and Medicine Among the Latter-day Saints: Science, Sense, and Scripture* (New York: Crossroad, 1993), 105.

18. Bush, *Health and Medicine Among the Latter-day Saints*, 106.

19. William F. May, "Religious Justifications for Donating Body Parts," *Hastings Center Report* 15 (February 1985), 38. The nature and composition of the resurrected bodily form has long been a source of speculation for Latter-day Saints. See Samuel Morris Brown, *In Heaven as It Is on Earth: Joseph Smith and the Early Mormon Conquest of Death* (New York: Oxford University Press, 2012), 58–63.

20. Wen, "Why Don't More People Want to Donate Their Organs?"; Fatemah Ghorbani et al., "Causes of Family Refusal for Organ Donation," *Transplantation Proceedings* 43 (March 2011), 405–406.

21. "Attitudes of the Church of Jesus Christ of Latter-day Saints Towards Certain Medical Problems," in "Mormon Medical Ethical Guidelines," ed. Lester E. Bush, Jr., *Dialogue* 12 (Fall 1979), 98.

22. Wayne A. Mineer, "Organ Transplants and Donation," in *Encyclopedia of Mormonism*, ed. Daniel H. Ludlow (New York: Macmillan Reference, 1992), 1051–1052, https://contentdm.lib.byu.edu/digital/collection/EoM/id/4024

23. No ecclesiastical policy has been issued on the determination of death for medical purposes, entailing de facto recognition of brain death criteria stipulated in the Uniform Determination of Death Act. See Nikolas T. Nikas, Dorinda C. Bordlee, Madeline Moriera, "Determination of Death and the Dead Donor Rule: A Survey of the Current Law on Brain Death," *Journal of Medicine and Philosophy* 41:3 (June 2016), 237–256.

24. Cecil O. Samuelson, Jr., "I Am Thinking of Donating Some Organs for Transplantation. Am I Wrong in Wanting to Do So?" *The Ensign*, February 1988, https://www.lds.org/ensign/1988/02/i-have-a-question/am-i-wrong-in-wanting-to-donate-organs-for-transplantation?lang=eng&clang=ase. Samuelson's commentary remains the most substantive exposition in LDS teaching on the theological, ethical, and medical aspects of organ donation.

25. The Church of Jesus Christ of Latter-day Saints, *General Handbook of Instructions* (Salt Lake City: 1989), 11–6; The Church of Jesus Christ of Latter-Day Saints, "Medical and Health Practices," in *Handbook 2: Administering the Church*, Section 21.3.7, https://www.lds.org/handbook/handbook-2-administering-the-church?lang=eng. See Appendix A for full statement.

26. Health Resources and Services Administration, "General FAQ: Bone Marrow Transplantation," https://bloodcell.transplant.hrsa.gov/about/general_faqs/index.html#1990%20number%20tx%20inUS

27. Organ Procurement and Transplantation Network, "An Evaluation of the Ethics of Presumed Consent," https://optn.transplant.hrsa.gov/resources/ethics/an-evaluation-of-the-ethics-of-presumed-consent/; I. Glenn Cohen, "Selling Bone Marrow: *Flynn v. Holder*," *New England Journal of Medicine* 366 (January 26, 2012), 296–297; Adam Martin, "You Can Now Sell Your Bone Marrow for $3,000," *The Atlantic*, December 2, 2011, https://www.theatlantic.com/national/archive/2011/12/you-can-now-sell-your-bone-marrow-3000/334788/

28. Abbas Rana, Elizabeth Louise Godfrey, "Outcomes in Solid Organ Transplantation: Success and Stagnation," *Texas Heart Institute Journal* 48:1 (2019), 75–76.

29. U.S. Department of Health and Human Services, "Organ Donation Statistics," https://www.organdonor.gov/statistics-stories/statistics.html, accessed August 3, 2020.

30. Jason Swensen, "The Church Has Partnered with the Red Cross Nearly 100 Years," *Church News*, March 26, 2015, https://www.lds.org/church/news/the-church-has-partnered-with-the-red-cross-nearly-100-years-?lang=eng; Richard A. Mimer, "Blood Transfusions," in *Encyclopedia of Mormonism*, ed. Daniel H. Ludlow (New York: Macmillan Reference, 1992), 131–132, https://contentdm.lib.byu.edu/digital/collection/EoM/id/5527

31. Donate Life America, "National Donor Sabbath," https://www.donatelife.net/nds/, accessed August 3, 2020.

32. May, "Religious Justifications," 38–42.

33. May, "Religious Justifications," 39.

34. May, "Religious Justifications," 41. May does not contend that commerce in the body, including buying and selling in an organ market, is ethically unacceptable but rather claims it is a position for which there are no religious warrants.

35. May, "Religious Justifications," 42.

36. Van Evans, Daniel W. Curtis, Ram A. Cnaan, "Volunteering Among Latter-Day Saints," *Journal for the Scientific Study of Religion* 52 (December 2013), 827–841.

37. The Book of Mormon, *Mosiah* 2:17.

38. The Book of Mormon, *Mosiah* 4:19–21.

6

Born Dying

Heaven Is Very Near

Landon and Malia, parents of a teenage daughter and three boys between ages four and eight, felt for some time that their family was "incomplete" and were trying to have a fifth child. While Malia had manageable complications with prior pregnancies, symptoms emerged early in this pregnancy as the fetal heartbeat was undetectable by ultrasound during the twelfth week. At nineteen weeks, Landon and Malia received distressing news from a perinatologist: their developing child, a boy, was three weeks delayed in growth based on a femur measurement. The physician recommended amniocentesis for further testing while informing the couple, "There's no chance you are having this baby." Medical testing led to a provisional diagnosis of intrauterine growth restriction (IUGR), a condition in which the placenta cannot provide nutrition and oxygen adequate for development. By twenty-three weeks, the child weighed 160 grams, less than one-half of normal. The perinatologist reiterated his earlier prognosis and asserted, "I don't understand why you are choosing to continue this pregnancy." Their supportive ob-gyn informed Landon and Malia to "prepare for a good fight but also prepare for your child not making it."

Landon and Malia felt "completely defeated and overwhelmed" as they processed this information and the medical recommendations. They acknowledged "what the reality is and [that] the chance of our son's survival rests in the Lord's hands." They prayed for "the doctors who watch over us" and for "the faith and strength to do whatever God asks so God can bless us." They also requested prayers on their son's behalf from their family, friends, and fellow church members, and their congregation fasted and provided meals, visits, and overnight care for the four children. Landon gave a blessing to Malia that promised their child, whom they had named Benjamin in the intervening weeks, would be born, though the blessing did not say anything about Benjamin's life span.

The pressure from specialists to discontinue the pregnancy came to a head during a conversation with a fourth perinatologist. The physician reiterated to the couple what was now familiar information: Any delivery with a prospect

Mormonism, Medicine, and Bioethics. Courtney S. Campbell, Oxford University Press (2021). © Oxford University Press.
DOI: 10.1093/oso/9780197538524.003.0007

for survival beyond birth required that Benjamin reach about 400 to 500 grams and Benjamin would likely need to be delivered at twenty-five to twenty-six weeks. These dire prospects meant the couple needed to make a decision on proceeding with the pregnancy. Infuriated, Malia thought, "How dare you ask me to end the life of my child?" After the couple had a brief private conversation, Malia replied, "We will do whatever we need to do to get this little guy here with the best stats we can and then you doctors will take over and we will deal with whatever happens then." When the physician inquired about whether the "go forward" conviction was rooted in religious reasons, Malia invoked parental commitment: "This is my son and I will do whatever I can to give him the best possible chance at life." A religious consideration did lie in the background of Landon and Malia's desire to continue the pregnancy: a clear threshold in the ecclesiastical policy of The Church of Jesus Christ of Latter-day Saints (LDS) for family membership is birth and conferral of a name and blessing on the infant.[1]

Following referral to a university hospital, the senior physician asked Landon and Malia about their goals for Benjamin to ensure their decisions were informed and within a realm of reasonable expectation. They indicated they wanted medical interventions to "give Benjamin a chance at birth," to which the physician replied, "We can certainly do that." Landon and Malia felt for the first time a mutuality of trust and a convergence of goals for care. This did not mean the difficult days were at an end. At twenty-eight weeks, Malia experienced a significant increase in blood pressure and upon hospital admittance was diagnosed with "mirror syndrome," meaning her heart was mirroring the blood-flow problems Benjamin was experiencing. A specialist indicated Malia's own health was now of paramount concern and requested consent for an immediate delivery with the knowledge that Benjamin would not survive. The couple privately considered this option and decided that as long as Malia remained stable, they would continue the pregnancy. They reasoned that, with Benjamin now weighing 335g, "we are so close and feel we need to give him just a bit more time."

Their refusal of the physician's recommendation initiated a lengthy conversation about family between Malia and the physician. Malia related her convictions that their temple marriage had conferred to Landon and Malia a promise of an eternal family. They had a firm faith that death does not break the eternal bonds of family and they would be parents to Benjamin whether in mortal life or in the post-mortal life. The physician, an adherent of a different religious tradition, commended their integrity but observed that Malia "should not be a martyr. You need to care for yourself because you have a family and children that rely on you."

In the thirty-first week of pregnancy Malia was diagnosed with severe pre-eclampsia. She authorized a cesarean section delivery and immediate admission of Benjamin (who now weighed 417 grams, just shy of 1 pound) into the neonatal intensive care unit (NICU). Benjamin was so small he could not be intubated for respiratory support of his undeveloped lungs. What transpired in the ensuing fifteen minutes was something the NICU staff related later to Malia was a "miracle" that "we have never seen before": Benjamin took a breath on his own and his oxygen saturation levels were initially at a normal range. Eventually, a respiratory tube was placed, and Landon gave Benjamin the family-defining name and blessing.

A subsequent echocardiogram diagnosed Benjamin with pulmonary hypertension and an opening between the blood vessels leading from the heart (patent ductus arteriousus or PDA) that posed risks of heart failure. Given his underlying conditions, the physicians assessed Benjamin's prospects for six-month survival at about 65%, with a 20 to 50% chance of two-year survival. Although Benjamin achieved small developmental steps in the NICU, his underdeveloped lungs and respiratory problems developed into a chronic lung condition (bronchopulmonary dysplasia or BPD) associated with chronic pulmonary hypertension. These conditions occasionally required resuscitation and caused secondary complications of a staph infection, flu, and edema. The issue of whether to continue life support loomed on almost a daily basis, and Landon and Malia dreaded the recurrent "When should we stop care?" conversations with nearly every attending physician. Malia acknowledged, "I know we may have to make the call to stop treatment. That terrifies me. What scares me more is that in the effort of trying to save him we are keeping him here longer than he wants to be." Strikingly, Landon and Malia attributed a kind of agency to Benjamin. Following one resuscitation, they reflected, "Benjamin is very much on the fencepost deciding whether or not he wants to stay with us or return home. . . . So many times Benjamin has stood on the threshold of both Heaven and Earth and weighed his options. We are there yet again." Even the divine will seemed subordinated to Benjamin's symbolic agency: "The Lord allowed Benjamin to choose. . . . Benjamin is in charge. We just keep going forward until Benjamin decides."

The medical issues, parental decisions, and Benjamin's agency converged after four and a half months in the NICU. Benjamin was experiencing severe edema and jaundice from early-stage liver failure attributable to continual

infusions of steroids. He was on high doses of opioids and was no longer con-
scious. These conditions led consulting cardiologists to decide Benjamin was
"too sick" to undergo a surgical procedure to repair his heart abnormalities.
Landon and Malia reflected that they had done everything they could, that
Benjamin in effect had "decided" to go to his heavenly home, and the time had
arrived to discontinue medical interventions. They asked each other: "Will we
be able to live with this decision later?" They brought their children to the hos-
pital for a last visit with their sibling; Malia requested that once the machines
were discontinued, she wanted to hold Benjamin "unattached" from all the
technologies. In these final moments, the room was filled not only with family
and NICU staff but a profound spiritual presence that Malia described as "the
closest thing to heaven." Saddened but not distraught, Landon and Malia re-
flected, "Benjamin's death was not and is not a tragedy. His birth and short life
was a divine miracle. By all statistical and medical knowledge, Benjamin was
never supposed to live and we had four and a half months of great memories
with him. We love him and will surely miss him, but we know our life here on
Earth must continue until the day comes when we can be reunited with him.
We are an eternal family."

It is a much safer medical world to bring a child into since the 19th-century LDS pioneer community experienced an infant mortality rate of 9%.[2] Advances in preventive medicine and biomedical technologies have substantially decreased infant and maternal mortality in the past century. The personal wrenching, parental distress, and moral anguish experienced by Landon and Malia are now a rare phenomenon relative to the vast majority of uncomplicated births of healthy children. Still, one in eight babies in the United States are born prematurely (prior to thirty-six weeks' gestation), the primary reason nearly a half million infants have some stay in a neonatal intensive care unit (NICU). The possibility that infants born at twenty-four weeks' gestation, and in some cases even earlier, can have their biological lives sustained through intensive medical interventions illustrates a remarkable expansion of medical progress. The survival rate for premature infants born between twenty-two and twenty-eight weeks is now at 79%, with increasingly higher rates of survival without long-term health ramifications for infants born at twenty-five weeks or later.[3]

The stories of infants who are born dying often push in very different professional, parental, and ethical directions. Health care professionals, especially nurses who have the most direct contact with the families and ongoing care of the infants, find themselves caught in the midst of the medical and emotional maelstrom and experience "moral distress."[4] Parents are confronted with existential questions about continuing a complicated pregnancy when the prognosis for a successful delivery or survival of the child beyond birth is uncertain or grim. They experience an almost unbearable burden of determining whether it may be better for a child born with severe complications to discontinue medical interventions. They may re-story their experience from tragedy to "miracle" and request high-risk interventions to provide some prospect of life for their child. Parents bear witness to the successes of medical interventions to save life and the limits of these interventions to prolong life.

The parental narratives of infants born dying present unique illustrations of how LDS convictions of the revealed reality—including the salvific value of embodied life, parental commitment and autonomy, the eternal family relationship, and medical futility—influence decisions regarding life endings at the beginning of life.[5] These experiences carry forward a historical bioethical debate about selective nontreatment of seriously impaired newborns, a controversy between families, professionals, and institutions that ultimately prompted regulatory intervention by the federal government.

Gray Zones

The question of appropriate care for premature infants or infants born with a physical impairment emerged in the late 1960s due to advances in mechanical ventilation and intravenous feeding (total parenteral nutrition [TPN]). As these technologies improved the prospects for successfully prolonging neonatal life, professional discourse formulated criteria for withholding or discontinuing treatments: "In the context of certain irremediable life conditions, intensive care therapy appears harmful. These conditions are . . . [the] inability to survive infancy, inability to live without severe pain, and inability to participate, at least minimally, in human experience."[6] These criteria reflected both medical and nonmedical (relational experience) judgments, which implied that decision-making required mutual collaboration of physicians and parents. The articulation of these thresholds for harmful treatment was important in light of reports in the medical literature describing parental and professional decisions to withhold treatment from infants born with Down syndrome or spina bifida.[7]

The catalyst for broad public discussion about selective nontreatment was a controversy in 1982 about an infant known as "Baby Doe," who had been born with Down syndrome and esophageal atresia, which obstructs nourishment from passing to the stomach. A routine and almost invariably successful surgical procedure can correct this obstruction, without which an infant will die from either starvation or pneumonia. Baby Doe's parents declined to authorize surgery, however, believing that "it was not in the best interests of the infant, their other two children, and the family entity as a whole for the infant to be treated."[8] The infant was instead provided with medications for pain and restlessness. The treatment refusal was immediately litigated, with a trial court determining that the parents had legal authorization to make the decision, and Baby Doe died at six days, prior to a hearing by the U.S. Supreme Court. In the context of publicized reports that infants with physical but non-life-threatening impairments, including Down syndrome, were subject to selective nontreatment in various hospital special-care nurseries, the Baby Doe case set off a firestorm of controversy. The case symbolized a decision-making pattern regarding impaired infants that stressed assessments of psychosocial quality of life rather than the benefits and risks of the medical intervention.

The federal government responded to the contentious issue by promulgating so-called Baby Doe regulations that deemed that selective non-treatment of newborns violated the state's responsibility to ensure equal treatment for persons and infants with disabilities. The regulations mandated necessary medical treatment for virtually all impaired infants regardless of their condition or its severity. Medical professionals agreed with the underlying moral sentiment to treat infants with non-life-threatening impairments but vigorously objected to governmental intrusion into medical decision-making and the pediatrician-parent-infant relationship. The American Academy of Pediatrics (AAP) recommended instead that health care institutions convene an "Infant Bioethics Committee" to address medical and ethical issues in the care of critically ill infants "when the forgoing of life sustaining treatment is being considered."[9] Subsequently, in a case in which parents had refused a surgical intervention for an infant born with spina bifida, the U.S. Supreme Court held that the Baby Doe regulations were not a valid application of federal antidiscrimination law.[10] Congress then passed amendments requiring states receiving federal funding for child protection services to develop procedures on reporting "medical neglect" of newborns. The concept of medical neglect consisted of withholding or withdrawing medically indicated treatment, including medication, food, and fluids, unless any of three conditions obtained: the infant was deemed by "reasonable medical judgment" to be (a) chronically and irreversibly comatose; or (b) the treatment would prolong dying, would not effectively treat the infant's life-threatening conditions, or would be "futile" in prolonging the infant's life; or, (c) the treatment would be "virtually futile" for survival and its provision would be "inhumane."[11] These criteria for nontreatment possessed important symbolic value as they reflected a "consensus statement among the federal government, the American Academy of Pediatrics, and various advocacy groups for the rights of the disabled."[12] The thresholds nonetheless presumed a level of medical knowledge about treatment outcomes and treatment futility not then available in neonatology.

Reflecting the AAP's more recent commitment to a treatment standard based on the "best interests" of the child,[13] neonatologists John Lantos and William Meadow have delineated several morally ambiguous or "gray zones" in decision-making for premature and low-birth-weight infants that permit professional and parental discretion regarding technological interventions.[14] A first gray zone is the status of birthweight as the primary indicator for delivery-room resuscitation and neonatal intensive care. The survival rates of infants born above 750 grams have increased so dramatically in the past two decades that treatment interventions are almost invariably provided; however, neonatologists, on grounds of medical futility, rarely provide resuscitation interventions when the infant weighs less than 400 grams. The gray zone of birthweight, as illustrated in the case of Benjamin, then pertains to infants born between 400 and 750 grams, customarily between twenty-two and twenty-five weeks' gestation. In such cases, neonatal intensive care following initial resuscitation involves a so-called trial of therapy for a limited duration to facilitate more informed prognostic assessments. Through this limited trial, infants with a poor prognosis due to conditions such as intraventricular hemorrhage, severe lung disease, or sepsis may "declare themselves" as unlikely to live.[15] Prognostic uncertainties following these limited-duration interventions often lead to further interventions, creates ethical questions, initiates professional moral distress, and, as with Landon and Malia, prompts protracted parental anguish.

A second zone of moral ambiguity concerns decision-making about premature infants for whom survival beyond birth will be accompanied by severe physiological or neurological impairments. These circumstances present difficult questions about whether it is obligatory or discretionary to provide life-extending treatment when an infant's quality of life is anticipated to be very burdensome. Lantos and Meadow propose a moral continuum in which the nature of the impairment is linked with judgments about whether medical interventions are morally required. One legacy of the original Baby Doe case is that a child born with Down syndrome with an esophageal blockage is now "a paradigm case example of a condition in which treatment is considered obligatory." The other end of the continuum is represented by a child born with the neural tube defect anencephaly, the absence of a forebrain and cerebral cortex. This condition is a "paradigm case of medical futility"[16] for which it would be wrong to provide medical interventions other than comfort care. Between these two paradigmatic conditions, numerous other impairments for which neonatal intensive care may not be morally obligatory or could be permissibly withdrawn if initiated provide a morally ambiguous zone for professional and parental discretion.

The moral landscape of neonatal intensive care presents ongoing issues about who should assume the roles of principal decision-makers and apply medical and ethical criteria. The AAP maintains that parents, in consultation with neonatologists and pediatricians, are best positioned to assess and articulate the child's best interests.[17] However, the intense nature of treatment decisions in the neonatal setting and the profound emotional bonds that arise between parents and their infant have led some scholars to advocate a "broad shoulder" approach in which physicians assume greater responsibilities in difficult decisions to continue or discontinue interventions.[18] Furthermore, continued medical advances have expanded the treatment possibilities of neonatal intensive care from infants with objective impairments to infants born at the fluid thresholds of viability. Infants born at what were once considered pre-viable gestational stages now receive aggressive intensive care interventions to maintain life. The ethical imperative that continuing life is virtually always in an impaired child's best interests is matched by a technological imperative embedded in the progressive narrative of medical advances to use all available technology. The convergence of ethical and technological imperatives means that parents experience an almost irresistible momentum for continued treatment and a do-everything mandate. This progressive narrative is both reinforced and challenged by a counternarrative about the intrinsic value of children in LDS teachings and practices.

Theological Consolation: "Too Pure, Too Lovely"

Scriptural passages in the LDS canon and ecclesiastical commentary indicate that children who die before a time period designated as the age of "accountability" are received into God's presence through the grace of salvation. The most direct scriptural teaching was conveyed to Joseph Smith in a vision addressing a personal decade-long question about the salvific status of his brother Alvin, who had died from an apparent medical misuse of the purgative calomel prior to the organization of the LDS Church.[19] Joseph Smith learned that his brother was in the presence of God and that "all children who die before they arrive at the years of accountability are saved in the celestial kingdom of heaven."[20] This was an extraordinary theology of consolation in an era of high childhood mortality.

LDS teaching on infant or childhood mortality re-stories the concept of "premature death" as pertaining to a child's physical age but not their spiritual development or maturation. The vulnerability of children to lethal diseases and death is frequently interpreted through the salvation narrative as a sign of a special, protected, and unique status: children who die prior to accountability displayed their faithfulness to God in the pre-mortal life and so need not be subjected to the vicissitudes of mortal life. Bruce R. McConkie, a prominent ecclesiastical leader and writer of the mid-20th century, appropriated this narrative in teachings on the salvation of young children. McConkie's exposition appealed to a statement voiced by Joseph Smith at the funeral of a two-year-old child: "The Lord takes many away, even in infancy, that they may escape the envy of man, and the sorrows and evils of this present world; they were too pure, too lovely, to live on earth."[21] McConkie extended this theology of consolation to persons who mature biologically but do not develop moral accountability due to a mental impairment: "They never arrive at the years of accountability and are considered as though they were little children." This narrative has significant implications for persons who do acquire moral agency: such persons "need the tests and trials to which we are subject and . . . our problem is to overcome the world and attain that spotless and pure state which little children already possess."[22]

The theology of consolation reiterated by the prophetic successors to Joseph Smith focuses on the spiritual and moral wholeness of infants and children whose death prior to accountability symbolizes divine deliverance from an evil world.[23] This theodicy is frequently cited in parental narratives that bear witness of their faithful anguish in caring for their children who are born dying. In one account, a family faced an arduous decision about having a risky heart surgery performed on their daughter who had been born prematurely and with a severe heart anomaly. The child would die without the operation, but the surgery provided no guarantee that the girl would either survive or live a reasonable quality of life. The parents ultimately opted for the surgery; the narrative relates "the baby lived for eight months, but did not thrive" and required further operations. In reflecting on the emotional and financial ordeal of the parents, their ecclesiastical leader observed, "The death of an infant, though emotionally wrenching, is perhaps the situation on which there is the most theologic comfort available."[24] This narrative concludes with the consoling theodicy that children born dying are "too pure" for mortal life.

The prospect of salvation and familial union in the next life does not obviate the necessity of grief and mourning. Parents who have lost a child at birth or infancy relate confusion, frustration, and anger intensified by a "lack of conversation" on such experiences within the Church community.[25] The narrative of hope does allow parents to re-story their experience from tragedy to a "blessing" or "gift." The narrative also does not imply theological fatalism and diminished agency on the part of parents in caring for their children. Even while acknowledging the reality of a divine plan for their child or symbolically imputing agency to their impaired infant, parents who often feel powerless are not choiceless. The primary calling of parents is to nurture their children through the passages and transitions of mortal life. The salvific conviction that eternal progress requires a passage from an unembodied spirit being to embodiment in mortal life provides a theological motivation for using medical interventions to continue complicated pregnancies to birth. This choice can be made even when such interventions pose some risk to the health of the mother (Malia's experience) or when the prospect of a postbirth survival with relational interaction for the infant is diminished. Furthermore, consonant with the healing vocation of medicine, parents may likewise request a judicious use of medical technologies, including appropriate aggressive interventions in neonatal intensive care settings, as an initial resort to prolong life.[26]

The soteriological narrative offers a threshold permitting the anguishing decisions of parents to forgo or withdraw medically futile interventions. The salvific necessity of embodiment means that so far as an infant has obtained bodily form in mortal life, the parents bear no ethical obligation to request treatment that will be futile relative to the infant's survival and that holds the likelihood of imposing unnecessary pain, distress, or suffering for the child. This appeal to futility and unnecessary suffering, which corresponds to the Baby Doe criteria, is embedded in LDS policy on the limits to life-prolonging treatment: "when dying becomes inevitable, it should be seen as a blessing and a purposeful part of eternal existence. Members should not feel obligated to extend mortal life by means that are unreasonable."[27] The concept of "unreasonable means" is not further explicated in ecclesiastical policy, permitting parental interpretation as informed by medical recommendations. The policy lacks comprehensiveness, however, for as illustrated in the narrative of Landon and Malia, it is medically difficult to determine when an impaired infant is "inevitably" dying.

Situated within the bioethics debate about selective nontreatment of impaired newborns, LDS teaching and practice dictate that the interests of a child born with severe, life-threatening impairments are invariably best served through life-prolonging medical interventions, but this presumption can be overridden in circumstances of medical futility through processes of parental agency, prayer, and medical consultation. One LDS parent framed well the values of alleviating suffering and best interests that informed a couple's anguished decision to discontinue aggressive medical interventions and respiratory support for their four-day-old son diagnosed with Ebstein's anomaly: "I really believe that there are worse things than death. I wouldn't want Charlie to suffer. . . . I wouldn't want to keep him here just for ourselves. . . . I feel like as a parent to let go is you giving up, but in a way you're doing . . . what is [in] their best interest and sometimes that decision is to let them pass away."[28]

Beyond the Infant?

The LDS ethical principles on ending life at the beginning of life offer a middle path between a vitalistic imperative of doing everything possible, a view that rejects the concept of medical futility, and a biblicist imperative to rely exclusively on divine healing and forgo medical interventions entirely on the grounds that they are invariably futile. This middle path presents its own ethical complications, however. Advances in neonatal resuscitative technologies have shifted the thresholds of viability for premature infants and of medical futility for children born with severe, life-threatening impairments. The fluidity of viability and futility is illustrated by current medical assessments on the duration of gestation required for resuscitation measures to possess a reasonable hope of addressing underdevelopment of a premature infant's lungs (as well as other complications of premature birth). In the 1990s, a threshold of 50% survival rate for resuscitation applied to infants born at twenty-five to twenty-six weeks' gestation. These infants now survive at close to 90%, while the 50% survival rate applies to infants born at twenty-three to twenty-four weeks'. Professional guidelines recommend against viability assessments and resuscitation efforts prior to twenty-two weeks' gestation.[29] However, a recent medical report indicated that a young girl born at twenty-one-weeks, four-days' gestation is now a thriving toddler.[30] The fact that a few infants born prior to twenty-two weeks have survived with minimal complications does not mean that the professional standard of care should make twenty-one weeks the threshold between medically efficacious and invariably futile treatments, but biomedical advances make it increasingly difficult to draw a hard line regarding viability and medical futility.

The fluidity of viability also entails that it is not possible to use specific medical criteria to give content to the ecclesiastic generalization that using "unreasonable" means to prolong life is nonobligatory. What is "unreasonable" shifts between cultures and time periods within a culture. Two criteria often invoked to differentiate between obligatory and optional medical treatments include (a) the probability of benefit to the patient and (b) the proportionate benefit to the patient relative to the risks, harms, or burdensomeness of successful treatment. Treatments that are minimally beneficial and substantially burdensome are ethically optional, and depending on the nature of burdensomeness and prospect of pain and suffering, the ethical obligation may reside in *not* providing such treatments. Though the criteria of "benefit," "proportion," and "burden" are susceptible to circumstantial interpretation, there is prudential moral wisdom in allowing these concepts to be somewhat fluid. For example, providing antibiotics to treat pneumonia is a reasonable intervention in most medical circumstances but can be unreasonable when dying is indeed "inevitable." The open-ended nature of "unreasonable means" of life prolongation entails that decisions about providing, withholding, or withdrawing medical interventions in neonatal intensive care will hinge on the ongoing progress and availability of medical technologies. This is an almost intractable ethical problem in end-of-life care generally since there is almost always "something else" contemporary medicine can do to prolong life.[31] It also presents an intractable theological problem since it may appear that the divine will for a child born dying is contingent on available technologies and their successful application. Ecclesiastical teaching is thus best positioned to present general principles and then defer to the professional expertise of physicians and the moral agency of families to govern contextual application of ethical concepts.[32]

A further ethical complication in applying a benefit/burden standard of care is the experiential reality that parents invest a substantial portion of their identity and emotional lives in their relationship with their child. Although decisions to withhold and to withdraw medical interventions are *ethically* equivalent, the parental binding of identities makes it *psychologically* difficult for parents, having initially authorized intensive life-saving interventions, to decline subsequent medical interventions. One implication of conferring responsibility for assessing the best interest of the infant or what constitutes an unreasonable means of prolonging life to the moral agency of the parents is that the parents' deep and moving commitments to their child will be entangled in all determinations of the child's interests. Parental aspirations for a meaningful and lasting connection with their infant motivate authorizing life-extending medical interventions and requests for subsequent interventions necessitated by complications from the underlying condition or the first intervention. It is emotionally anguishing for parents to feel that in declining ongoing interventions they are "giving up" on their child (or on God's healing). Nevertheless, parents should be the presumptive primary decision-makers, as guided by considerations of the best interests of their child, as the parents will live with and bear the lifelong repercussions and relationships of decisions to authorize or decline treatment. The principle of justice is violated when a third party, such as the government or even the medical profession, imposes a decision on the parents in an emotionally and ethically fraught situation when the parents will ultimately live with the burdens of the decision.

Ethical responsibility requires attentiveness not only to parental distress but also to the moral distress experienced by professionals, especially nurses, in neonatal intensive care.[33] Moral distress refers to a circumstance in which professionals committed to ethically responsible practices encounter institutional limits or decision-making structures that render them powerless to act on what is right or to prevent moral wrongs.[34] Parental decision-making can generate moral distress or compromised professional integrity for caregiving staff. An LDS NICU nurse experienced intense moral distress related to continued parental requests for surgical interventions on a child born with trisomy 18, a chromosomal disorder responsible for abnormalities throughout the body's organ systems and that is typically fatal within the first year of life. The infant had displayed continual signs of physical discomfort, pain, and suffering, frequently requiring sedation, through the course of numerous invasive interventions. When a new surgery was proposed to address the infant's jaw displacement, the nurse considered talking to a nursing supervisor about the overall care plan for the child. The nurse, invoking the proportionate-burden criterion, sought understanding regarding "why are we [professional staff] doing this?," when "this" constituted a burdensome intervention with minimal long-term medical benefit for the child even from a successful procedure. The nurse had no desire to contravene parental wishes but felt morally compromised, perceiving their nursing skills were contributing to prolonging the baby's suffering. This led the nurse to express a quintessential sentiment of moral distress: "I feel like I'm a bad person no matter what I do." The nurse did not feel empowered to request an ethics consult, observing that in several years of neonatal intensive care nursing, an infant bioethics consultation had never occurred nor had any member of the institutional ethics committee visited the NICU. Effectively, the institution cultivated an environment of ethical neglect toward the nurses. Ultimately, the nurse determined that personally withdrawing from the case wouldn't change the interventions and outcomes for the infant and that they were best positioned to provide the quality nursing care the infant should receive.

A further question is whether advocacy of the child's interests should be insulated from societal interests. The costs of NICU care are substantial whether in an individual case (the hospital care for Benjamin exceeded $1.5 million) or cumulatively. The aggregate cost of care for premature infants has been estimated at 1 to 2% of total health care costs,[35] a figure of around $35 billion in 2020 health care costs. While scholars have emphasized that comprehensive prenatal care can prevent premature births and diminish costs, this presumes prospective knowledge of the many causes of prematurity and the imposition of paternalistic restrictions on behaviors that influence prematurity. Lantos and Meadow contend that advocacy of either prevention or neonatal intensive care is a false choice: "the best thing for babies is not to choose between one and the other, but to build systems that seamlessly provide both. Such systems do not reduce costs but they do improve outcomes."[36] Furthermore, Lantos argues that NICU care embodies a morally potent symbol: "NICUs make money only because they implicitly make a compelling moral claim upon society. . . . NICUs stand for our society's moral commitment to children, our excellence in caring for them, and even for our moral progress over time in recognizing that our tiniest citizens have rights."[37] The societal interest in containing costs and utilizing intensive care can conflict with the societal expression of moral commitment to vulnerable infants. The integrity of parental decisions about care for their newborn must be insulated from social questions about constrained resources.

Re-Storying Quest: Tragedy to Blessing

LDS parental stories of their experiences in the culture of neonatal intensive care can be illuminated through a typology of illness narratives developed by medical sociologist Arthur Frank.[38] The conventional story of illness told by the medical profession is a *restitution* narrative in which medical interventions enact a professional promise to restore a person to wholeness and their familiar place in their world. This narrative frame is inadequate for infants born dying as there is no baseline of wholeness for the particular infant that medicine can restore. An alternative narrative frame bearing greater authenticity for the parental experience of world disruption and identity dissonance is a *chaos* narrative. Serious, unexpected, and life-threatening conditions render the customary familiarity of the world unintelligible; parental devotion and care are subject to powers and forces beyond parental control, occasionally to such an extent that words, let alone stories, cannot express the experience. The chaos narrative holds potency for LDS parents who ask, "How can the Lord deprive me of raising this child?"[39] LDS families who undergo the shattering ordeal of their child born dying encounter the basic question of theodicy: how can a loving God permit identity-disrupting experiences to afflict good persons who live with fidelity to God's teachings? It is not possible for parents who have entered a covenantal relationship of an eternal family to experience the birth of a child with significant medical complications that threaten their survival without wondering what purpose the ordeal has within the divine plan. This is the essential issue in the chaos narrative.

Parents who experience such mortal wrenching nonetheless display remarkable resilience attributable to their devotion and re-storying their experience through the theological and ethical resources of the community, including the eternal significance of family life, the importance of mortality, and the necessity of embodiment. LDS parents who bear witness to a premature birth of an impaired child who is at risk of severe developmental abnormalities or death give voice to a third narrative form; a quest is born from the ashes of an implicit chaos narrative. The *quest* narrative manifests a parental experience of finding their way through their ordeal to a new identity, a new relationship with their child, and ultimately to new sources of meaning. In contrast to the restitution narrative, the quest for place, identity, and meaning is markedly different from a "going back" or restoring of familiar patterns of life; in contrast to the chaos narrative, the quest represents a "going forward" of parents to discover the new and unfamiliar.

The narrative of re-storying quest is illustrated in the repeated characterization by LDS parents of their brief relationship with their dying child not as a "tragedy" but as a "miracle," "blessing," "mission," or "a gift, a gift to us, and a gift to other people."[40] The re-storied parental experience opens a path from an encounter with chaos, meaninglessness, and ethical irrationality to an ongoing discovery of a "new normal" and of healing the brokenness in the parental world. The soteriological narrative of the loving character of the divine, the giftedness of human life, and the "pure" spiritual state of a child born dying delivers parents from chaos not by restoring their common world but rather by initiating a journey to a new identity, relationship with their child, and understanding of God.

The characteristics of a re-storying quest can be expressly delineated. A first element is characterized by parental realization of *human finitude* and limited understanding about ultimate meaning to these arduous experiences. This is coupled with an expression of faith in and submission to *providential care and purposiveness* for experiences that defy human comprehension. As articulated by a mother whose child died six months after birth following repeated surgical procedures, "I felt that God was in control regardless of what I could or couldn't see in those moments."[41] The re-storying quest also contains a *narrative witness*, a public sharing of the child's and the family's ordeal, which opens to a new understanding of parenting identity, that parenting does not end but continues in a different form: "I can continue to be her parent [as] . . . I can share her story and hopefully help other people to maybe find more faith or trust or hope."[42] The parents find meaning through a new identity as storytellers.

Re-storying further resides in parental aspirations for a *meaningful relationship* with their child no matter how transient its duration. Every narrative of a child born dying ritualizes this relationship through a *naming* of the child accompanied by a father's blessing. The spiritual practice of naming reflects a view that, though disempowered regarding the life or health of their child, parents retain the salvific and morally potent power to confer identity and relationship. Naming symbolizes the family's welcoming embrace to the infant. The quest enables parental expression of convictions that the relationship will be *renewed* in an eternal fullness in post-mortal life. As articulated by Benjamin's parents, "We are an eternal family," and in the eternities, "We can be reunited with Benjamin."

The re-storied quest reflects a view that death does not have the final word about human life but is a *passage* in the eternal progress of the child and parents. Scholars have maintained that early LDS teaching and practice—embedded in a social context of high rates of infant, child, and maternal mortality—manifested a theology of "conquering death" such that the disruption of familial relationships, however anguishing, is provisional, not final.[43] Salvific narratives that children who die prior to accountability are saved in heaven and ritualizing ethics through spiritual practices of conferring a name and blessing on a child born dying do not supplant chaos, grief, and mourning but they offer a way for parents and the community to appropriate death in their common identity and look beyond death to a promised time when God will wipe away all tears.

Notes

1. LDS ecclesiastical policies about the family status of a stillborn child allude to the possibility "that a stillborn child may be part of the family in the eternities. Parents are encouraged to trust the Lord to resolve such cases in the way He knows is best." The Church of Jesus Christ of Latter-Day Saints, "Church Policies and Guidelines, Medical and Health Policies: Stillborn Children (Children Who Die Before Birth)," in *General Handbook: Serving in the Church of Jesus Christ of Latter-day Saints,* Chapter 38.7.11, February 19, 2020, https://www.churchofjesuschrist.org/study/manual/general-handbook/38-church-policies-and-guidelines?lang=eng#title_number91

2. Melvin L. Bashore et al., "Mortality on the Mormon Trail, 1847–1868," *BYU Studies Quarterly* 53 (2014), 109–123, https://byustudies.byu.edu/content/mortality-mormon-trail-1847-1868

3. Barbara Stoll et al., "Trends in Care Practices, Morbidity, and Mortality of Extremely Preterm Neonates, 1993–2012," *Journal of the American Medical Association* 314 (2015), 139–151, https://jamanetwork.com/journals/jama/fullarticle/2434683

4. Andrew Jameton, "Dilemmas of Moral Distress: Moral Responsibility and Nursing Practice," *AWHONN's Clinical Issues in Perinatal and Women's Health* 4 (1993), 542–551.

5. See Joshua J. Perkey, "Loss and Childlessness: Finding Hope amid the Pain," *The Ensign*, October 2012, https://www.lds.org/ensign/2012/10/loss-and-childlessness-finding-hope-amid-the-pain?lang=eng
6. Albert R. Jonsen et al., "Critical Issues in Newborn Intensive Care: A Conference Report and Policy Proposal," *Pediatrics* 55 (1975), 756–765.
7. Raymond S. Duff, A. G. M. Campbell, "Moral and Ethical Dilemmas in the Special-Care Nursery," *New England Journal of Medicine* 289 (October 25, 1973), 890–894, https://www.nejm.org/doi/full/10.1056/NEJM197310252891705; Anthony Shaw, "Dilemmas of 'Informed Consent' in Children," *New England Journal of Medicine* 289 (October 25, 1973), 885–890.
8. Alan Meisel, *The Right to Die* (New York: John Wiley and Sons, 1989), 436–437.
9. American Academy of Pediatrics, Infant Bioethics Task Force and Consultants, "Guidelines for Infant Bioethics Committees," *Pediatrics* 74 (August 1984), 306–310.
10. *Bowen v. American Hospital Association*, 476 US 610 (1986).
11. Michael White, "The End at the Beginning," *Ochsner Journal* 11 (Winter 2011), 309–316, https://www.ncbi.nlm.nih.gov/pmc/articles/PMC3241062/
12. John D. Lantos, William L. Meadow, *Neonatal Bioethics: The Moral Challenges of Medical Innovation* (Baltimore: Johns Hopkins University Press, 2006), 74.
13. White, "The End at the Beginning."
14. Lantos, Meadow, *Neonatal Bioethics*, 109–111.
15. Lantos, Meadow, *Neonatal Bioethics*, 109.
16. Lantos, Meadow, *Neonatal Bioethics*, 110.
17. American Academy of Pediatrics Committee on Fetus and Newborn, "Non-Initiation or Withdrawal of Intensive Care for High-Risk Newborns," *Pediatrics* 119 (February 2007), 401–403.
18. Christopher Meyers, "Cruel Choices: Autonomy and Critical-Care Decision-Making," *Bioethics* 18 (2004), 104–119.
19. Samuel Morris Brown, *In Heaven as It Is on Earth: Joseph Smith and the Early Mormon Conquest of Death* (New York: Oxford University Press, 2012), 22–25.
20. *Doctrine and Covenants* 137:10.
21. Joseph Fielding Smith, ed., *Teachings of the Prophet Joseph Smith* (Salt Lake City: Deseret Book, 1976), 196–197.
22. Bruce R. McConkie, "The Salvation of Little Children," *The Ensign*, April 1977, https://www.lds.org/ensign/1977/04/the-salvation-of-little-children?lang=eng
23. In the words of current LDS president, Russell M. Nelson, "Every now and then, the Lord takes these special spirits back home to him early, as if to spare them some of the weighty trials that mortality would have brought." Russell M. Nelson, *The Gateway We Call Death* (Salt Lake City: Deseret Book, 1995), 40.
24. Val D. MacMurray, Kim Ventura, "Decision Models in Bioethics," in *Perspectives in Mormon Ethics: Personal, Social, Legal, and Medicine*, ed. Donald G. Hill, Jr. (Salt Lake City: Publishers Press, 1983), 269; cf. Sharon Belknap, "Don't Let My Baby Die," *The Ensign*, December 2001, https://www.lds.org/ensign/2001/12/dont-let-my-baby-die?lang=eng

25. Danielle Christensen, "Finding Hope: When Babies Return to Their Heavenly Home Before Ever Coming to Their Earthly One," *LDS Living*, January 16, 2020, http://www.ldsliving.com/Knowing-Heaven-How-2-Latter-day-Saint-Families-Who-Didn-t-Take-Their-Baby-Home-From-the-Hospital-Hope-for-an-Eternal-Family/s/92240?utm_source=ldsliving&utm_medium=email

26. The ambiguous soteriological status of stillborn children raises its own distressing challenges. See Melinda E. Jennings, "Our Stillborn Baby," *The Ensign*, February 2006, https://www.lds.org/ensign/2006/02/our-stillborn-baby?lang=eng; Jeanne B. Inouye, "Stillborn Children," in *Encyclopedia of Mormonism*, ed. Daniel H. Ludlow (New York: Macmillan, 1992), 1419, https://eom.byu.edu/index.php/Stillborn_Children.

27. The Church of Jesus Christ of Latter-Day Saints, "Church Policies and Guidelines, Medical and Health Policies: Medical and Health Practices," in *General Handbook: Serving in the Church of Jesus Christ of Latter-day Saints*, Chapter 38.7.7.

28. Christensen, "Finding Hope."

29. Hannah C. Glass et al., "Outcomes for Extremely Premature Infants," *Anesthesia and Analgesia* 120 (June 2015), 1337–1351.

30. Kaashif A. Ahmad et al., "Two-Year Neurodevelopmental Outcome of an Infant Born at 21 Weeks' 4 Days' Gestation," *Pediatrics* 140 (December 2017), http://pediatrics.aappublications.org/content/140/6/e20170103

31. Atul Gawande, *Being Mortal: Medicine and What Matters in the End* (New York: Henry Holt and Company, 2014), 149–190.

32. The Church of Jesus Christ of Latter-day Saints, *Teachings of the Presidents of the Church: Joseph Smith* (Salt Lake City: Intellectual Reserve, Inc., 2011), 281–291. Cf. Mark L. McConkie, R. Wayne Boss, "'I Teach Them Correct Principles and They Govern Themselves': The Leadership Genius of the Mormon Prophet Abstract," *International Journal of Public Administration* 28 (2005), 437–463.

33. A. Janvier et al., "Moral Distress in the Neonatal Intensive Care Unit: Caregiver's Experience," *Journal of Perinatology* 27 (2007), 203–208, https://www.nature.com/articles/7211658.pdf?origin=ppub

34. Andrew Jameton, "What Moral Distress in Nursing History Could Suggest About the Future of Health Care," *AMA Journal of Ethics*, June 2017, https://journalofethics.ama-assn.org/article/what-moral-distress-nursing-history-could-suggest-about-future-health-care/2017-06

35. Jonathan Muraskas, Kayhan Parsi, "The Cost of Saving the Tiniest Lives: NICUs Versus Prevention," *AMA Journal of Ethics* 10 (October 2008), 655–658, https://journalofethics.ama-assn.org/article/cost-saving-tiniest-lives-nicus-versus-prevention/2008-10

36. Lantos, Meadow, *Neonatal Bioethics*, 142–143.

37. John D. Lantos, "Hooked on Neonatology," *Health Affairs* 20 (September/October 2001), https://www.healthaffairs.org/doi/full/10.1377/hlthaff.20.5.233

38. Arthur W. Frank, *The Wounded Storyteller: Body, Ethics and Illness*, 2nd ed. (Chicago: University of Chicago Press, 2013), 75–136.

39. Jennings, "Our Stillborn Baby."

40. Christensen, "Finding Hope"; Sharon Belknap, "Don't Let My Baby Die," *The Ensign,* December 2001, https://www.lds.org/ensign/2001/12/dont-let-my-baby-die?lang=eng

41. Perkey, "Loss and Childlessness."

42. Christensen, "Finding Hope."

43. Brown, *In Heaven as It Is on Earth*.

7

Dying Well

Life Endings and Medical-Assisted Death

Vigil

New-age music emanated softly from a CD player on a chair in a darkened hospital room. A male adult sat on the floor writing by the illumination of lights from the nurses' station outside the room. A serenity seemed present in the room, disrupted by the ever-present beeping of the machines monitoring Sarah's physiological status as she lay motionless in the hospital bed. The beeps and statistical data from the machines relayed information to Sarah's professional caregivers and to her three adult children that their mother was still alive despite her present unconsciousness, a condition that had persisted for a week due to a car accident and an intracranial hemorrhage. Every two hours, a nurse entered the room, recorded vital signs data, monitored the life-sustaining fluids, and then left the adult children to the shadows of the room and of life.

The adult children had experienced helplessness and a depersonalized dissonance: their experience of what was happening to "Mother" was manifestly different from the staff's professionalized treatment of another intensive care "patient." Sarah had authorized her children to act as collective proxy for medical decisions through an advance directive; short conversations between the attending physicians and the family had addressed authorization for interventions to reduce swelling in the brain and provide medication administration. Still, the children experienced something missing from interactions with professional staff. The requisite institutional protocols, information exchange, and legal documents seemed rather hollow as they found themselves encountering ultimate boundaries of human life. The prospect of a lingering or persistent unconsciousness or even death evoked for the children matters about the meaning of their mother's mortality that were unacknowledged by professional caregivers.

As family and professional staff contemplated several medical alternatives over several weeks—including to "wait and see" if Sarah would regain

Mormonism, Medicine, and Bioethics. Courtney S. Campbell, Oxford University Press (2021). © Oxford University Press.
DOI: 10.1093/oso/9780197538524.003.0008

consciousness, additional medical interventions to stabilize regulating bodily systems, and neurosurgical procedures—an ethical chasm opened: if it came to a choice about continued interventions, risk-filled surgical procedures, or discontinuing life-prolonging treatment, what decision should be made, on what basis, and what would that decision mean? These questions could not be answered by institutional procedures; they resided in value convictions and constructed meanings of mortality about which there was no consensus among the family members despite their shared religious faith, let alone between the family and professional staff. Familial disagreements occurred because the children situated their mother's current condition and prognosis differently within the context of Sarah's prior life history, her spiritual and relational values, her prospects for recovery and recuperation, and their barely vocalized personal concerns about assuming responsibilities for familial caregiving. Disagreements with the professional staff occurred because the staff had no narrative of their patient as a mother and a person outside the hospital setting. Sarah was "another stroke victim" who would receive the care, treatment, and professional attention any ICU patient should receive. For the children, however, their mother was not "any ICU patient."

Although neurological specialists could not make a confident prognosis until (or if) Sarah recovered consciousness, they raised the prospect that a prolonged unconscious stupor would likely mean very significant limits for Sarah's recuperation of her physical, cognitive, and relational capacities. Even with a successful surgical procedure to relieve pressure on the brain (an outcome that could not be guaranteed), Sarah's prognosis included some permanent paralysis, loss of mobility, compromised memory, and speaking difficulty. These impairments cumulatively could entail a substantially compromised quality of life for a woman who had been quite independent. Two of the children believed it was important to try any surgical procedure that would maintain their mother's life; the other sibling questioned the benefits of the procedure relative to its burdens. Physicians and family alike experienced their fallibility and finitude in this encounter with mortality.

The most meaningfully charged ethical choices confronted by members of The Church of Jesus Christ of Latter-day Saints (LDS) and by LDS medical professionals concern decision-making at the end of life. The successes of modern medicine in prolonging life present difficult questions about continuing or stopping medical treatment, requesting full life-support interventions, or authorizing a do-not-resuscitate order. More recently in some jurisdictions with significant LDS populations and professionals, end-of-life options have expanded to encompass requests by terminally ill patients to hasten death by medical means. This chapter examines ethical considerations confronting the LDS community in decisions regarding both forgoing medical life support and in medically assisted dying.

Dying Well: Death as Eschatological Passage

As with other bioethical questions, LDS ethics regarding end-of-life decisions are contextualized by the soteriological narrative that affirms birth and death as "passages" in the eternal progress of all persons.[1] Birth marks the passage of the self from pre-mortal life into an embodied life, with the "soul" comprising the integration of spirit and body, a condition necessary for persons to acquire a fullness of joy.[2] Death marks a temporary separation of spirit and body, a provisional experience of dis-embodiment preparatory to an eventual literal resurrection of the person and the reuniting of the embodied soul in a perfected and glorified form.[3] The salvation narrative re-stories ethical and familial decisions about requesting medical treatments to prolong life or declining further medical interventions to allow a terminally ill person to die. As articulated by one scholar, "Belief in everlasting life after mortal death should allow faithful Latter-day Saints to make wise and rational decisions regarding artificially prolonging life when medical means to restore useful and functional existence have been exhausted."[4] Three illustrative theological assumptions are embedded in this claim: death is a natural part of mortal experience that is temporarily deferred by "artificial" medical interventions, ethical decisions about end-of-life treatments fall within the scope of wise personal stewardship and agency, and medical interventions can be legitimately discontinued when they are futile in furnishing a "useful and functional existence." Although the content of "useful" and "functional" life needs greater specification, LDS end-of-life ethics rejects a vitalistic theology holding that biological life is to be preserved and prolonged at all costs.

Narratives by LDS patients and families emphasize the value of conversations about planning for end-of-life care, utilizing advance directive instruments, and enacting covenantal commitments for caregiving responsibilities to bear with the burdens of fellow church members, including presence to the dying and shared mourning with the survivors.[5] A poignant essay by a cancer victim published posthumously challenged the LDS religious community "to confront death and dying better [and] . . . to find a way to help [dying persons] that consists of genuine acts of love and compassion."[6] The author was critical of insensitive comments and common platitudes that discount or ritualize the profound existential anguish and practical ordeals experienced by dying persons and their families. A well-meant invitation to "call if you need anything" can be experienced as shallow and isolating when a dying person needs the presence of another to "walk with me through my memories." The constructive realism of the author manifested a communal critique and call to live the covenantal responsibility to "be with" dying persons.[7]

There is no distinctive LDS ideal of a "right" way of dying or a "good death." The perspective of hospice physician Ira Byock that "dying well is fundamentally about people experiencing something that has meaning and value for them" is certainly resonant with the communal context of dying and caring for the dying.[8] The hopes embedded in the moral culture for dying persons and their families include (relative) freedom from physical pain and emotional suffering, opportunities for remembering, and conversations of familial legacy and memory. The integrity of the dying person throughout their life may culminate in a final "testimony" and admonition to survivors to live the gospel of Christ. These are not aspects of dying unique to the community. The narrative interpretation of the dying process corresponds with various professional and psychological understandings of dying as an occasion for finding new perspectives, making meaning of mortality, healing of self and of relationships, and experiencing growth in personal character for the dying person and for their covenantal companions.[9]

LDS ecclesiastical teaching about end-of-life medical treatment decisions has paralleled developments in bioethical discourse about "the right to die."[10] The initial ecclesiastical policy was formulated in the same year as a landmark case that served as a catalyst for widespread public discussion. In 1975, a young female college student, Karen Ann Quinlan, was diagnosed in a persistent comatose condition subsequent to a drug and alcohol overdose. Ms. Quinlan's parents, devout Roman Catholics, acted as proxy decision-makers for their daughter's medical care and eventually requested that the ventilator sustaining their daughter's life be withdrawn. The New Jersey Supreme Court upheld this request by recognizing that patients had rights of privacy and bodily integrity to be free of unwanted invasive medical treatment.[11] Fifteen years later, the U.S. Supreme Court heard arguments in its first right-to-die case, brought by the parents of another young woman, Nancy Cruzan, who had suffered a similar neurological injury from an automobile accident and whose biological life was sustained through feeding tubes. The Court held that feeding tubes providing nutrition and hydration constituted a medical procedure that patients (or their proxies) had legitimate rights to refuse or have withdrawn. However, the Court agreed that a state could require patients to meet an evidentiary standard of "clear and compelling" documentation regarding their treatment preferences, a standard they did not believe Ms. Cruzan met.[12] These groundbreaking cases established a *negative* right to die—that is, a right of patients to *noninterference* from invasive, life-prolonging medical treatments—and prompted greater public education about completing advance directives to authorize end-of-life treatment decisions.

Patient requests and professional practices also initiated controversies about a *positive* right to die, a patient claim to *assistance* from medical professionals in hastening death. This right received legal recognition through the 1994 passage of the Oregon Death with Dignity Act, a citizen referendum permitting a decisionally capable terminally ill patient to request and self-administer a medication from an attending physician to bring about their death. Variations of legalized physician-assisted dying have subsequently been adopted in nine states and the District of Columbia.[13] The limited scope of physician-assisted dying has generated further questions regarding the applicability of a positive right to die for terminally ill patients who have lost decision-making capacity but have requested physician assistance in an advance directive or are incapable of physical self-administration, and for non–terminally ill patients who experience irremediable psychological distress and suffering, as well as whether physicians (or nurse practitioners) have professional moral authority to administer lethal medication to patients. The practices extending a positive right to die beyond patient self-administration of a death-hastening medication to clinician-administered medication for terminally ill patients who have lost decision-making capacity or physical function or for nonterminal patients suffering from an unacceptable quality of life are now commonly referred to as *medically assisted dying.*[14]

LDS ecclesiastical policy has sought to present a middle ground for exercising negative rights to decline treatment that contrasts with both vitalistic pursuit of biological life prolongation and with legalized positive rights to medical-assisted death (or what ecclesiastical policies define as "euthanasia"). The current policy affirms that the sanctity of life is the primary religious value at stake in decisions about ending life, which can be honored in decisions either to continue or refuse treatment but is deemed violated by social policies permitting euthanasia. The relevant portion of the policy regarding forgoing life-prolonging medical interventions states, "The Church of Jesus Christ of Latter-day Saints does not believe that allowing a person to die from natural causes by removing a patient from artificial means of life support, as in the case of a long-term illness, falls within the definition of euthanasia. When dying from such an illness or an accident becomes inevitable, it should be seen as a blessing and a purposeful part of eternal existence. Members should not feel obligated to extend mortal life by means that are unreasonable. These judgments are best made by family members after receiving wise and competent medical advice and seeking divine guidance through fasting and prayer."[15]

Several aspects of this statement are noteworthy. First, the moral realities of the salvation narration lead to re-storying dying as a "blessing" and thereby imbued with purposiveness rather than tragedy or fatalism. Second, the organizing decision-making concepts are illustrative of reasoning by negation. The cessation of life support is not considered the moral equivalent of euthanasia, while the claim that there is no obligation to prolong life by unreasonable means presumes some conception of "reasonable" treatments. Theological purposiveness and inferential reasoning seem intended to allow maximal breadth for responsible moral agency. However, these intentions are complicated by the indeterminacy of "unreasonable" interventions and appeals to an "inevitability" threshold. The determination of "inevitable dying" is in large measure a professional assessment constrained by the fallibility and finitude of physicians.

Inevitability and Finitude. Physician Atul Gawande has observed that determining when a person *is* dying is a professional puzzle for which medicine no longer has clear answers. In the context of technologies that expand the possibilities of life prolongation, Gawande expresses uncertainty about just "what the word 'dying' means anymore. In the past few decades, medical science has rendered obsolete centuries of experience, tradition, and language about our mortality and created a new difficulty for mankind: how to die."[16] As medical technology expands the boundaries of life prolongation and transforms circumstances of "inevitable" dying into a reversible or chronic condition, ascertaining when thresholds of inevitability and medical futility are crossed is increasingly difficult. The concept of "inevitable" dying is marked with the same fluidity and susceptibility to technological expansion that characterizes "viability" at birth. This reality of medical finitude and fallibility risks misjudgments of both undertreatment and overtreatment.

Since determinations of dying rely on fallible medical judgments, an error that presumes a person is inevitably dying when that is not the case can lead to undertreatment and a premature demise of a patient. To avoid this possibility, the logic of medical finitude and fallibility intimates trying the "next thing," a posture conducive to overtreatment. Physician fallibility and finitude, professional evasion of dying, and expanding research provide an irresistible cultural and professional momentum to continue treatment as though everyone, not just an occasional fortunate survivor, will experience increased longevity.[17]

A central difficulty is that the medical profession imbibes a professional agnosticism about the meanings of mortality that parallels social and personal agnosticism. Gawande observes, "The problem with medicine and the institutions it has spawned for the care of the sick and the old is not that they have had an incorrect view of what makes life significant. The problem is that they have had almost no view at all."[18] In the absence of substantive visions of dying well, the practical necessity for decisions is filled by imperatives of technology, professional ethos, and culture: "we fall back on the default, and the default is: Do Something. Fix Something."[19] If the society and profession give negligible attention to the question of a meaningful life, choices regarding end-of-life care that raise such a question create existential perplexity. This agnosticism about dying well within a narrative of a life well-lived confers significance to the ecclesiastical construction of "inevitable" dying. The purposiveness of dying is informed by the values of the dying person and their family members and enacted by the covenanted community. The professional finitude and fallibility lurking within determinations of when dying becomes inevitable, or of judgments that medical treatments will be futile, are effectively supplanted by a religious re-storying that is confirmed through spiritual practices of prayer and fasting. Medicine need not go where it ethically should not.

Unreasonable Means of Life Prolongation. The implicit LDS distinction of reasonable/unreasonable medical measures of prolonging life parallels analogous conceptual categories in bioethics, such as natural/artificial, ordinary/extraordinary, customary/heroic, and obligatory/optional treatments. The ethical issues embedded in these categories concern whether any medical treatments are *always* morally required and whether these categories support judgments of both when it is permissible to *forgo* a life-prolonging medical intervention and when it is obligatory to *not* provide such an intervention. Any medical intervention must have some prospect of medical benefit to the patient—whether improving health, offering a better quality of life, or prolonging life—and prospective benefits must be weighed against burdens the patient experiences from undergoing the intervention; for example, prolonging life should be balanced against increased pain. Within this framework of benefit and proportionality, no medical treatments as such are always morally required and virtually every medical intervention—from ventilators, to chemotherapy, to dialysis, to transfusions, to feeding tubes, to antibiotics—can be considered an "unreasonable" means of prolonging life in the circumstances unique to each patient.[20] When a medical treatment is anticipated to provide minimal medical benefit to the patient, even if conferring psychological benefit for family members, and the patient considers the burdens more substantial than the anticipated benefits, it can be morally *wrong* to provide such treatment.

The inchoate concept of "unreasonable" means in LDS ecclesiastical policy could be specified through these ethical concepts. The policy recommends a procedural method to differentiate between reasonable and unreasonable means of life prolongation: consulting medical professionals, spiritual practices of fasting and prayer, and requesting divine guidance. This procedure manifests LDS principles of respecting the ethical integrity of medicine and honoring moral agency. However, the deliberations of medical professionals, the patient, and the family invariably invoke considerations of prospective benefit and proportionate burdens. These criteria are patient-centered. As emotionally fraught as treatment termination decisions can be, physicians have a professional responsibility to refrain from recommending end-of-life interventions designed more to treat family distress than patient welfare. Familial distress and anticipatory grief about the impending loss of a loved one can be addressed in ways other than providing "everything possible" and by healing professionals rather than technologies, including counseling, bereavement support, and the presence of a community of covenantal caring. Caregivers bear witness to the dying person that they are remembered, a valuable part of the community, and that they will not be abandoned.

The patient-centered focus of what comprises unreasonable means of life prolongation similarly precludes appeals to the economic primacy of societal interests. The issue here is not trivial. Studies indicate that about 30% of Medicare expenditures, or about $170 billion annually, is devoted to medical care during the last six months of life.[21] The symbolism of end-of-life care makes it morally immune from implicit, bedside rationing of resources. It can be a noble and even altruistic action for a dying person to decline life-prolonging treatment out of concern to diminish the overall costs of care for family members. However, if economic rather than patient-centered considerations are embedded in determinations of unreasonable means of prolonging life, decisions about the negative right to die risk being transformed into a *duty* to die for the sake of the greater good.

LDS teaching on prolonging life and ending medical treatments displays the communal pattern of articulating generalized principles and conferring responsibility for ethical applications of these principles to the exercise of moral agency and the ethical commitments of the healing professions. These principles are reoriented, however, when the issue concerns a patient's positive right to request self-administered life-ending medication or medical-assisted death. On these controversial practices, the LDS Church has presented both ecclesiastical teaching and a public moral witness.

Euthanasia and Medical-Assisted Death

Dr. Savay is an LDS physician in internal medicine with three decades of practice in his local community, where he is well recognized among his colleagues for his professionalism and care for elderly patients as they approach life's ending. His relationship-centered philosophy of medicine includes commitments to keep his patients comfortable, ensure meaningful relationships for patients and family, provide medications to relieve pain or suffering, refrain from medical treatments that would prolong life unnecessarily, and avoid being the cause of death. His patients and families choose an array of legal alternatives in their experience of dying, from patient refusal of chemotherapy for readily treatable lymphoma to family insistence on years-long life prolongation of a loved one in a persistent vegetative condition. These alternatives can become professionally complicated as Dr. Savay lives in a state that has legalized physician-assisted death, a procedure in which a terminally ill patient requests that their physician write a life-ending medication to hasten their death. A few of Dr. Savay's patients have asked for his help in dying. Ms. Abby, an elderly woman with Parkinson's diagnosed with a life-ending lung disease, faced what she considered to be a poor quality of life and informed Dr. Savay she was "ready to die." Mr. Eason, diagnosed with congestive heart failure and immobility because of the strain from physical exertion, also requested Dr. Savay's assistance.

Dr. Savay has no legal duty to participate in a patient's request for life-ending medication. His philosophy of care means that providing medication to relieve

pain, even when the patient is terminally ill and death is foreseen, is a legitimate
medical practice for patient comfort; however, he believes writing a prescription
for life-ending medication places a physician like himself in a moral compro-
mise as it reflects participating in a patient's suicide. Dr. Savay is professionally
comfortable referring patients who request his assistance in dying to hospice
programs and patients' rights organizations that have more expertise with the
assisted-death process and can provide the medications that patients request.

While determinations about *prolonging life* are circumstantial, LDS
policy on using medicine for *hastening death* is restrictive and absolutist.
The social ferment about the positive right to die and euthanasia was intimated in the 1974 LDS medical ethical guidelines: "The Church does not
look with favor upon any form of mercy killing. It believes in the dignity of
life and that faith in the Lord and medical science should be appropriately
called upon and applied to reverse conditions that are a threat to life."[22] Two
elements of this statement are striking in historical perspective. First, cultural discourse on the right to die was constructed as "mercy killing." This
necessarily means ecclesiastical rejection of a positive right to die: killing a
patient on grounds of mercy presumes a moral logic of compassion rather
than respecting patient claims of entitlement. Second, the ecclesiastical assessment that mercy killing is not viewed with "favor" seems understated, as
illustrated by comparison with other ecclesiastical positions articulated in
the 1974 compilation: "the church opposes," "the church does not approve,"
and "the church does not object." An assessment that mercy killing is "not
favored" is a less prohibitionist directive than conveyed by the language of
"opposition" or "not approved," but presents a modest moral critique relative to the language of "not opposed." Nonetheless, an act constructed as
"killing" would seem to demand a rather strenuous objection. As public
sentiments, bioethical arguments, and public policy in some countries have
since shifted to permit euthanasia and medical-assisted dying, subsequent
ecclesiastical policies have sought greater conceptual clarity and more definitive judgments.

A revised policy articulated in 1976 (the year following the *Quinlan* case)
that was subsequently (1983) included in the church administrative handbook presented an ecclesiastical position on "euthanasia" similar to that
formulated for elective abortion. The language of "right to die," "does not
favor," and "mercy killing" was supplanted by a specific restrictive assessment: "Because of its belief in the dignity of life, the Church opposes euthanasia."[23] The use of euthanasia terminology invited practical confusion

because treatment refusal decisions, exemplified in *Quinlan* and *Cruzan*, were often designated as "passive euthanasia." Some members presumed this statement meant that the LDS Church did not support advance directives or other methods for discontinuing burdensome medical interventions. A 1989 policy revision reaffirmed ecclesiastical opposition to euthanasia but also provided needed conceptual clarity by defining euthanasia—"deliberately putting to death a person suffering from incurable conditions of disease"—and differentiating euthanasia from forgoing unreasonable means of prolonging life. The ecclesiastical opposition to euthanasia also was grounded in (unspecified) divine commands rather than the value of life's dignity: "[A] person who participates in euthanasia [as defined] violates the commandments of God."[24] The most recent revisions of ecclesiastical policy on euthanasia[25] have brought further clarity:

- The moral rationale for opposition to euthanasia shifts from the dignity of life or divine commandments to the "sanctity of life," as indicated in a prefatory statement: "The Church of Jesus Christ of Latter-day Saints believes in the sanctity of human life, and is therefore opposed to euthanasia."
- The language of "incurability" in the definition of euthanasia is specified. Rather than pertaining to disease "conditions," the phrasing now refers to an "incurable condition *or* disease." This implies a broader range of conditions, encompassing not only cancer, amyotrophic lateral sclerosis (ALS), or Alzheimer's as incurable diseases but also chronic pain or severe depression as incurable conditions. This expands the scope of ecclesiastical opposition insofar as these conditions are deemed legal grounds in some jurisdictions for a patient request for medically administered dying.
- Concepts of "deliberateness" and temporality are embedded in the definition of euthanasia: "Such a deliberate act [of euthanasia] ends life immediately through, for example, frequently-termed assisted suicide." The intentionality behind euthanasia is thereby construed not as merciful relieving of the pain or suffering of the dying person, as intimated in the 1974 guidance, but rather to "end life immediately." This conjoining of intentionality and outcome makes no room for appeals to mercy or compassion as grounds for hastening death. The statement also affirms that physician-assisted "suicide" legislation in numerous U.S. states is the conceptual and moral equivalent of euthanasia.

- The statement clearly differentiates forgoing medical interventions that "allow a person to die from natural causes" from euthanasia. Ending life by withholding or withdrawing medical interventions and providing necessary comfort care is neither conceptually nor morally equivalent to ending life through euthanasia, which remains "a violation of the commandments of God."

This ongoing effort to establish conceptual clarity and more definitive positions is rather unusual for LDS ecclesiastical policy. It displays attentiveness to shifts in public discourse regarding the positive right to die especially as these rights are enacted in various legislative statutes. The revisions have provided rationales for a prophetic moral witness against legalizing self-administered or medically-assisted dying in several U.S. states and Canada. I first examine the bioethical complexities of medical assisted dying before turning to the public ecclesiastical moral witness.

Ethical Dissonance and Equivalence

The disagreement between LDS ecclesiastical policy and arguments advocating legalized physician-prescribed medications self-administered by a patient to hasten their death takes place in two different realms. A primary issue, the very terms of the public debate, discloses a value-laden discursive dissonance. LDS teaching has consistently framed actions in which a physician provides or administers medications to a terminally (or chronically) ill patient to hasten death as an act of "euthanasia," a subset of which are acts of "assisted suicide." These classifications convey moral evaluations, which is why such nomenclature has been vigorously rejected by legalization proponents. Legalization advocates have intentionally adopted language of "physician assisted dying" to refer to the process by which a decisionally capable terminally ill patient self-administers lethal medication prescribed by a licensed physician as well as "death with dignity" to designate the outcome and nature of the death achieved through this process.

Legalization advocates have argued that portraying these processes as the moral equivalent of "suicide" (as stipulated in LDS ecclesiastical policy) misrepresents the nature and ethics of the actions of both patient and physician and imposes cultural stigmas that further burden the life-ending ordeals of terminally ill patients. Conceptually, physician "assistance" in dying comprises five features differentiating it from "suicide" assistance: A person with (1) a diagnosed terminal illness and (2) decision-making capacity (3) intends to hasten their inevitable death by (4) self-administering a medication prescribed by (5) a professionally qualified and licensed physician. This framework of the action contrasts with cultural and ecclesiastical constructions of suicide as an action by a person (1) who is suffering but has no terminal illness and (2) possesses presumptively impaired decision-making capacity with (3) the aim of self-inflicting (preventable) lethal harm through (4) means—guns, knives, poisons, gas—that (5) entirely exclude physician involvement.[26]

A similar conceptual invention informs rejection by legalization advocates of "euthanasia" language, which is seen as ethically, professionally, and legally prejudicial. In the first instance, state statutes that permit legalized physician-assisted dying have retained legal prohibitions on euthanasia. To avoid advocating a process that is directly prohibited by law, legalization advocates adopted the less-charged, rather bland phraseology of "medical assisted dying" to designate those circumstances in which a physician directly administers life-ending medication. Second, public forums indicate that respondents, such as citizens voting on such laws, are less supportive when "euthanasia" language is used to designate the proposed legal revision.[27] The recourse to medical-assisted dying phrasing is best illustrated in 2016 legislation passed by the Canadian Parliament (and opposed by the LDS Church)[28] that ratified two forms of such assistance: (a) "clinician-administered medical assistance in dying," wherein a physician or nurse practitioner directly administers a death-causing medical substance to an eligible patient, and (b) "self-administered medical assistance in dying," wherein the authorized professional prescribes a drug ingested by the eligible patient to hasten death.[29]

"Medical assisted dying" as constructed by advocates not only expands the scope of physician participation from prescription to administration but also expands the scope of eligible patients who may make a legally valid request. Advocates have long voiced friendly criticism of current physician-assisted dying statutes as overly restrictive (and potentially discriminatory) since their eligibility requirements exclude patients who (a) had previously been decisionally capable but lost decision-making capacity as their illness progressed (e.g., a person with end-stage Alzheimer's), or (b) were not physically capable of self-administration (e.g., a person with advanced ALS), or (c) were nonterminal but experienced severe, unbearable suffering from physical or mental illness. That is, the safeguards to ensure an in-formed patient decision in physician-assisted dying statutes are erased by medical-assisted dying as persons who are not terminally ill, or are physically incapable of self-administration, or no longer are mentally capable of making a contemporaneous choice have access to a medical procedure to end life.

In a religious moral culture, concepts, categories, definitions, and "naming" are soteriologically and morally potent, leading to skepticism about language that rationalizes or obfuscates the moral nature of actions. It is prophetically significant that LDS ecclesiastical policies retain the language of "assisted suicide" and "euthanasia" in ongoing statements of opposition. The prophetic critique is that the language of dignified death or (medical) assistance in dying is not intended as neutral moral terminology by which to conduct public dialogue on controversial issues, but rather seeks to resolve the ethical and policy questions by definitional fiat. What reasonable person, after all, would oppose a "dignified" death?

The second realm of dispute between LDS ecclesiastical policy and dis-course advocating assisted-dying legislation concerns the ethical (and legal) equivalence advanced by legalization advocates between the negative right to die through treatment refusals with the positive right to hastened death through physician-prescribed or -administered medications. For advocates and many bioethics scholars, a clear moral continuity exists between a pa-tient refusing futile medical treatments, a patient ingesting prescribed lethal medication, and a professional administering lethal medication, since the outcome and implicit intent in all three circumstances are the patient's death. The practical culmination of this assumption of equivalence is manifested in arguments that patients should be able to stipulate their requests for medical assistance in dying through their advance directive, a document heretofore reserved exclusively for treatment refusals.[30] The challenge the equiva-lence argument presents to LDS teaching on end-of-life decision-making is whether it is ultimately coherent and defensible to permit forgoing "unrea-sonable" life-prolonging measures while denying professional and moral au-thority to self-or clinically administered assistance in dying.

The broad scope of permitted actions constructed as medical assis-tance in dying does rely on established precedents for refusing medical interventions (such as advance directives and patient agency) and patient self-administration of lethal medications. The moral logic of medically as-sisted dying builds from the negative and positive rights to die implicit in these practices. The equivalence argument is supported by a purely pragmatic assessment of outcomes. Each of these practices—forgoing treatment, physician-prescribed medication, and clinically administered medication—culminates in the person's death. If all that matters ethically are direct, immediate consequences, then these three life-ending procedures seem equivalent. This purported equivalence makes ethical and policy reasoning largely a matter of instrumental rationality: the end of death both dictates and justifies the means of achieving death.[31]

The implicit claim of LDS ecclesiastical policies is that the ends of experiencing dying and death as a blessing do not justify any and all means. The difficulty with the consequentialist argument for moral equivalence is that it neglects (1) moral intentionality (what ecclesiastical policy characterizes as "deliberately" putting another person to death), (2) professional responsibility in causation of death, and (3) the moral character of actions independent of consequences. Moral assessments cannot be reduced entirely to outcomes, especially when, given the realities of human finitude and fallibility, consequences must be assessed prospectively rather than retrospectively. Religious traditions incorporate human finitude and fallibility in emphasizing that the intentionality underlying actions matters morally, and especially so regarding the ending of a human life. While matters of intentionality in ending human life are very complicated to parse out,[32] the concept of right intention illuminates a moral difference between decisions to decline life-prolonging medical interventions and requests for hastened death by medical assistance. The core intentions underlying patient refusal of medical interventions are to cease medically futile treatments; avoid invasive, burdensome prolongation of life; and offer palliative care for symptom relief. The core intention in advocacy of self-administering medication or medically assisted dying is the patient's death. The subsidiary intention of relieving pain and suffering is achieved by ending the life of the person that suffers.

A second distinction pertains to professional responsibility for causing or hastening death. While physicians can almost always do more to prolong life, withdrawing or withholding life-prolonging interventions does not make the physician complicit in either intending or causing patient death. Discontinuing interventions based on medical futility and the patient-centered criteria of minimal prospective benefit and substantial burdensomeness means death results from the underlying disease process, what LDS teaching frames as "natural causes." By contrast, in self-administered and clinically administered assistance in dying, a person dies not from disease but by medical and pharmacological interventions. The extent of professional responsibility is less direct when a physician writes a life-ending prescription as patients can (and often do) choose not to use the prescription. Since the patient makes this decision, the physician is situated at a psychological and moral distance from the actual death. This diminished professional responsibility for causing death has made medical professionals more amenable to participating in a prescription procedure than with clinically administered life-ending medications, which unquestionably make the physician or nurse practitioner the agent of death.[33] In either circumstance, however, the dying process is medicalized.[34]

The vocation of healing and caring central to the ethical integrity of medicine is diminished when medicine is sought solely for technical expertise. When medicine cannot save life, a default to more technology that prolongs biological life is morally inadequate, if not immoral. When patient options at the end of life are limited to life prolongation at some physical burden or to a protracted dying with comfort care, patient preference for assistance in dying from a medication or lethal injection is understandable. However, these modes of hastening death, inevitably reliant on the professional's technical expertise, neglect the moral reality that medicine can always carry out its core mission with integrity and heal even if there is no prospect for cure. The ethical equivalence argument neglects the manner in which medically assisted dying deforms the physician-patient relationship into transactional technique rather than holistic caring.[35] Medicine is not a morally neutral profession; the commitment to caring and healing even when curing is not possible is a distinctive feature of professional integrity.

The moral reality of bearing witness to compassionate caring and healing of terminally ill persons when curative medicine is precluded is envisioned in hospice care. The hospice philosophy of care—encompassing values of patient self-determination, the highest quality of life while dying, symptom relief, and companionship and nonabandonment of the patient—exemplifies a profound commitment of caregiver solidarity with the terminally ill.[36] Physicians as healers have alternatives to facilitate dying well, such as palliative care, caregiving presence, sufficient pain control, and professional responsibility to bear embodied witness to patient needs for relationship, touch, re-storying, and nonabandonment. The healing relationship embodies a moral narrative of medicine constituted by transforming moral identities rather than transacting technical expertise.

My principal critique of the purported moral equivalence between forgoing life-prolonging medical interventions and patient self-administered medications to hasten death and the more recent purported equivalence with medically assisted dying relies on this interpretation of medicine as a vocation of healing. The LDS objection to "euthanasia" (including "assisted suicide") is premised on two religious appeals: the sanctity of human life and a violation of divine commandments. This divergence displays that different principles of ethical reasoning can arrive at similar concrete positions. Moral convergence despite pluralism in values is an important consideration given the frequent public moral witness of the LDS Church opposing legalization of physician- and medical-assisted dying.

Epistolary Ethics

The LDS Church expressed opposition to legalizing physician prescriptions to hasten death by patient self-administration on at least two occasions: the Oregon referendum in 1994 and a Colorado referendum in 2016. In addition, the LDS Church endorsed an interfaith statement in 2016 that opposed passage of the Canadian legislation on medically assisted dying.[37] In each instance, the legalization measures were approved by citizens or by legislators. This discloses how bearing a public moral witness—of going on public record when there is a significant ecclesiastical value at issue, such as the sanctity or dignity of life—is important for moral accountability and integrity even if the preferred policy position does not prevail.

The Oregon process to legalize "death with dignity" included vitriolic attacks against religious views opposed to legalization. LDS ecclesiastical leaders joined with Catholic leaders (the state's largest religious denomination) to mobilize opposition to the citizen referendum among members of religious communities. A letter distributed to LDS leaders and members in Oregon congregations expressed unequivocal ecclesiastical opposition: "A fundamental doctrine of the Church of Jesus Christ of Latter-day Saints is that each person is a child of God. Consequently, life—a gift from God—is sacred and to be cherished. The Lord has commanded that man should not kill, 'nor do anything like unto it' (*Doctrine and Covenants* 59:6). One who assists the suicide of another violates God's commandments. The Church of Jesus Christ of Latter-day Saints opposes the judicial or legislative legalization of such assistance."[38] The letter urged LDS members to study the referendum, participate in voting, and if so inclined, offer volunteer support, including financial contributions, to a broad-based citizens' committee organized to defeat the initiative.

A form of epistolary ethics, the letter contains several statements whose interrelationship is rather ambiguous. The first two assertions—affirming human beings as God's children and the giftedness of human life—presume familiarity with the salvation narrative. The second set of assertions presents both conceptual and theological claims. Physician prescriptions written for the purpose of ending the life of a terminally ill patient are, for reasons discussed previously, portrayed as acts of assisting "killing" and "suicide." This categorization permits application of the scripturally sanctioned prohibition on killing (the same appeal in ecclesiastical objections to elective abortion) to support the ecclesiastical conclusion that such actions are contrary to divine commands. The letter then engages in a remarkable shift from commandments binding for members of the religious community to a claim about public policy, with negligible explication about why the ethics of the religious community are relevant to policy deliberation.

The pattern of epistolary ethics was displayed two decades later when Colorado citizens voted on the state's "Medical Aid in Dying" act. A month prior to the Colorado vote, the LDS First Presidency distributed a letter on "assisted suicide" to "Church leaders and members in Colorado."[39] The content of the letter reiterated teaching from the ecclesiastical policy on "euthanasia and prolonging life" with the principles of the sanctity and gift of human life affirmed as bookends for ecclesiastical opposition. However, the Colorado epistle invoked shared *human* values about protecting vulnerable human life without citing scriptural imperatives of a *religious* community. The argument from common moral concerns relied on interpretations of cultural empirical experience with legalized physician-assisted suicide, which "suggests that such legalization can endanger the vulnerable, erode trust in the medical profession, and cheapen human life and dignity." These anticipated consequences of legalization reflect broader social, cultural, and professional issues for all citizens irrespective of religious commitments. This broader dimension is augmented by an appeal to a familial and communal consideration: "the decision to end one's life carries a lasting impact far beyond the person whose life is ending."

The complication with relying on nontheological, experientially informed critiques is that the feared consequences are not substantiated in the empirical reporting provided by various U.S. states that have legalized physician-assisted death. Professional adherence to statutory requirements of decision-making capability and self-administration has minimized risks to vulnerable persons from legalization. Some studies do indicate increased risks to vulnerable persons, especially persons suffering from mental illness, under the regimes of legalized euthanasia in European countries.[40] The ethical concern to protect vulnerable persons from familial or medical hegemony at the end of life provides grounds for critique of medically assisted dying since the safeguards of decision-making capacity and self-administration are rescinded in the kind of legislation enacted in Canada. However, in the American setting of legalized physician prescriptions for patient self-administration to hasten death, the question of heightened risks to vulnerable persons is more speculative than empirically warranted.

The epistolary ethics of preserving societal trust in the medical profession and affirming the dignity of human life raise matters of professional and societal ethos that are very difficult to assess empirically.[41] The issue of diminished trust in medicine manifests the professional and ethical conflicts encountered by LDS physicians such as Dr. Savay who receive requests for participation in life-ending prescriptions and subsequently refer the patients to other physicians, patient rights organizations, or hospice care, should the patient pursue a life-ending medication. The critique of diminished societal trust also runs contrary to the general ecclesiastical ascription of medicine's ethical integrity. The epistolary emphasis on the responsibilities of citizens to participate in political processes neglects the practical issues that professionals encounter when physician-assisted death is legalized.

The Colorado letter concluded by encouraging church members to "let their voices be heard in opposition to measures that would legalize physician-assisted suicide." This admonition reflects a general ecclesiastical pattern on public policy issues. The letter never once makes reference to LDS *Church* opposition to the specific Colorado initiative. The "opposition" is to be vocalized by church members as citizens rather than the Church as a distinct moral voice. The moral pattern is that ecclesiastical leaders articulate fundamental ethical principles while adherents assume responsibility for exercising moral agency and self-governance on matters that concern their specific practical, medical, and civic experience. This ethical distancing allows ecclesiastical leadership to maintain that advocacy of a position on a policy question facing church *members* does not overstep norms and boundaries of religion and politics in civic discourse. Ecclesiastical leaders can avow, albeit in an awkward and indirect manner, that the Church has not presented a position on a *specific* public policy or legislative proposal. Members as citizens are conferred with the responsibility of wise use of moral agency regarding legalizing physician-prescribed medications to end life.

A 2020 Utah legislative proposal designated "The End of Life Options Act" to permit physician-prescribed medications for patient self-administration has not received full consideration by the state legislature.[42] A 2015 polling analysis conducted when a death-with-dignity bill was first introduced in the legislature indicated a majority of Utahans (58%) supported what was portrayed as "right to die legislation," while 35% of the respondents were "somewhat" or "definitely opposed" to such a law.[43] The overall support for legalization is important even though approval levels in Utah are significantly less than what is represented in broader national studies.[44] Rates of approval differ depending on whether a survey invites responses on the general concept of physician assistance in a patient's death (national studies) or to a specific policy initiative (the Utah context).

An overview of the Utah survey claimed "there's not much of a religious divide on the question" but this seems an oversimplification. While the proposed legislation received strong support from persons with no religious identification (94%), and from Protestant (80%) and Catholic (76%) respondents, noticeable differences were reflected among LDS respondents according to a self-identified level of church participation. Persons who self-described as "non-active" church members supported the proposed legislation at a level of 87%; persons identifying as "somewhat active" also expressed a substantial level (79%) of support. This level of support can be attributed to the emphasis in the moral culture on personal agency. However, a majority of persons identifying as "active" LDS members opposed the legislation by a ratio of 54% to 38%, with 40% indicating they were "definitely opposed." As neither the Utah "Death with Dignity Act" nor the "End of Life Options Act" has received full legislative or public consideration, ecclesiastical leadership has not issued any statement specific to the Utah context. However, should a physician-assisted death proposal be seriously debated as a public policy initiative, an ecclesiastical response similar to the Oregon and Colorado contexts (states where the LDS population is about 3%) can be anticipated. Both general and local ecclesiastical leadership will likely engage in "epistolary ethics," distributing letters to congregations expressing opposition to such legislation, encouraging LDS citizens to engage in policy discussion, and recommending voting against legalization.

The strategy of ethical distancing provides immunity for church leadership from critiques that it is influencing public policy, an issue much more germane in Utah, where the population is around 70% Mormons, than in states like Oregon or Colorado. Moral distancing seeks to respect legal and cultural traditions mandating separation of church and state as they apply to the church as an organized religious community while not muting the moral witness and civic engagement of ethical principles enacted by church membership. The focus on civic responsibility and personal moral agency can neglect professional responsibilities of LDS physicians faced with the real-life practical consequences of legalization. The patterns of moral witness illustrated in ecclesiastical epistolary discourse about end-of-life ethics can be extended by examining ecclesiastical teaching and policy involvement regarding legalized elective abortion and medical marijuana.

Notes

1. The necessity of understanding death through the LDS salvation narrative is displayed in the teachings of the current president, Russell M. Nelson: "Before we can comprehend the reason for the gateway and the purpose of death, we must first understand the purpose of life." Russell M. Nelson, *The Gateway We Call Death* (Salt Lake City: Deseret Book, 1991), 1.
2. *Doctrine and Covenants* 88:15; 93:33–34.
3. L. Kay Gillespie, "Death and Dying," in *Encyclopedia of Mormonism,* ed. Daniel H. Ludlow (New York: Macmillan, 1992), 364–366, https://eom.byu.edu/index.php/Death_and_Dying. The ontological and soteriological equivalence of birth and death is conveyed in the title of a popular devotional book. See Paul H. Dunn, Richard M. Eyre, *The Birth We Call Death* (Salt Lake City: Aspen Books, 1999).
4. Donald B. Doty, "Prolonging Life," in *Encyclopedia of Mormonism,* ed. Daniel H. Ludlow (New York: Macmillan, 1992), 1159–1160, https://eom.byu.edu/index.php/Prolonging_Life
5. Nancy Madsen-Wilkerson, "When One Needs Care, Two Need Help," *The Ensign,* March 2016, https://www.lds.org/ensign/2016/03/when-one-needs-care-two-need-help?lang=eng; Arla Halpin, "Finding Answers for Family Caregiving," *The Ensign,* June 2018, https://www.lds.org/ensign/2018/06/finding-answers-for-family-caregiving?lang=eng
6. William M. Timmins, "On Death and Dying," *The Ensign,* April 1989, https://www.lds.org/ensign/1989/04/on-death-and-dying?lang=eng
7. Cf. Joan Halifax, *Being with Dying: Cultivating Compassion and Fearlessness in the Presence of Death* (Boston: Shambhala Publications, 2009).

8. Ira Byock, *Dying Well: Peace and Possibilities at the End of Life* (New York: Riverhead Books, 1997), 217.

9. Kory Scadden, "Caring for Loved Ones in Their Final Hours," *LDS Living*, http://www.ldsliving.com/Caring-for-Loved-Ones-in-Their-Final-Hours/s/75170

10. "Attitudes of the Church of Jesus Christ of Latter-day Saints Towards Certain Medical Problems," in "Mormon Medical Ethical Guidelines," ed. Lester E. Bush, Jr., *Dialogue* 12 (Fall 1979), 98.

11. *In re Quinlan*, 70 N.J. 10, 355 A.2d 647 (NJ 1976); Peter G. Filene, *In the Arms of Others: A Cultural History of the Right-to-Die in America* (Chicago: Ivan R. Dee, 1998), 11–93. Ms. Quinlan remained on a feeding tube until her death from pneumonia in 1985.

12. *Cruzan v. Director, Missouri Department of Health*, 497 U.S. 261 (1990); Filene, *In the Arms of Others*, 161–186. The Cruzan family subsequently obtained further evidence about their daughter's treatment preferences, and Ms. Cruzan died six months after the Court decision following withdrawal of the feeding tube.

13. National Academy of Medicine, *Physician Assisted Death: Scanning the Landscape* (Washington, D.C.: National Academies Press, 2018). Physician-assisted death, in which a patient self-administers a prescribed medication, is now legal in Oregon, Washington, Montana, Vermont, California, Colorado, Hawai'i, New Jersey, Maine, and the District of Columbia. Medical-assisted dying, in which a physician administers lethal medication to a patient, is now lawful in Canada, Belgium, Germany, the Netherlands, Switzerland, and Victoria, Australia.

14. Thaddeus Mason Pope, "Medical Aid in Dying: When Legal Safeguards Become Burdensome Obstacles," *ASCO Post*, December 25, 2017, https://www.ascopost.com/issues/december-25-2017/medical-aid-in-dying-when-legal-safeguards-become-burdensome-obstacles/

15. Newsroom, "Euthanasia and Prolonging Life," https://newsroom.churchofjesuschrist.org/official-statement/euthanasia-and-prolonging-life; cf. The Church of Jesus Christ of Latter-Day Saints, "Church Policies and Guidelines, Medical and Health Policies: Prolonging Life," in *General Handbook: Serving in the Church of Jesus Christ of Latter-day Saints*, Chapter 38.7.9, February 19, 2020, https://www.churchofjesuschrist.org/study/manual/general-handbook/38-church-policies-and-guidelines?lang=eng#title_number91

16. Atul Gawande, *Being Mortal: Medicine and What Matters in the End* (New York: Henry Holt and Company, 2014), 158.

17. Gawande, *Being Mortal*, 169–171.

18. Gawande, *Being Mortal*, 128.

19. Gawande, *Being Mortal*, 174.

20. Determinations of reasonable and unreasonable means of prolonging life do not apply to comfort care or palliative care, which are directed to symptom relief and always morally required.

21. Kaiser Health News, "End-of-Life Care: A Challenge in Terms of Costs and Quality," June 4, 2013, https://khn.org/morning-breakout/end-of-life-care-17/; T. R. Reid,

"How We Spend $3,400,000,000,000," *The Atlantic*, June 15, 2017, https://www.theatlantic.com/health/archive/2017/06/how-we-spend-3400000000000/530355/

22. "Attitudes of the Church of Jesus Christ of Latter-day Saints Towards Certain Medical Problems," ed. Lester E. Bush, Jr., *Dialogue* 12 (Fall 1979), 98.

23. Lester E. Bush, Jr., *Health and Medicine Among the Latter-day Saints: Science, Sense, and Scripture* (New York: Crossroad, 1991), 38.

24. Courtney S. Campbell, "Sounds of Silence: The Latter-day Saints and Medical Ethics," in *Theological Developments in Bioethics: 1988–1990*, ed. B. Andrew Lustig (Dordrecht, The Netherlands: Kluwer Academic Publishers, 1991), 34–36.

25. Newsroom, "Euthanasia and Prolonging Life."

26. Courtney S. Campbell, *Bearing Witness: Religious Meanings in Bioethics* (Eugene, OR: Cascade Books, 2019), 222–247.

27. Robert P. Jones, *Liberalism's Troubled Search for Equality: Religion and Cultural Bias in the Oregon Physician-Assisted Suicide Debates* (Notre Dame, IN: University of Notre Dame Press, 2007), 55–96.

28. Newsroom Canada, "Euthanasia and Prolonging Life," https://canada.lds.org/euthanasia-and-prolonging-life

29. Government of Canada, "Medical Assistance in Dying," https://www.canada.ca/en/health-canada/services/medical-assistance-dying.html

30. David Gibbes Miller, Rebecca Dresser, Scott Y. H. Kim, "Advance Euthanasia Directives: A Controversial Case and Its Ethical Implications," *Journal of Medical Ethics* 45:2 (2019), 84–89.

31. Some bioethical argumentation maintains that, once physician and patient concur that death is the inevitable outcome of a disease process, a clinically administered euthanasia procedure is more humane and ethical. James Rachels, "Active and Passive Euthanasia," *New England Journal of Medicine* 292 (January 9, 1975), 78–80.

32. Neil M. Gorsuch, *The Future of Assisted Suicide and Euthanasia* (Princeton, NJ: Princeton University Press, 2006), 48–75, 181–218.

33. William F. May, *Testing the Medical Covenant: Active Euthanasia and Health Care Reform* (Grand Rapids: Eerdmans, 1996).

34. Courtney S. Campbell, "Mortal Responsibilities: Bioethics and Medically-Assisted Dying," *Yale Journal of Biology and Medicine* 92:4 (2019), 733–739.

35. Edmund D. Pellegrino, "Doctors Must Not Kill," in *Euthanasia: The Good of the Patient, The Good of Society*, ed. Robert I. Misbin (Frederick, MD: University Publishing Group, 1991), 30.

36. Courtney S. Campbell, "Moral Meanings of Physician-Assisted Death for Hospice Ethics," in *Hospice Ethics: Policy and Practice in Palliative Care*, ed. Timothy W. Kirk, Bruce Jennings (New York: Oxford University Press, 2014), 223–249.

37. Newsroom Canada, "Euthanasia and Prolonging Life."

38. "Statement of The Church of Jesus Christ of Latter-day Saints on the Question of Physician-Assisted Suicide," May 29, 1997 (letter in my possession).

39. The Church of Jesus Christ of Latter-day Saints, Office of the First Presidency, "Letter on Assisted Suicide," October 12, 2016, https://www.churchofjesuschrist.org/church/news/first-presidency-asks-members-to-oppose-recreational-marijuana-assisted-

suicide?lang=eng. See also Tad Walsh, "LDS Leaders Ask Mormons to Oppose Legalization of Assisted Suicide, Recreational Marijuana," *Deseret News*, October 13, 2016, https://www.deseretnews.com/article/865664777/LDS-leaders-ask-Mormons-to-oppose-legalization-of-assisted-suicide-recreational-marijuana.html

40. Rachel Aviv, "The Death Treatment," *The New Yorker*, June 22, 2015, http://www.newyorker.com/magazine/2015/06/22/the-death-treatment; National Academy of Medicine, *Physician Assisted Death*, 66–72.

41. Roman Catholic teaching on physician-assisted death has voiced a similar concern, contending that such practices are symptomatic of a "culture of death." John Paul II, "*Evangelium Vitae* (The Gospel of Life)," March 1995, http://w2.vatican.va/content/john-paul-ii/en/encyclicals/documents/hf_jp-ii_enc_25031995_evangelium-vitae.html

42. Utah State Legislature, "HB 93: End of Life Prescription Provisions," https://le.utah.gov/~2020/bills/static/HB0093.html; NOLO, "Death with Dignity in Utah," January 10, 2020, https://www.nolo.com/legal-encyclopedia/death-with-dignity-utah.html

43. Bryan Schott, "Majority of Utahans Support Right-to-Die Legislation," UtahPolicy.com, December 10, 2015, https://utahpolicy.com/index.php/features/today-at-utah-policy/7910-poll-majority-of-utahns-support-right-to-die-legislation

44. A 2018 study indicated that 72% of Americans believed "doctors should be legally allowed . . . to end a terminally ill patient's life using painless means." Megan Brenan, "Americans' Strong Support for Euthanasia Persists," Gallup, May 31, 2018, https://news.gallup.com/poll/235145/americans-strong-support-euthanasia-persists.aspx

8

Public Bioethics

Abortion and Medical Marijuana

In March 1997, in the wake of scientific reports of the first mammalian cloning, the National Bioethics Advisory Commission (NBAC) was charged with examining the scientific, ethical, religious, and legal issues on human reproductive cloning. I was invited to prepare a scholarly manuscript on religious perspectives.[1] Shortly thereafter, the Public Affairs Office of The Church of Jesus Christ of Latter-day Saints (LDS) contacted me to ask if I would provide my thoughts on whether the Church should make a public statement about cloning, and on the content of such a statement. Church president Gordon B. Hinckley had already encountered media questions on the subject.[2] The public affairs representative took pains to remind me that the ecclesiastical leadership typically did not take public stands on policy questions, unless the matter was directly connected to the salvific mission of the Church or otherwise implicated a deep-rooted moral value. What was not going to happen was an official statement along the lines of a pastoral letters: rather, any statement should follow the customary ecclesiastical pattern of a couple of sentences.

I responded that there were valid reasons for both issuing an ecclesiastical position and providing a substantive rationale. On theological grounds, reproductive human cloning implicated values of profound importance to the LDS Church, including convictions about the family, human sexuality, and parenting that do not sit comfortably with human cloning. I also proposed pragmatic reasons for issuing a statement: I would soon be calling the Church, as with virtually every other U.S. denomination, on behalf of NBAC. Particularly in the midst of a controversial public issue, it is difficult to control the interpretation of silence. Silence can be interpreted as indifference or as tacit consent or as implicit endorsement of a practice. Finally, I invoked a political argument, that the Church had an opportunity to exemplify its teachings about responsible citizenship by contributing to public discourse. I subsequently provided to the Public Affairs Office an outline of items a

Mormonism, Medicine, and Bioethics. Courtney S. Campbell, Oxford University Press (2021). © Oxford University Press.
DOI: 10.1093/oso/9780197538524.003.0009

statement might address, including the ecclesiastical rationale for the statement, an endorsement of the general legitimacy of scientific inquiry, theological issues raised by reproductive human cloning and an evaluation of
cloning based on these theological values, and a recommendation to NBAC.
Not quite the desired two sentences, but these elements could briefly reflect
the complexities of the issue while avoiding the reactionary condemnations
of cloning science issued by other denominations.

I do not know what became of this exchange of ideas and messages. After
an initial five-day flurry, I was not contacted again. As the deadline for my
NBAC report drew near, and I had not received any statement from the
LDS Church on human cloning, I made my dutiful phone call to the Public
Affairs Office and received the following statement: "The First Presidency
and Quorum of the Twelve Apostles of The Church of Jesus Christ of Latter-
day Saints have declared that: God has commanded that the sacred powers
of procreation are to be employed only between man and woman, lawfully
wedded as husband and wife. We declare the means by which mortal life is
created to be divinely appointed."[3] This statement was taken verbatim from
the 1995 "Proclamation on the Family" and its most notable aspect is that
it met the two-sentence requirement. As a statement aimed at addressing
human cloning issues confronted by NBAC, however, it was a failure.[4]

Some two decades following this ecclesiastical evasion, I question whether
my experience would be repeated in the 2020s. As discussed in the preceding chapter, the issue of legalized physician assistance in a patient's
death reveals limits to peaceful compromise between ecclesiastical policy
with both professional morality and public policy of the secular state. Even
on this very charged issue, the ecclesiastical witness is best characterized as
"epistolary": letters encouraging church membership to engage in public
deliberations and the policy process do not present a prophetic stance that
positions the LDS Church as a public moral agent. Ecclesiastical policy
presumes that the primary audience for ethical wisdom on health, medical, and moral issues is the LDS community, as church members, citizens,
and professionals. This teaching has seldom addressed itself to the healing
professions or to questions of public policy, an ethical insularity coherent
with the principles of respect for moral agency and trust in the healing vocation of the professions. Ecclesiastical deference to the relationship between
patient and physician relies on a prior societal and professional consensus
that bioethical issues can be largely resolved within the relationship free of ecclesiastical influence. However, some issues have prompted the LDS Church

to forgo ecclesiastical silence and present a public witness of its values and positions on a policy question to a broader civic audience. Two questions especially illustrative of this religious presence in public reasoning concern legalization of elective abortion and legalization of medical marijuana.

This chapter presents an exposition of the conflicting ways that ecclesiastical teaching and bioethics professionals construct and negotiate a religious presence in the common square of public bioethics discourse. The core issue is whether the pluralistic realm in which controversial bioethical issues are contested should necessarily be a "religion-free" zone. I examine why this dispute emerges and the consensus breaks down for the LDS community on issues such as physician-assisted death, elective abortion, and medical marijuana.

Criteria for Public Reasoning

The public square of health and medical issues is constructed in LDS teaching through three core criteria. The first concept, grounded in the ethical principles of respect for moral agency and justice, supports *civic engagement and responsibility* by individual members. The rationale for participation in rather than withdrawal from civic life is embedded in a scriptural call that church membership be "anxiously engaged in a good cause, and do many things of their own free will. . . . For the power is in them, wherein they are agents unto themselves."[5] The call to civic engagement encompasses relevant LDS professionals, such as physicians, nurses, public health workers, educators, and other policy-makers, who contribute expertise to health issues in their professional roles. The issue for either citizen or professional is whether the concept of a "good cause" is subject to personal interpretation and moral agency (e.g., communal coordination of blood donation) or is informed by an ecclesiastical "epistolary" ethic (e.g., voting recommendations regarding legalizing physician assistance in dying).

The call to civic engagement can be communally organized, as most recently illustrated by a grassroots forum, "Mormon Women for Ethical Government" (MWEG). Concerned about declines in public civility and political accountability, MWEG organized in early 2017 as a "pro-ethics" organization affirming "defense of the fundamental values of democracy and constitutional law."[6] MWEG advocates for various ideals in political governance, including decency, honor, accountability, transparency, and justice, as a way to address "a tidal wave of corruption and self-interest in the political landscape."[7] MWEG articulates principles for peacemaking that exemplify moral continuities between scriptural teachings and narratives and the philosophies of Mahatma Gandhi and Martin Luther King, Jr. This communal instantiation of commitments and values of the moral culture offers a distinctive but collaborative LDS religious presence in public reasoning that may otherwise go unrecognized in individual civic engagement.

A second criterion for ecclesiastical public engagement is general adherence to *separation* of church and state.[8] This position reflects long-standing LDS resistance to governmental intervention in religious practices as well as the historical experience of state-sponsored persecution that ultimately drove the early Mormon community out of the United States. This experience informs ecclesiastical critiques about the overreaching nature of governmental policies to promote civil rights that contravene constitutional protections of freedom of religion and conscience, protections that secure the sacred principle of moral agency.[9] The separationist posture acknowledges the historical realities so concerning in the American founding that religion is often used as a guise for obtaining political authority, can be culturally divisive, and can lie behind political oppression of minority religions or of nonreligion.[10] Partitioning religion and politics avoids corrupting the integrity of religious witness with political concerns and imposes barriers to theocratic pretensions of dominant religions.[11]

The principle of separation is not absolute, however. As a religious community whose integrity is constituted in part by moral values, the LDS Church affirms a responsibility to address public policy questions that contain moral dimensions, particularly when specific policies may *compromise the core moral values* that constitute the community moral culture.[12] Although the core moral values have not been systematically articulated, precedent and principle suggest a cluster of interrelated values, including religious liberty, moral agency, family integrity, embodiment and the sanctity of human life, the inherent dignity and equality of persons, the covenantal health ethic, and care for the poor and stranger. Such values, which correspond to my account of LDS ethical principles, are embedded in recent ecclesiastical statements on religious liberty, marriage equality, nondiscrimination, refugee assistance, and immigration, and of direct pertinence for bioethics discourse, on matters of legalized elective abortion and medicinal marijuana.

The meaning of these values, all of which are re-storied by the salvation narrative, construct an issue as a "moral" question rather than solely a political, economic, or professional consideration. They also provide justification for a public moral witness, although they do not necessarily prescribe specific policy *solutions*. In this regard, a public policy that failed to account for or conflicted with these values would be morally inadequate even if politically defensible. A prominent example of this ecclesiastical moral witness is presented in "The Family: A Proclamation to the World," which affirms the indispensable role of the family in the salvation narrative, in religious communities, and in secular societies. The proclamation makes good on its witness to "the world" by concluding with an explicit policy implication: "We call upon responsible citizens and officers of government everywhere to promote those measures designed to maintain and strengthen the family as the fundamental unit of society."[13] This principle does not, however, explicitly designate the specific measures that could be enacted to sustain the family role in society. The question to be pursued now is how the criteria of civic engagement, separation, and core moral values bear on health and bioethical public policy issues, particularly given advocacy within bioethics for a religious-free realm of public policy.

Privatizing Religion

Although scholars argue that bioethics has important intellectual roots in religious-based thinking,[14] over time bioethics forged its own acquiescence with what philosopher Richard Rorty designated as "the Jeffersonian compromise" that the rationalistic Enlightenment made with religion in the 17th and 18th centuries.[15] This sociopolitical compromise formulates core philosophical assumptions of a liberal democratic society: (a) a secular, or nonreligious, justification of government grounded in popular consent and convictions of "we the people" rather than a divine mandate;[16] (b) an extensive zone of personal liberty restricted by prohibitions against harm or injury of others; (c) governmental neutrality between different religions, as well as between religion and nonreligion; and (d) nonreligious, secular justifications for laws and policies, even if the policy advances a secondary religious end.[17] A liberal democratic society committed to protecting individual rights and securing the common welfare will (e) embrace a pluralistic public philosophy and (f) be agnostic about the human good or the good society. This pluralistic proposition consists of three interrelated features bearing on the issue of the public presence of religiosity: privatized religion, a secular public square, and religious restructuring.

Privatizing Religion. The social compromise permits expression of religious conviction within various private and nonpublic forums, including families and religious communities, and religious-based educational or health care institutions. This privatized realm for religious discourse secures guarantees of religious liberty and protections from political or theocratic authority for minority religious traditions (including early Mormon communities). Societally, a privatized religion helped to advance the revolutionary aspiration to establish a viable democratic, pluralistic society unthreatened by wars of religion that historically devastated Europe (or surrogate "culture wars" that undermine civility and public trust in the contemporary era).[18]

A Secular Public Square. The social corollary of privatized religion is that the public realm of discourse and policy should be open to rationally accessible perspectives and free from religious authoritarianism and values—ideally, free from religious influence altogether. This provides social space for the lifeblood of a democratic system, civic discourse regarding balancing personal rights and the common good. The insight of the American founding is that demarcating boundaries for the religious and secular facilitates both the vibrancy of civil society and the integrity of religion.[19] This philosophic claim was verified in the following generation by French political theorist Alexis de Tocqueville, who observed that the American experiment with religious liberty illustrated that religion has greater societal influence insofar as it *refrains* from claims to political authority and focuses on cultivating moral norms and virtues in adherents.[20] This understanding is likewise reflected in LDS views on the responsibility of moral agency for engaged citizenship and in the assumption of the separation principle that walling off religion and politics from each other to some extent diminishes the potential for corruption of both.

Religious Restructuring. In policy formulation oriented toward a common good, the sources of moral reasoning must be accessible to all citizens. As Rorty contends, "The only test of a political proposal is its ability to gain assent from people who retain radically diverse ideas about the point and meaning of human life."[21] A civic discourse oriented by publicly accessible rational argument consequently entails that interposing religious values into public discussions of ethics or policy violates the boundaries necessary for the political compromise to succeed. Religious communities that seek to claim a public voice must "restructure" or "translate" their values or positions into secular terms—for example, the biblical image of God is recast as human dignity or the ethic of love is translated into beneficence. The process of moral restructuring involves avoiding appeals to scripture, religious authority, or historical communal experience as sources of moral authority because such appeals are not publicly shared or accessible. As indicated in the preceding chapter, the LDS epistolary ethic of the 2016 Colorado letter opposing legalized assisted death engaged in restructuring by appealing to shared values, such as societal trust in medicine and protection of the vulnerable, rather than relying on explicitly religious values.

Rorty claims that, for faith traditions, the political compromise of privatized religion, rationally accessible public reasoning, and restructuring argumentation is "a reasonable price to pay for religious liberty."[22] As a historically persecuted religious minority, LDS political theology embraces protections of religious liberty and freedom of conscience as part of the pluralistic commitment to a realm of extensive personal liberties. The moral culture also has a significant stake in establishing boundaries around political ideologies that oppress religious minorities or risk compromises to religious integrity. However, a totalizing secularization of a civic life immune from religious influence is less amenable to LDS teaching. The implication for religious communities of the claim that public life should be a religion-free realm of reasoning is that when a community believes a public policy infringes on its core values, the values must meet a threshold test of public accessibility. This threshold test holds regardless of whether the religious witness is a prophetic critique (opposition to assisted death) or serves as a potent motivation for advancing secular aims (e.g., religious argumentation for civil and human rights).

The principles of liberal democracy and philosophically cogent arguments on moral compromise and sustaining a peaceful social order support a religion-free realm of bioethics discourse on civic and policy issues.[23] The principle of religious liberty confers social permission for religious expression in private settings of family, religious congregations, and religion-based educational or health care institutions but does not extend to explicitly religious argumentation about bioethical policy issues. This means that occasional ecclesiastical statements that interpose religious considerations into the policy realm of bioethics confront an embedded cultural, political, and philosophical resistance. The integrity-defining questions then become whether such cultural resistance entails a diminished responsibility of religious communities to address public policies that implicate or compromise core values, and how any public ecclesiastic moral witness can make an argument accessible to a pluralistic moral culture. The issues of legalized abortion

and medical marijuana present illuminating examples of how the LDS tradition negotiates a public witness to core moral values in a secular public
square.

An Evolving Ethic of Abortion

During the founding years of Mormonism, U.S. states largely followed
English common law permitting abortion prior to quickening (the mother's
experience of fetal movement within her womb), while criminal sanctions
could be imposed for abortions post-quickening.[24] This gradualist approach
was radically altered by the latter part of the 19th century. The profession
professionalization of medicine, exemplified by the establishment of the
American Medical Association, aimed to ensure safer medical procedures
for childbirth in the context of high neonatal and maternal mortality. The
professionalization of medicine and medicalization of pregnancy coincided
with various cultural, philosophical, and religious paternalistic controls over
women's reproductive lives.[25] Advancing medical knowledge about fetal development also increased religious interest in the moral status of unborn
human life, as illustrated by a radical departure from seventeen centuries of
Roman Catholic teaching in an 1869 declaration of Pope Pius IX that en-
soulment occurred at conception, meaning that any direct abortion was the
moral equivalent of homicide.[26] By the end of the 19th century, the common-
law approach to abortion had been supplanted by restrictive laws on abortion in all U.S. states, with provisions for legal pregnancy termination limited
primarily to a mortal threat to the mother from continuing the pregnancy.

   Early LDS teaching on abortion reflected this prohibitionist approach.
19th-century leaders portrayed abortion as a form of "murder" and a sin
of "shedding innocent blood" for which there was no forgiveness. In contrast to Catholicism, LDS condemnation of abortion was formulated in the
absence of definitive teaching on the timing of the union of the enlivening
spirit with the developing physical body. While other religious traditions increasingly affirmed conception as the threshold for moral status and legal
protection of human life, early-20th-century LDS leaders gave credence to
the criterion of quickening: "the body of man enters upon its career as a tiny
germ or embryo, which becomes an infant, quickened at a certain stage by
the spirit whose tabernacle it is, and the child, after being born, develops into
a man."[27] The ontological issue of the beginnings of human life has never

been resolved in LDS teaching. The perennial ecclesiastical pattern resists identifying a specific biological criterion and instead re-stories the question within the salvation narrative. As articulated by Dallin H. Oaks, "individual identity began long before conception." Hence, "[the LDS] attitude toward abortion is not based on revealed knowledge of when mortal life begins for legal purposes. It is fixed by our knowledge that according to an eternal plan all of the spirit children of God must come to this earth for a glorious purpose."[28] This ecclesiastical agnosticism about the beginnings of mortal life creates complications for policy realms where definitive criteria for moral and legal protections are essential.

The national movement to expand reproductive rights during the early 1960s, including expanded contraceptive access and legalization of elective abortion, culminated in the 1973 U.S. Supreme Court decision in *Roe v. Wade* that overturned all restrictive abortion laws on the basis of a woman's right to privacy and bodily integrity.[29] In the decade preceding *Roe*, professional and legislative advocacy of legalization provided a backdrop for renewed ecclesiastical teaching. The American Legal Institute proposed expanding the historical therapeutic provision for abortion when continuing pregnancy posed risks to the mother's life to include provisions for abortion when pregnancy endangered the mother's physical or mental health, or the pregnancy was the result of rape or incest, or the child would be born with a severe, and potentially fatal, physiological anomaly. These expanded provisions for therapeutic abortion were enacted in reforms to abortion law in several states and introduced in the Utah legislature in 1969.[30] Shortly thereafter, the ecclesiastical leadership issued a statement asserting that following "careful consideration," the leadership was "opposed to any modification, expansion, or liberalization of laws on these vital subjects."[31] This statement initiated the ecclesiastical pattern of public engagement with a policy *issue* by articulating basic principles without directly addressing the merits of a specific *law*. The Utah legislature rejected the reform bill, which meant that, as with most U.S. states prior to *Roe v. Wade*, legal abortion in Utah was restricted to circumstances of a mortal threat to a pregnant woman's life.

In June 1972, the First Presidency offered what remains the most elaborate statement on the ecclesiastical rationale for opposition to elective abortion.[32] An absolutist critique of abortion as a social practice—"The Church opposes abortion"—was qualified by permissible exceptions reflecting the broadened scope of therapeutic abortion in the 1969 Utah legislative proposal that the Church had previously rejected: "The Church . . . counsels its members not to submit to or perform an abortion except in the rare cases where, in the opinion of competent medical counsel, the life or good health of the mother is seriously endangered or where the pregnancy was caused by rape and produces serious emotional trauma in the mother." The maternal physiological and psychological criteria provided necessary but not sufficient conditions for an abortion. A decision on pregnancy termination required procedures of consultation with a local ecclesiastical leader and "divine confirmation through prayer." The statement did not contain a permissible exception for fetal indications.

While the policy expanded the scope of therapeutic abortion, it also made the ecclesiastical prohibition of abortion virtually absolute, insofar as the cases of permissible abortions constitute approximately 1% of all abortions (about ten thousand annually) of all abortions in the United States.[33] Even if rare *acts* of therapeutic abortion are permissible, the insistent question is the rationale for ecclesiastical opposition to a social *practice* of elective abortion, especially as the policy offers no theological threshold for the beginnings of human moral status. The first rationale, reflected in the statement's designation of abortion as a "crime," presumes a legal context for assessing elective abortion rather than an exclusively religious or moral evaluation. Irrespective of the moral status of unborn human life or the legal status of abortion, the claim is that abortion is *analogous to killing* another person. The statement concluded with a scriptural admonition: "Thou shalt not steal, neither commit adultery, nor kill, nor do anything like unto it."[34] This scriptural-based analogy of abortion as "like" killing is the *only* rationale articulated in current ecclesiastical policy.[35]

The construction of abortion as a crime analogous to killing raises an ecclesiastical issue about communal membership. Significantly, the statement rejects historical ecclesiastical teaching that forgiveness is not available for abortion. An attestation of ecclesiastical humility errs on the side of understanding abortion as not equivalent to murder and presumptively in the class of forgivable actions: "No definite statement has been made by the Lord one way or another regarding the crime of abortion. So far as is known, he has not listed it alongside the crime of the unpardonable sin and shedding of innocent blood. . . . [Abortion] will be amenable to the laws of repentance and forgiveness." The morally serious act of abortion does not then imply that the *agent* is morally incorrigible.

An oppositional argument framing abortion as analogous to killing is hard to sustain without some presupposition about the theological status of unborn human life that may be difficult to defend in forums of publicly accessible reasoning. The analogy to killing is supplemented in the ecclesiastical policy by a more explicit theological conviction: abortion is contrary to the *procreative imperative* delineated in the salvation narrative. The moral wrong of elective abortion in part resides in its disruption of "the processes in the procreation of offspring" and thereby symbolizes violating "one of the most sacred of God's commandments—to multiply and replenish the earth." A third rationale situates abortion within the narrative of *moral decline* and as a symptom of deeper moral and spiritual evils in society, particularly as manifested in sexual promiscuity: "Abortion must be considered one of the most revolting and sinful practices in this day, when we are witnessing the frightening evidences of permissiveness leading to sexual immorality." The implicit concern is that legalized abortion will too readily be used as a form of birth control and a resort for the unintended consequences of sex outside the marital context.

Ecclesiastical statements in the post-*Roe* era have introduced some substantive and some terminological changes. The evaluative and restrictive language of the 1972 policy providing therapeutic exceptions for abortion to preserve maternal "good health" or in circumstances of "emotional trauma" has been discarded and the therapeutic allowance for abortion in nonconsensual sex is now extended to pregnancy resulting from incest. A 1989 revision provided the first exception based on fetal indications.[36] The current policy continues to frame abortion as an action "like unto" killing but clarifies that ecclesiastical opposition is directed at "elective abortion for personal or social convenience." The three "possible exceptions" to this general critique are then identified: (1) pregnancy resulting from forcible rape or incest; (2) a competent physician determines that continuing the pregnancy will place the mother's life or health in serious jeopardy; (3) a competent physician determines that the fetus has severe defects that will not allow the baby to survive beyond birth.[37] The LDS Church website statement on abortion invokes a fourth rationale for opposition to elective abortion that reflects an explicitly theological claim: "Human life is a sacred gift from God."[38] The language of "gift" of life suggests a *responsive* and *creative* rationale for continuing pregnancy (such as embodiment and adoption), even while recognizing the justifiability of abortions in the three exceptional circumstances.

Agency and Abortion

LDS ecclesiastical opposition to elective abortion then appeals to legal concepts (a "crime" analogous to killing), cultural concerns (societal moral decline and sexual permissiveness), and theological considerations (gift of life, procreative imperative, and embodiment). This means that legalized elective abortion implicates core moral values in LDS teaching, including protecting human life, procreative sanctity, the purposes of family life, and the sacredness of sexual intimacy. Such values confer on abortion the status of a moral issue rather than solely a medical, health, political, legal, or personal issue. Insofar as these core moral concerns may not be instantiated in public policies and court decisions on legalized abortion, an ecclesiastical moral witness on abortion law may be warranted. However, the posture of moral opposition and policy critique presumptively runs contrary to another core value of the tradition, respect for moral agency, which in its secular correlate of respect for autonomy and personal privacy is the primary grounding for abortion rights.[39]

How can the core value of moral agency be reconciled with teaching that opposes elective abortion?[40] The LDS understanding of respect for choice on abortion has been addressed directly by two senior ecclesiastical leaders. Dallin H. Oaks acknowledges the resonance of freedom of choice within LDS moral culture: "Pro-choice slogans have been particularly seductive to Latter-day Saints because we know that moral agency, which can be described as the power of choice, is a fundamental necessity in the gospel plan."[41] Latter-day Saints are for Oaks necessarily theologically (not politically) advocates of choice but in a manner that constructs moral agency and "choice [a]s a method, not the goal." The theology of agency means affirming not only "choice" but also "right" choices. Put another way, Oaks's exposition entails that not all choices are morally equal. An ethical critique of public policy on elective abortion presumes a distinction between the legal right to choose (choice as a goal) and morally right choices (choice as a method).

This distinction allows Oaks to challenge the privatized religiosity of the political compromise: "If [Latter-day Saints] say we are anti-abortion in our personal life but pro-choice in public policy, we are saying that we will not use our influence to establish public policies that encourage righteous choices on matters God's servants have defined as serious sins." In this construction of agency, it is then morally problematic, if not hypocritical, to support a privatized religiosity on abortion while affirming that civic discourse on policy matters is to be religion-free. There is an implicit claim in this interpretation about the theological purposes of law: law is not simply a restraint on wrongdoing but instantiates a pedagogical purpose of promoting good values, or alternatively, law is not positivistic and morally neutral but should encourage right choices. However significant and distinctive this theology of law, moral influence on public policy can and should expand beyond advocating for restrictive laws on elective abortion. Policy changes to encourage right choices should encompass education on sexuality, contraceptive access, mandated insurance coverage for maternal care and neonatal care, adequate maternal and paternal leave, promoting adoption as a valid and feasible choice, and advocating a cultural ethos that displays no tolerance for sexual violence against women.

The current LDS prophetic leader, Russell M. Nelson, likewise has formulated a theological construction on moral agency in the abortion context. Nelson portrayed decisions about abortion as framed by a prior choice "to embark on a journey that leads to parenthood." Moral agency and freedom of choice pertain to this basic decision "to begin or not to begin" a path to parenthood. When conception occurs, "that choice [of parenting] has already been made." That is, the exercise of agency prior to *conception* restrains a woman's right to choose *during* pregnancy: "a woman is free to choose what she will do with her body. . . . [But], her choice to begin the journey [of parenting] binds her to the consequences of that choice. She cannot 'unchoose.'"[42]

Nelson's critique is directed against elective abortion and clearly would not apply to circumstances of involuntary pregnancy from sexual assault. However, his theology of agency addresses something more fundamental than philosophical advocacy of self-determination that has legalized rights to abortion. It is a multivalent critique regarding both agency and a privatizing culture that separates sexuality and procreation. However, the claim that moral agency applies prior to but not during pregnancy presumes an idea of a moral agent who is rationally deliberate, intentional, and analytic about all reproductive choices, free from either internal or external influences, and unmoored from relationships and cultural narratives. The construction of moral agency as inherently binding to consequences, even if unintended, unanticipated, and undesired, seems narrowly constricted to decisions regarding sexual intimacy and abortion and is seldom applied to other situations of moral choice.

The culture of privatized choices and a religion-free zone of public policy in a pluralistic democratic society offer a minimalistic social ethic constituting the interrelationship of personal freedom and the philosophical "harm" principle: liberty of choice and actions of morally mature individuals are to be respected so long as they do not impose harms to others.[43] Furthermore, the culture of personal autonomy, including contraception and reproductive technologies as matters of privacy, has successfully separated the sexual from the procreative such that it makes no sense in the public religious-free realm to claim that a choice for sexual intimacy is *necessarily* a choice to become a parent. The secular nonharm ethos permits extensive personal liberty to continue or terminate a pregnancy with the pre-viable fetus constructed as a biological other not capable of experiencing harm. The four rationales for ecclesiastical opposition identify other-regarding harms of abortion that are unlikely to be compelling for a secularized moral culture that already endorses the immunity of private choices and of public policy from religious influence. Notably, despite his very harsh indictment that societies permitting legalized elective abortion engage in a "war on the defenseless," Nelson does not issue a prophetic call for policy change or more restrictive abortion laws. The argument is theological and ethical, not legal, and appeals to convictions shared within the religious community but unlikely to be persuasive in the religious-free zone of the secular public square.

Given the continual ecclesiastical criticism of legalized elective abortion in the wake of the *Roe* decision,[44] the LDS Church issued a rather remarkable statement in the context of a 1991 Utah legislative proposal to narrow the range of legally permissible abortion. While condemning "the devastating practice of abortion for personal or social convenience," the statement maintained the church "as an institution has not favored or opposed specific legislative proposals or public demonstrations concerning abortion." LDS members were encouraged to "let their voices be heard in appropriate and legal ways that will evidence their belief in the sacredness of life."[45] This distinction between Church and membership[46] reflects the three orienting criteria of public ecclesiastical moral witness: the personal agency of members to undertake civic responsibility; the "separationist" principle of refraining from advocacy of "specific" policies or laws, thereby preserving policy immunity from direct religious dictates; and social space for an ecclesiastical witness on the core moral value (sanctity of life) at stake. The long-standing position of unequivocal opposition to elective abortion exhibits a pronounced prophetic moral witness: liberal abortion laws are morally flawed and contrary to formative values of a life-oriented society, including reverence for the gift of life, responsible sexuality, and respect for choice as a moral method. The efforts of Oaks and Nelson to take seriously arguments about freedom of choice on abortion reflect an approach of moral suasion according to shared values rather than values that are specific to a faith tradition and hence alien in public discourse. It follows from these commitments that religious values should not be the *basis* for public policy or law except as those values are part of a societally shared moral legacy, but religious communities can articulate public moral positions for the *process* of civic engagement and public reasoning.

Legal scholar Lynn Wardle has argued that ecclesiastical teaching reflects the historical leadership's responsibility to teach correct principles rather than dictate specific choices, which provides both ecclesiastical clarity and policy efficacy: "the Church has taken a clear position on the foundational issue (elective abortion should not be legal) and avoided the bramble bush of political battles on the many lesser issues that seem to divide even the most sincere pro-life groups and persons."[47] The consistent moral witness on the "foundational" issue leaves to the policy process, including engaged civic responsibility, subsidiary issues about the scope of abortion rights, such as procedures for informed consent, parental notification requirements, waiting periods, and public funding.

Still, the ecclesiastical focus on legalized abortion can neglect opportunities for civic engagement on numerous policy issues that could both preserve freedom of choice and limit the frequency of abortion without imposing draconian legal resolutions. This includes education regarding responsible sexuality and contraceptive literacy. It is rather remarkable that one-half century into the contraceptive revolution nearly 45% (2.8 million) of the pregnancies in the United States on an annual basis are unintended.[48] Forty-two percent of the unintended pregnancies end in abortion. Education on responsible sexuality does not mean discarding church teachings on sexual abstinence prior to marriage, but it does mean that insofar as elective abortion is framed as the moral equivalent of a killing act, other educational practices, health care interventions, and social policies can be advocated without moral compromise. The values underlying ecclesiastical promotion of adoption as a preferable recourse in the case of unintended pregnancy[49] can be extended to various integrity-preserving measures that prevent unintended pregnancies. This includes advocacy that the United States follow the path of other countries that have substantially lower rates of abortion by providing a comprehensive commitment of insurance coverage for maternity care, neo-natal care, and comprehensive parental leave. In addition, civic engagement that affirms women's equality in education and in work life (including eq-uitable compensation) and expresses intolerance for cultural practices that encourage sexual objectification and physical and structural violence against women should be part of a more comprehensive ecclesiastical moral witness on elective abortion. Ecclesiastical resources exist to re-story abortion as a matter of social justice rather than choice.

Legalizing Medical Marijuana

The LDS public witness on elective abortion has parallels with but also diverges from the more recent question of legalized medical marijuana. The catalyst for this discussion was a 2010 Colorado citizen referendum to legalize medical marijuana. In response to an inquiry at a conference of Colorado church leaders, the current prophetic leader, Russell M. Nelson, commented that there was no church position on medical marijuana. The implication of ecclesiastical silence is that use of medical marijuana is a matter of personal medical needs, consultation with physician, and respon-sible moral agency. Some years later, a physician overseeing LDS missionary health services offered the following response to a parental inquiry about use of medical cannabis for their daughter who was suffering from a rare degen-erative disease: "In states where the medicinal use of marijuana is approved legally and given under the guidance of a licensed physician, the church has [no] position. . . . The use of medical marijuana can be a great blessing to individuals who are suffering with these diseases in which other treatments have failed."[50] Within appropriate legal and professional parameters, use of medical marijuana is guided by an ethical imperative, relief of suffering, rather than ecclesiastical mandates. The difficulty with ecclesiastical silence is that as medicinal marijuana is legalized in thirty-one states, access could depend on the jurisdiction in which a person resides, or to put the point di-rectly, LDS members in Colorado would have more latitude using medical marijuana than in a state where it remained illegal, as in Utah.

This initial re-storying of medical marijuana as a "blessing" reflects the principles of medicine as a healing vocation, respect for professional in-tegrity, and responsible moral agency. However, the emergence of a Utah Medical Cannabis Act (UMCA), a citizen referendum on the fall 2018 ballot permitting persons with specified illnesses to "acquire and use medical cannabis for personal use" from private dispensaries as well as to grow a limited number of cannabis plants for "personal medical use,"[51] prompted a different ecclesiastical response. The UMCA effectively proposed privatizing medicinal marijuana without requiring a physician prescription. The UMCA presented ethical complications for ecclesiastical teaching in that the ecclesiastical position seemed to hinge on the status of legalization: the commitment to professional integrity does not make provision for political processes that change the status of a medical practice from illegal to legal, and the initial UMCA provisions failed to place access to medicinal marijuana under the professional oversight of physicians and pharmacists.

In April 2018, the LDS First Presidency issued a statement concurring with the perspectives articulated by the Utah Medical Association in its statement of professional opposition to the ballot initiative. The First Presidency statement unfortunately confused the issue by twice referring to the "Utah marijuana initiative" rather than as an initiative to legalize "medical" marijuana. The pattern of ecclesiastical nonengagement with specific policies was reflected in the statement's deference to professional expertise and avoidance of explicit church opposition to the UMCA: "We respect the wise counsel of the medical doctors of Utah. . . . The public interest is best served when all new drugs designed to relieve suffering and illness and the procedures by which they are made available to the public undergo the scrutiny of medical scientists and official approval bodies."[52] The Church subsequently engaged the services of a law firm to provide an analysis of the prospective legal issues raised by passage of the UMCA. This thirty-one-point analysis presented what the Church portrayed as "grave concerns" and "serious adverse consequences" were the initiative to be enacted.[53] The ecclesiastical leadership implicitly appealed to moral agency by inviting all persons to read the legal analysis and "make their own judgment."

By August 2018, this invitation to personal agency was supplanted when the LDS Church joined a coalition formally opposed to the UMCA. An ecclesiastical leader, Jack N. Gerard, indicated that ecclesiastical opposition to the UMCA was not a matter of principle; indeed, the core principle was that medical marijuana manifested the healing ethic of Christ and covenantal caring: "From the church's standpoint, our number-one priority as followers of Jesus Christ is to follow his example, to assist those that are afflicted, to try to relieve human pain and suffering."[54] The relief of pain and suffering, embedded in principles of love and covenantal care, became the core moral issue warranting public ecclesiastical engagement. The problem with the UMCA was instead procedural: inadequate regulation specific to the Act. While Gerard affirmed that "[t]he church does not object to the medical use of marijuana if doctor prescribed in dosage form through a licensed pharmacy," the privatization provisions of the UMCA failed to provide professional controls for medical marijuana use and thereby heightened the prospect of "serious consequences to health and safety." The issue of inadequate oversight—a claim that presumes the principle of the professional integrity of medicine—and the potential for health risks were combined in a church-distributed email message to all LDS members in Utah encouraging a "no" vote on the UMCA, which was portrayed as a "serious threat to health and public safety, . . . by making marijuana generally available with few controls."[55]

Within a span of months, the LDS position on medical marijuana shifted from "no position," to "opposition," to "does not object," to "support" grounded in the ethic of covenantal caring and professional integrity: "The Church supports the use of medicinal marijuana when it is distributed correctly to people enduring chronic pain and suffering."[56] These shifting views on medical marijuana in general in the context of simultaneous retention of opposition to the UMCA reflect an ecclesiastical moral witness carried out on two levels. The evolving position of "support" reflected ecclesiastical responsibilities to teach correct principles focused primarily on professionally authorized methods for the healing purposes of relieving pain and suffering. A second principle involved societal trust in the integrity of the medical profession to responsibly oversee a "medical" rather than a privatized practice. This principled position established the second realm of moral witness in which ecclesiastical statements emphasized a need for a procedural solution, primarily through a legislative process to develop the provisions of professional oversight. Professionalized regulation provided safeguards by which the overriding moral imperative, the relief of pain and suffering, could be enacted without generating adverse consequences.

Notwithstanding the complex and nuanced LDS opposition, the UMCA received 53% of the vote. The LDS Church was involved in negotiations with various groups and legislators to draft a post-vote compromise legislative bill that modified the UMCA and drew the ire of advocates of legalization.[57] This legislative process delayed the effective date of medical marijuana until 2020. The question is why church leadership decided medical marijuana was a moral issue of such import that it displaced the ecclesiastical principle of separation with ecclesiastical advocacy in the public square.[58]

Comparing the ecclesiastical approach to legalizing medical marijuana with that embedded in the critique of legalized elective abortion allows for meaningful lessons to be drawn. The LDS position on legalizing elective abortion, including critiques of permissive abortion laws, has been consistently and unequivocal in opposition. Elective abortion has been presented as a threat to values core to the Church identity, such as the sanctity of life, the integrity and responsibility of family life, covenants of chastity and marital sexuality, and refraining from killing, and as ultimately contrary to the divine plan for human salvation. These values have supported a public moral witness that, while informed by specific court cases or legislative proposals, has focused on articulating the substantive ethical issues of elective abortion rather than delineating procedural remedies (or, it must be added, constructive resolutions to the causes of abortion).

The LDS position on legalizing medical marijuana by contrast was equivocal and publicly confusing. It is difficult to square the initial "no position" teaching, which left matters to professional integrity and personal moral agency, with the extensive ecclesiastic engagement on the UMCA or even with the evolving posture of principled support for medical marijuana coupled with procedural opposition to the UMCA. The ecclesiastical position also seemed to imply that the morally responsible choice about medical marijuana was defined by jurisdiction. "Why, from the LDS Church's perspective," one writer editorialized, "should the science be compelling enough to permit the use of cannabis in one state but not another?"[59]

Second, the LDS Church struggled to find its *moral* voice on medicinal marijuana. The principled imperatives for supporting regulated medical marijuana, the Christian injunctions to care for the needy and relieve pain and suffering, were not articulated until months into the policy discussion. This moral articulation followed earlier ecclesiastical statements that framed the issue first as a matter of medical *science* and subsequently as a *legal* issue. The ecclesiastical reliance on other professional analyses regarding the benefits, risks, and legal implications of medicinal marijuana is understandable but also meant it was difficult to ascertain why medical marijuana was an ecclesiastical concern in the first place. Only when ecclesiastical leaders situated medical marijuana within an ethic of imitating the healing ministry of Jesus did the Church meet its own criteria for addressing the moral dimensions of a public policy issue and for speaking on terms other than as a political special-interest group: advocating correct principles when core values are compromised (or advanced) by public policy. A more coherent engagement with the UMCA would have framed the issue first as a matter of how to best relieve the pain and suffering of persons with chronic illnesses, which is not a scientific or legal matter, but a shared professional, religious, and human commitment.

The question of how ecclesiastical policy addresses matters of the changing legal status of a medical practice, such as assisted death, or a treatment, such as medicinal marijuana, is likely to continue to be a vexing consideration. The professional and public discussion about using psychedelic therapeutics, such as psilocybin, is following the path of medicinal marijuana.[60] The LDS commitment to reliance on professional medical judgment about therapeutic benefit would seem to support a professionally regulated approach to psychedelic therapeutics that historically have been ecclesiastically prohibited through the provisions of the Word of Wisdom.

Religious Ethics and Policy

The question of religious influence on policy questions, including bioethical and health care issues, is a perplexing question for liberal pluralistic cultures. Such cultures preserve religious integrity by recognizing the autonomy of religious faith and community in personal and congregational life and preserve political integrity by requiring a principle of general separation of religion from politics. However, it can be philosophically problematic to construct public policy as a "religion-free" zone. It manifests either an odd lack of or a selective historical moral consciousness insofar as the societal movement for civil rights for minorities and for women clearly relied on the involvement of religious traditions and churches, charismatic religious leaders such as Martin Luther King, Jr., and advocacy based on religious values and narratives. The ethical resources of faith communities can be potent sources for moral progress. When faith communities speak to questions of public policy, or when they speak truth to power, they have a *civic* responsibility to articulate the moral grounds and principles for their participation. LDS ecclesiastical leadership has presented these values consistently on elective abortion, and more recently, on immigration reform. Ecclesiastical statements on medical marijuana appeared reactionary and inconsistent because the moral values warranting ecclesiastical participation were not clearly formulated. The efficacy of any religious-based witness in the public square will ultimately depend on the resonance of these moral values with those of the civic community.

The ethical and policy questions presented by legalized abortion and legalized medical marijuana are embedded in broader structural and cultural issues about the social determinants of health and the moral responsiveness of the health care system to the human experience of disease, disability, and suffering. The LDS commitment to teach correct principles should convey a moral witness about universal access to essential health care.

Notes

1. Courtney S. Campbell, "Religious Perspectives on Human Cloning," in National Bioethics Advisory Commission, *Cloning Human Beings*, vol. 2 (Rockville, MD: NBAC, 1997), D1–D64.
2. "Leader of Mormon Church Looks to Future," *Sunstone* 20:2 (1997), 72.
3. Don Lefevre, "LDS Church Public Affairs Office: Statement on Human Cloning," March 21, 1997. See Campbell, "Religious Perspectives on Human Cloning."
4. Courtney S. Campbell, "Prophecy and Citizenry: The Case of Human Cloning," *Sunstone* 21:2 (1998), 11–15.
5. *Doctrine and Covenants* 58:27–28.
6. Mormon Women for Ethical Government, *The Little Purple Book: MWEG Essentials* (Salt Lake City: Common Consent Press, 2018), 23.
7. MWEG, *The Little Purple Book*, 1–2.
8. This is sometimes formulated as a principle of "anti-establishment." See Dallin H. Oaks, "Religious Values and Public Policy," *The Ensign*, October 1992, https://www.lds.org/ensign/1992/10/religious-values-and-public-policy?lang=eng&_r=1.
9. Governmental critiques of religious liberty as a cloak for discrimination are presented in U.S. Commission on Civil Rights, *Peaceful Coexistence: Reconciling Nondiscrimination Principles with Civil Liberties* (Washington, DC: USCCR, 2016). The LDS Church urged President Barack Obama to reject the recommendations and their characterization of religious liberty. Kelsey Dallas, "Faith Leaders, Including LDS Presiding Bishop, Call on Obama, Congress, to Reject Controversial Religious Liberty Report," *Deseret News*, October 12, 2016, https://www.deseret.com/2016/10/12/20598089/faith-leaders-including-lds-presiding-bishop-call-on-obama-congress-to-reject-controversial-religiou#president-barack-obama-gestures-while-saying-its-important-to-think-about-the-why-when-choosing-a-goal-or-profession-as-he-speaks-to-young-african-leaders-initiative-event-at-the-omni-shoreham-hotel-wednesday-aug-3-2016-in-washington. More recently, the LDS Church issued a statement advocating "fairness for all" in conflicts between religious

liberty and LGBT rights prompted by consideration of the Equality Act by the U.S. Congress (Newsroom, "Church Expresses Support for 'Fairness for All' Approach," May 13, 2019, https://newsroom.churchofjesuschrist.org/article/church-expresses-support-fairness-for-all-approach).

10. See Thomas Jefferson, "Virginia Statute for Religious Freedom," and James Madison, "Memorial and Remonstrance against Religious Assessments," in *The Sacred Rights of Conscience: Selected Readings on Religious Liberty and Church-State Relations in the American Founding*, ed. Daniel L. Dreisbach, Mark David Hall (Indianapolis: Liberty Fund, 2009), 250–251, 528.

11. W. Cole Durham, Jr., "Church and State," in *Encyclopedia of Mormonism*, ed. Daniel H. Ludlow (New York: Macmillan, 1992), 282–283, https://contentdm.lib.byu.edu/digital/collection/EoM/id/3541. The separation principle is a product of 20th-century Mormonism and the changed relationship between the LDS Church and the U.S. government. The principle is not always adhered to in local or state politics, particularly in Utah, where the LDS Church retains pronounced cultural influence. See Oaks, "Religious Values and Public Policy," 10.

12. Boyd K. Packer, "Our Moral Environment," The Church of Jesus Christ of Latter-day Saints, April 1992, https://www.lds.org/general-conference/1992/04/our-moral-environment?lang=eng

13. The Church of Jesus Christ of Latter-day Saints, "The Family: A Proclamation to the World," September 1995, https://www.lds.org/topics/family-proclamation?lang=eng&old=true

14. LeRoy Walters, "Religion and the Renaissance of Medical Ethics in the United States: 1965–1975," in *Theology and Bioethics: Exploring the Foundations and Frontiers*, ed. Earl E. Shelp (Dordrecht, The Netherlands: D. Reidel, 1985), 3–16; John H. Evans, *The History and Future of Bioethics: A Sociological View* (New York: Oxford University Press, 2012), 3–66.

15. Richard Rorty, *Philosophy and Social Hope* (New York: Penguin Books, 1999), 168–174; cf. Thomas Jefferson, "Letter to Danbury Baptists," in *The Sacred Rights of Conscience: Selected Readings on Religious Liberty and Church-State Relations in the American Founding*, ed. Daniel L. Dreisbach, Mark David Hall (Indianapolis: Liberty Fund, 2009), 528.

16. LDS scripture and teaching offer both liberal democratic and theological rationales for government. *Doctrine and Covenants* 134; The Pearl of Great Price, *Articles of Faith*: Articles 11, 12.

17. Kent Greenawalt, *Religious Convictions and Political Choice* (New York: Oxford University Press, 1988), 21.

18. James Hunter, *Culture Wars: The Struggle to Define America* (New York: Basic Books, 1991).

19. Jefferson, "Virginia Statute for Religious Freedom," 250–251.

20. Alexis de Tocqueville, *Democracy in America*, ed. J. P. Mayer (New York: Doubleday & Company, 1969), 287–301.

21. Rorty, *Philosophy and Social Hope*, 173.

22. Rorty, *Philosophy and Social Hope*, 173.

23. Vincent Barry, *Bioethics in a Cultural Context* (Boston: Wadsworth, Cengage Learning, 2012), 67; Jonathan D. Moreno, "The End of the Great Bioethics Compromise," *Hastings Center Report* 35:1 (January–February 2005), 14–15.

24. James C. Mohr, *Abortion in America. The Origins and Evolution of National Policy* (New York: Oxford University Press, 1978).

25. Emily Martin, *The Woman in the Body: A Cultural Analysis of Reproduction* (Boston: Beacon Press, 2001).

26. Lisa Sowle Cahill, "Abortion: Roman Catholic Perspectives," in *Bioethics,* 4th ed., vol. 1, ed. Bruce Jennings (New York: Macmillan Reference USA, 2014), 37–41.

27. First Presidency, "The Origin of Man," *Improvement Era,* November 1909, 77–80, https://www.lds.org/ensign/2002/02/the-origin-of-man?lang=eng. The context for this statement was an exposition on human nature in light of theological challenges posed by evolution.

28. Dallin H. Oaks, "The Great Plan of Happiness," *The Ensign,* October 1993, https://www.lds.org/general-conference/1993/10/the-great-plan-of-happiness?lang=eng. As noted in Chapter 2, fluidity about the beginnings of life also informs LDS views on the moral status of embryos created through in vitro fertilization.

29. *Roe v. Wade,* 410 U.S. 113 (1973).

30. Lynn D. Wardle, "Teaching Correct Principles: The Experience of the Church of Jesus Christ of Latter-day Saints Responding to Widespread Social Acceptance of Elective Abortion," *BYU Studies Quarterly* 53 (2014), 107–140.

31. "Church Opposes Abortion Bill," *Deseret News,* January 23, 1969.

32. This policy, with some minor organizational changes, was reissued ten months later in the wake of *Roe v. Wade.* Gilbert W. Scharffs, "The Case Against Easier Abortion Laws," *The Ensign,* August 1972, https://www.lds.org/ensign/1972/08/the-case-against-easier-abortion-laws?lang=eng; First Presidency, "Policies and Procedures: Statement on Abortion," *New Era* 3, April 1973, https://www.lds.org/new-era/1973/04/policies-and-procedures-statement-on-abortion?lang=eng

33. Guttmacher Institute, "Induced Abortion in the United States," September 2019, https://www.guttmacher.org/fact-sheet/induced-abortion-united-states

34. *Doctrine and Covenants* 59:6.

35. The Church of Jesus Christ of Latter-Day Saints, "Church Policies and Guidelines, Policies on Moral Issues: Abortion," in *General Handbook: Serving in the Church of Jesus Christ of Latter-day Saints,* Chapter 38.6.1, February 19, 2020, https://www.churchofjesuschrist.org/study/manual/general-handbook/38-church-policies-and-guidelines?lang=eng#title_number91

36. Courtney S. Campbell, "Embodiment and Ethics: A Latter-day Saint Perspective," in *Theological Developments in Bioethics: 1990–1992,* ed. B. A. Lustig (Dordrecht, The Netherlands: Kluwer Academic Publishers, 1993), 52.

37. The Church of Jesus Christ of Latter-Day Saints, "Church Policies and Guidelines, Policies on Moral Issues: Abortion," in *General Handbook: Serving in the Church of Jesus Christ of Latter-day Saints,* Chapter 38.6.1. I have not been able to find the rationale for the puzzling inclusion of the term "forcible" in the therapeutic allowance for abortion in cases of rape or incest.

38. "Abortion," The Church of Jesus Christ of Latter-day Saints, https://www.lds.org/topics/abortion?lang=eng (accessed August 4, 2020).

39. Peggy Fletcher Stack, "Surprise! The LDS Church Can Be Seen as More 'Pro-Choice' Than "Pro-Life' on Abortion," *Salt Lake Tribune*, June 2, 2019, https://www.sltrib.com/religion/2019/06/01/surprise-lds-church-can/

40. Carol Kuruvilla, "Mormon Mom Has an Abortion Story That Donald Trump Needs to Hear," *Huffington Post*, October 21, 2016, https://www.huffingtonpost.com/entry/mormon-mom-abortion-donald-trump_us_580a565ce4b000d0b1566cf0; Caroline Kee, "Here Is What An Abortion at 22 Weeks Is Actually Like," *Buzzfeed,* October 20, 2016, https://www.buzzfeed.com/carolinekee/woman-shares-late-term-abortion-story; Danielle B. Wagner, "Abortion: How President Nelson and Other Church Leaders' Teachings Provide Hope, Compassion, and Direction," *LDS Living,* January 28, 2019, http://www.ldsliving.com/Abortion-How-President-Nelson-and-Other-Church-Leaders-Teachings-Provide-Hope-Compassion-and-Direction/s/90173

41. Dallin H. Oaks, "Weightier Matters," *The Ensign*, January 2001, https://www.lds.org/ensign/2001/01/weightier-matters?lang=eng

42. Russell M. Nelson, "Abortion: An Assault on the Defenseless," *The Ensign*, October 2008, https://www.lds.org/ensign/2008/10/abortion-an-assault-on-the-defenseless?lang=eng. Cf. Russell M. Nelson, "Reverence for Life," *The Ensign*, November 1985, https://www.lds.org/general-conference/1985/04/reverence-for-life?lang=eng

43. John Stuart Mill, "On Liberty," in *The English Philosophers: From Bacon to Mill*, ed. Edwin A. Burtt (New York: Modern Library, 1939), 949–1041.

44. Legal scholar Lynn D. Wardle notes that in every LDS semiannual general conference between 1970 and 1995, ecclesiastical leaders expressed criticism or condemnation of elective abortion. Wardle, "Teaching Correct Principles," 122. The audience, however, was communal, not legal or political.

45. Lester E. Bush, Jr., *Health and Medicine Among the Latter-day Saints*: *Science, Sense, and Scripture* (New York: Crossroad, 1993), 165.

46. This formulation is reiterated in current ecclesiastical teaching: Newsroom, "Abortion," https://newsroom.churchofjesuschrist.org/official-statement/abortion

47. Wardle, "Teaching Correct Principles," 131.

48. L. B. Finer, M. R. Zolna, "Declines in Unintended Pregnancy in the United States, 2008–2011," *New England Journal of Medicine* 374 (2016), 843–852.

49. The Church of Jesus Christ of Latter-Day Saints, "Church Policies and Guidelines, Policies on Moral Issues: Single Expectant Parents," in *General Handbook: Serving in the Church of Jesus Christ of Latter-day Saints*, Chapter 38.6.18, https://www.churchofjesuschrist.org/study/manual/general-handbook/38-church-policies-and-guidelines?lang=eng#title_number113

50. Robert Gehrke, "LDS Church's Stance on Medical Marijuana Doesn't Make Sense," *Salt Lake Tribune*, April 13, 2018, https://www.sltrib.com/opinion/2018/04/12/gehrke-lds-churchs-stance-on-medical-marijuana-doesnt-make-sense-why-is-it-ok-for-a-nevada-mormon-but-not-a-utah-mormon/

51. Marijuana Policy Project, "Summary of Utah's Medical Cannabis Law," https://www. mpp.org/states/utah/summary-of-utahs-medical-cannabis-law/; Ben Lockhart, "13 Things Voters Should Know About Utah's Medical Marijuana Initiative," *Deseret News*, September 23, 2018, https://www.deseretnews.com/article/900033184/13-things-voters-should-know-about-utahs-medical-marijuana-initiative.html

52. Newsroom, "First Presidency Statement on Utah Marijuana Initiative," https://newsroom.churchofjesuschrist.org/article/first-presidency-statement-on-utah-marijuana-initiative; Luke Ramseth, "LDS Church Issues Statement Opposing Medical Marijuana Ballot Initiative," *Salt Lake Tribune*, April 12, 2018, https://www.sltrib.com/news/health/2018/04/10/lds-church-issues-statement-opposing-medical-marijuana-ballot-initiative-which-a-majority-of-utah-voters-supports/

53. Newsroom, "Utah Medical Marijuana Initiative," May 11, 2018, https://newsroom.churchofjesuschrist.org/article/marijuana-analysis.

54. Taylor W. Anderson, Benjamin Wood, "LDS Church Announces Opposition to Utah Medical Marijuana Initiative," *Salt Lake Tribune*, August 23, 2018, https://www.sltrib.com/news/politics/2018/08/23/lds-church-announces/

55. Ben Lockhart, "The Church of Jesus Christ of Latter-day Saints Joins Utah Coalition," *Deseret News*, August 23, 2018, https://www.deseretnews.com/article/900029171/mormon-church-joins-utah-coalition-saying-no-to-marijuana-initiative

56. Jason Swensen, "Church Says Yes to Regulated Medical Marijuana but No to Utah Initiative," The Church of Jesus Christ of Latter-day Saints, September 20, 2018, https://www.lds.org/church/news/church-says-yes-to-regulated-medical-marijuana-but-no-to-utah-initiative?lang=eng

57. Kelsey Foreman, "The Church Gets Involved in Re-Writing Utah's Medical Marijuana Laws," *Utah Business*, September 3, 2019, https://www.utahbusiness.com/prop2-medical-marijuana/

58. Legalization advocates filed a lawsuit in December 2018 contending church involvement violated state constitutional provisions of separation of church and state. The argument regarding church involvement was dropped in a May 2019 filing. See Nicole Nixon, "Patient Advocates May Sue over LDS Church Involvement in Utah Medical Marijuana Laws," *NPR Utah*, November 15, 2018, http://www.kuer.org/post/patient-advocates-may-sue-over-lds-church-involvement-utah-medical-marijuana-laws#stream/0; Morgan Smith, "Lawsuit Drops Claim Mormon Church Swayed Medical Pot Changes," *AP News*, May 3, 2019, https://apnews.com/bac195211378427080568acd03f14954

59. Gehrke, "LDS Church's Stance on Medical Marijuana Doesn't Make Sense."

60. Dustin Marlan, "Beyond Cannabis: Psychedelic Decriminalization and Social Justice," *Lewis and Clark Law Review* 23 (2019), 851–892; Michael Pollan, "The Trip Treatment," *The New Yorker*, February 9, 2015, https://www.newyorker.com/magazine/2015/02/09/trip-treatment

Epilogue

An LDS Case for Universal Health Care

In November 2019, Brigham Young University–Idaho, owned by The Church of Jesus Christ of Latter-day Saints (LDS), announced it would no longer accept Medicaid as a payment option from its students as of January 2020.[1] As students are required to have health insurance coverage, the policy change was anticipated to prompt greater enrollment in the university's student health plan, a program that requires a significantly higher annual premium, imposes ceilings on annual benefits, and provides no insurance coverage for birth control. The latter two provisions are contrary to benefit protections under the federal Affordable Care Act. The Brigham Young University system, including campuses in Utah and Hawai'i as well as Idaho, has criticized the Affordable Care Act as paternalistic and intrusive: "There are numerous government-imposed requirements that we don't believe are necessary to provide good health care to our students."[2] However, following national publicity, criticism by students, and a refutation by local health care institutions of the university's claim that continuing Medicaid coverage would overburden health care organizations, BYU-Idaho reversed its decision and apologized for any "turmoil" generated from the initial policy change.[3]

The brief controversy is illustrative of broader societal concerns about national health care coverage, cost-effective delivery of health care, and government directives that may conflict with principles of religious liberty.[4] It also highlights the general disarray of health care delivery in the United States, which has simultaneously managed to be both remarkably expensive and inefficient in securing good health outcomes. The prominent journalist and public intellectual Walter Cronkite once commented, "America's health care system is neither healthy, nor caring, nor a system."[5] A brief elaboration on Cronkite's observation would include the following points. American health care is better framed as "sick care": a substantially greater portion of resources, including critical and intensive care medicine, is dedicated to

Mormonism, Medicine, and Bioethics. Courtney S. Campbell, Oxford University Press (2021). © Oxford University Press.
DOI: 10.1093/oso/9780197538524.003.0010

treatments for persons who are seriously ill than to preventive measures that keep persons healthy and out of institutional care settings in the first place. Groundbreaking innovations in biomedical research involve an expensive quest to cure *diseases* that is prioritized over caring for *patients*. American medical education is so predominantly oriented by scientific learning about the pathologies of the bodily organism that the levels of empathy and care for patients among medical students actually decline.[6] These institutional and educational orientations foster criticisms of "impersonal medicine" even in a purportedly patient-centered medicine. And as illustrated in the BYU-Idaho example, American health care is delivered through a patchwork of systems and organizations, including federal, state, and private programs, in which efficiency rather than quality care are structural priorities. Health care becomes another commodity and market economy that conditions access through ability to pay.[7]

The LDS "welfare" system that seeks to assure that basic needs of persons and families are met is often touted as a distinctive success story within the Church and by national political leaders. The organizing principles of welfare—including self-reliance, personal responsibility, familial and communal support, and recipient service contributions—aim to promote personal and familial empowerment, diminish "free riding" at public expense, and minimize bureaucratic inefficiencies.[8] The moral culture of a community can be assessed by how it treats the most vulnerable of its members. At the same time, embedded within many of the bioethical questions discussed in the preceding chapters is a fundamental issue of social justice: the lack of access to basic or essential health care for millions of persons. LDS ecclesiastical teaching has not addressed this social inequity directly. However, long-standing concerns about federally imposed mandates or government-coordinated medicine that in practice default to a presumption of the efficiency of private, marketplace health care delivery can be resisted by a counternarrative in the moral culture. The welfare principles on "health" include a stipulation that LDS members should "practice good sanitation and hygiene and obtain adequate medical and dental care."[9] The LDS Church has extended the concept of "adequate care" in articulating criteria for health care reform at a local level and in establishing health priorities in programs of international humanitarian assistance. This epilogue draws on these examples to develop an LDS case for a system of universal access to basic or adequate health care services.

Health Care Reform

In December 2014, Utah governor Gary Herbert proposed the "Healthy Utah" program to expand health insurance coverage for an estimated ninety-five thousand lower-income persons, provide mechanisms for containing health care costs, and promote personal responsibility for good health.[10] The LDS Church soon issued a precedent-setting statement on health care reform:

> "We recognize that providing adequate health care to individuals and families throughout Utah is a complex and weighty matter. It deserves the best thinking and efforts from both the public and the private sectors. While the economic and political realities are being debated, we hope the discussion and decisions taken in this matter will be consistent with the God-given principles regarding care for the poor and the needy that in the end benefit all of His children. We reaffirm the importance for individuals and families to be as self-sufficient as their particular circumstances allow and recognize that the lack of access to health care can impair a person's ability to provide for self and family. We commend public officials for their efforts to grapple with these difficult issues and pray for their success in finding solutions that reflect the highest aspirations of society."[11] Framed as a "principled approach" to health care reform, the statement follows the pattern of public ecclesiastical witness by articulating communally valid principles that could support a variety of public policies, while refraining from endorsing a particular policy alternative. The principled approach discloses a moral concern that the health care coverage crisis not be resolved solely through economic efficiency and political expediency.

A first ethical principle embedded in the statement is that achieving health care adequacy requires collaboration between public and private sectors. In the wake of populist backlash against the federal Affordable Care Act, numerous congressional efforts at repeal, and judicial review, affirming the valued role of the public sector rather than deferring to private commercial market approaches is morally significant. The principle presents an implicit critique of a prevalent libertarian philosophy that the primary role of government is to protect citizens' civil liberties, and consequently, the provision of health care is deemed beyond governmental jurisdiction.[12] The ecclesiastical statement clearly doesn't encourage or defend a particular system of health care delivery, be it market-coordinated, government-coordinated, or a mixed system. Some scholars have advocated an entirely market-based, consumer-driven approach to provide basic health care for all persons.[13] The overriding ethical principle is that adequate health care is a *public* matter, a moral conviction reinforced by the concluding sentence that expresses hope for resolutions reflecting societal aspirations.

A second principle of adequate health care, articulated as part of the call for public deliberation to not be limited to economic and political realities, is that policy manifests divinely bestowed covenantal *care for the socially vulnerable*. The claim is not that these "God-given principles" be the foundation for policy or even integrated in policy deliberation but rather that sound public policy on health care reform should *correspond* with communal commitments of not abandoning persons in need. A threshold test of consistency between policy and religious ethical values is that equity has primacy over economic efficiency. Although caring for the poor, vulnerable, and needy is embedded in the religious community's commitments to love, hospitality, justice, and covenantal solidarity, an ethic of caring for the vulnerable should have meaningful resonance in civic discourse as it reflects moral continuities between religious and nonreligious values.

A pair of substantive principles are presumed in the claim that caring for the vulnerable will, "in the end," benefit all persons. The implication that health care reform should promote the interests of all "[God's] children" is a prophetic reminder of the divine identity (*imago Dei*) of each person. A third threshold of moral adequacy in access to health care is a commitment to the *moral equality* of persons irrespective of financial capabilities, or racial, sexual, ethnic, or other social categories. Preferential concern for the health care needs of the vulnerable can benefit the entire community in two direct ways that exert very different moral responsibilities than the "invisible hand" ideology esteemed in private, libertarian approaches to health care. Responsiveness to the health care needs of vulnerable persons provides opportunities for caregiving service. Insofar as human givers are first and always beneficiaries and recipients of gifts, expressing care and service for vulnerable persons manifests the gift-responsiveness of covenantal calling. Second, health care needs evoke a shared human condition of vulnerability— all persons at some point will experience their health vulnerabilities—that provides moral ground to enact interdependency and empathic solidarity.

A further principle is personal and familial self-sufficiency, which draws on core values of LDS welfare and personal responsibilities of stewardship and moral agency. As moral agents, persons assume stewardship responsibilities for their own health and well-being to the extent they can control. This commitment issues in a fourth criterion for adequate health care: accessible health care should be empowering rather than enabling and provide opportunities for vulnerable and ill persons to exercise *personal responsibility* for their own health and wellness. The corollary of the threshold test is that, consistent with principles of justice and fairness, persons should not be deprived of benefits of their participation in civic life for reasons they have *minimal control* over, such as their genetic legacy, the arbitrariness of disease, and the existential fragility of human life. It is difficult for a person (or caregivers) to assume full moral responsibility when experiencing the limitations imposed by disease. The interrelationship of agency, self-reliance, stewardship, and justice is presupposed in the ecclesiastical assertion that "lack of access to health care can impair a person's ability to provide for self and family." A deprived familial ability to access basic health care means that communal, social, and other resources must supplement family resources to meet the minimal stewardship standard of sufficiency.

The public ecclesiastical witness affirms that health care reform is a *moral* question that should not be addressed solely by economic efficiency or political expedience but rather should correspond with principles of the moral culture, including commitments of caring for the vulnerable, moral equality, responsive empathy, communal solidarity, and empowering self-reliance. It may not be possible for any policy to fully realize these values, but these principles articulate tests to differentiate between an ethically acceptable and relatively just health care system and a system that, however economically efficient or politically feasible, is morally inadequate, unhealthy, and uncaring. Ultimately, the threshold of moral adequacy in health care is an ongoing quest to realize what is designated as the "highest aspirations of society." These aspirations include emphasizing health care rather than sick care, establishing a level of essential health care for all citizens based on need rather than ability to pay, and ensuring that the most vulnerable persons are not abandoned by the healing professions and the religious community.[14]

Principles of Humanitarian Assistance

These principles for assessing health reform proposals on adequate health care did not specify what services would be included in an "adequate" level

of health care. The precepts and programs of international humanitarian assistance that Latter-day Saint Charities develop for communities in the absence of a health care infrastructure provide some direction. These programs
have provided $2.3 billion in public health education, health literacy
interventions, and access to basic subsistence and health care services in
197 countries, including food production, clean water, emergency response,
immunizations, maternal and newborn care, vision care, wheelchairs for
persons with disabilities, refugee assistance, and local community projects.[15]
A brief exposition of seven principles underlying humanitarian assistance
can give meaning to "adequate" health care and establish priorities for basic
health care services.

Donative Love. The humanitarian assistance programs are funded by
donations from LDS church members as well as by other persons and organizations. Implementing these programs requires humanitarian service
missions by LDS members and collaboration with non-LDS humanitarian
organizations. The humanitarian initiatives then express an ethic of self-
sacrificial, donative love of neighbor and hospitality to the stranger.[16]

Self-Reliance. The humanitarian assistance initiatives emphasize educational and health literacy, particularly with respect to food production, clean
water, maternal and newborn care, and other community-based programs.
These efforts of health literacy mirror the emphasis on public health and disease prevention programs responsible for the significant increases in life
expectancy in industrialized nations in the last century. The educational
emphasis reflects the maxim embedded in medical progress that knowledge empowers persons and families to assume responsibility for their own
health. Humanitarian assistance programs thereby share with the ecclesiastical statement on health care reform a priority commitment to cultivate
stewardship, self-reliance, and moral agency.

Community Coordination. The efficacy of the humanitarian initiatives
requires coordination with several communities, including relief organizations and the communities served. Organizational collaborations enable education to reach more persons and expand the impact of assistance
programs. The local communities served by the initiatives display agency
and stewardship by assuming responsibilities for defining their own needs
and values that best promote communal sufficiency. The humanitarian assistance programs require a coordinated pooling of resources and expertise for
community-formulated purposes, a structural matter of great significance
for health care reform.

Prevention. The ethic of preventive health care is central to humanitarian assistance and essential given the absence of a basic health care infrastructure in many communities. The food production program seeks to ensure that families and communities have a sustainable level of nutrition for good health, reflecting priorities embedded in the Word of Wisdom health ethic. The clean water program prevents the emergence of waterborne contagious diseases through constructing wells, establishing drinking water systems, and family hygiene education. The immunization program provides vaccines for measles, diarrhea, pneumonia, and maternal and neonatal tetanus.

Relieving Pain and Suffering. Prevention is never a complete panacea because many health conditions are attributable to genetic inheritance, accidents, and embodied limitations. Some initiatives focus on alleviating chronic impairments and disabilities that cause pain and suffering. The vision care program provides education, equipment, and supplies to community eye care professionals to enhance quality vision care and diminish the frequency of vision impairment and blindness. The wheelchair program assists local organizations in improving the services they provide to persons with physical disabilities based on the person's need and circumstances. These programs symbolize inclusiveness and nonabandonment of vulnerable persons with physical impairments and disabilities.

Covenantal Solidarity. Two initiatives address crisis interventions in emergency situations and reflect moral imperatives to save human life and avoid premature death. The maternal and newborn care initiative provides education and training for birth attendants as both a crisis-prevention and crisis-response measure. Education in maternal care, essential care for newborns, and neonatal resuscitation addresses life-threatening respiratory distress of babies at birth and the needs of mothers with serious postbirth medical conditions. The emergency response program, a crisis-response initiative, ensures that food and other subsistence needs are met in the aftermath of a natural disaster. These programs display the covenantal commitment of solidarity with persons experiencing unanticipated physical or sustenance afflictions over which they have no control.

Hospitality to Strangers. The refugee response initiative has perhaps the most explicit scriptural rationale, embedded in the injunction of Jesus to his disciples—"I was a stranger and you took me in"—and in biblical teachings requiring the covenanted community to care for the stranger "as one born among you, and thou shalt love him as thyself."[17] Refugee assistance assumes heightened symbolic ethical meaning through ecclesiastical discourse that re-stories contemporary refugee crises within the communal narrative identity as refugees from historical religious persecution. In relating that the LDS Church partnered with seventy-five humanitarian agencies in seventeen countries to address ongoing refugee crises in Europe, ecclesiastical leader Patrick Kearon invoked empathic moral imagination in observing that "members of the Church, as a people, . . . don't have to look back far in our history to reflect on times when we were refugees, violently driven from homes and farms over and over again."[18] While community engagement in refugee assistance has been endorsed by the LDS First Presidency,[19] its most compelling exposition was presented by then-president of the women's Relief Society program, Linda K. Burton. Burton situated responsiveness to refugee crises within the program's historical mandate to address "pressing calls and extraordinary occasions" with compassionate care and charitable service. Burton likewise invoked empathic and narrative imagination and the golden rule ethic in encouraging congregants to ask, "What if their story were my story?"[20] Although refugee assistance presumes a crisis context, an interpretation of shared vulnerability and shared identity is part of the moral case for health care reform and universal access.

A Moral Witness on Adequate Health Care

These seven principles underlie LDS humanitarian assistance in areas of the world that lack the minimal medical and communal health infrastructure to promote the capabilities of millions of persons to care for their basic subsistence and health needs. By analogy, these principles underlie ten propositions that can serve as guidelines to determine what constitutes an essential level of health care services for persons who have been failed by the morally inadequate health care system in the United States. A morally adequate health care system will:

- Cultivate personal responsibility for good health and wellness;
- Encourage health literacy through education about hygiene, nutrition, and fitness;
- Ensure that all persons have access to basic preventive health care services promoting personal and public health, including annual physical exams, maternal and newborn care, and immunizations;
- Prioritize health care rather than sick care;

- Guarantee care for functions critical to physical well-being such as dental and eye care;
- Balance preventive, chronic, and crisis medical interventions;
- Provide medical interventions, including rehabilitative and chronic care, for persons with various disabilities that affirm their inclusion as valued members of the moral community;
- Ensure that all persons can access critical care medical interventions that have a high likelihood of making a difference between saving life and premature death;
- Prioritize palliative care and symptom relief to alleviate unnecessary pain and suffering from illness, especially in the context of dying; and
- Affirm a societal responsibility to provide access for essential health care services.

These ten guidelines represent priorities for health care reform rather than an exhaustive list of health care interventions. As resource constraints invariably impose limits on making every medical intervention available to every person, specific health care interventions can be prioritized based on five criteria: the frequency of a disease condition, its communicability, cost-effective treatments, severity of pain and suffering, and the prospects for recovery or rehabilitation. Resource scarcity should not erode the core commitment to provide a basic level of essential or adequate health care services for everyone.

My reliance on LDS principles for humanitarian service as a precedent for community advocacy of universal access to morally adequate health care is intertwined with several moral assumptions. First, health care is a public or common good rather than a private commodity. Health care should be considered a shared public good because health insecurity is often beyond a person's control. Health crises are often undeserved, random, unpredictable, and of overriding importance. Persons should not be deprived of a fundamental good of civic life and responsibility due to circumstances that they cannot control. A health care system responsive to the needs of vulnerable persons ultimately redounds to the common good because all persons eventually will experience their own vulnerability, frailty, and finitude. This means the moral condition for receiving health care is need, not ability to pay.

Second, needs for health care are analogous to a person's basic needs for subsistence and security, including food and clean water, shelter, police and fire protection, and education.[21] These subsistence and security needs must be met for persons to participate in, contribute to, and receive benefits from civic engagement. In modern industrial societies, a collective societal infrastructure to coordinate care and beneficence must complement works of communal charitable assistance or familial resources to meet these core human needs. The financial burdens for health care and disease should be shared relatively equally between the healthy and the ill, rather than stratifying access to health care according to the rich ill and poor ill.

Third, a health care system that provides morally adequate care is warranted by principles of justice, reciprocity, and fair play. Since health care is a public good, all persons have responsibilities to contribute (through taxation, insurance premiums, or service) to the common financial and logistical resources that support the institutions, education, and professions that mediate morally adequate health care. Public trust in a system's effectiveness presupposes commitments of personal responsibility, self-reliance, and minimization of "free riders" (persons who could contribute to the common good but choose not to do so). One area of common ground in societal debates over health care reform is that "the rules should apply to everyone."[22]

At one point in our nation's history, it was possible for families and communities to be relatively self-sufficient in meeting their own health needs. Principles of self-reliance and personal responsibility are necessary features of ecclesiastical and bioethical discourse about health care reform. However, as good health is not simply a matter of personal choice, but is more strongly correlated with genetic predispositions, social and cultural determinants, and the broader environment,[23] access to morally adequate health care requires coordination beyond what individual families or communities can supply. LDS ethical principles, health practices, and coordinated programs of assistance provide a foundation for a moral claim that extends beyond current ecclesiastical emphases, namely, that universal access to adequate health care is a moral requirement of a relatively just society.

Ethical Constraints in the COVID-19 Pandemic

The moral challenge of providing adequate health care was heightened significantly in the context of the global coronavirus pandemic of 2020, which constituted over33 million confirmed cases and 1 million deaths from COVID-19 globally as well as over 205,000 deaths in the United States as of October 2020.[24] In addition to the evident impacts on physical and emotional health, public health and civic responses to the coronavirus pandemic required unprecedented changes for the worldwide LDS community in worship, missions, and humanitarian services that both challenged and displayed the resiliency of the church organization. LDS ecclesiastical leaders reaffirmed the commitment of church members to be "good global citizens" and "good neighbors" in the context of directives from government and public health officials to restrict public gatherings, including religious services.[25] Communal worship services were cancelled for several weeks, with religious instruction occurring within the home and family settings. Male priesthood holders were authorized to administer the core ritual of the sacrament to family members within their homes and provide the sacrament to other families as part of ministering responsibilities. LDS evangelizing missionaries serving in far-flung regions on the world returned to their homes for several months and subsequently received new callings to serve in regions deemed to be at lower risk for contracting the coronavirus. LDS temples, emblematic of the most sacred communal sites of worship, were closed for weeks to months depending on geographic location.

These forms of contraction of organization mission were coupled with an expansion of humanitarian assistance commitments. This assistance coordinated the provision of food and essential medical supplies to numerous countries as well as organizational and local mask-making initiatives to provide respiratory protections for both medical professionals and citizens.[26] Latter-day Saints also participated in two worldwide days of prayer and fasting in March and April 2020 in shared petition for divine assistance and providential relief and healing such that "the present pandemic may be controlled, caregivers protected, the economy strengthened, and life normalized."[27]

As ecclesiastically challenging as the coronavirus pandemic continues to be, the ethical perplexities of COVID-19 were chilling. The pandemic revealed underlying structural problems in health care delivery and health care access that contributed to increased risk of contagion and increased mortality. Four interrelated ethical issues have occupied professional and public discussion: (1) the responsibility of health care providers to treat infected patients in the absence of adequate personal protective equipment (PPE), including masks, shields, gloves, and gowns; (2) the binding nature of patient or familial requests for resuscitation measures; (3) the necessity of priority setting of crisis care interventions due to a scarcity of ICU beds and ventilators; and (4) racial and ethnic disparities in mortality that reflect underlying issues of stratified access to care.

Professional Responsibilities to Treat. Latter-day Saint communities and leaders echoed the public commendations of health care professionals, including physicians, critical care nurses, and respiratory therapists, who assumed heightened personal risk to themselves and their own family members in the course of treating patients infected with SARS-CoV-2. The risks were real and sometimes fatal as over 66,000 health care workers had been infected, with 318 reported deaths, in the first three months of the pandemic in the United States. The heightened risk assumption and increasing infection rates were due not simply to the contagiousness of SARS-CoV-2 but to an inadequate supply of personal protective equipment for what were often termed the "front line" professions in fighting the pandemic. The scarce supplies of equipment to diminish professionals' risk of infection led the church-owned Beehive Clothing facilities to temporarily shift their processes for making clothing for religious rituals to the production of over 1.5 million masks and two hundred thousand medical gowns.[28]

The perils of exposure for professionals prompted professional and ethical reconsideration about the principle of professional integrity relative to the level of permissible risk in professional responsibilities to treat patients. The distinctive ethical integrity of the healing professions has historically resided in a self-sacrificial commitment to give primacy to the interests of patients.[29] However, this professional commitment to care and treat the ill has never required professionals to become martyrs for their calling. For many professionals, the lack of the essential protective equipment required to provide care for infectious patients raised serious questions about the tenuous nature of the societal contract with the healing professions. A professional "ought" or responsibility to care presupposes the availability of basic resources so that care can be enacted. The reconceptualization of professional integrity and responsibility was articulated by an emergency room physician in a biting indictment of structural failures: "If health-care providers are going to risk their life, then there is a reciprocal obligation—*the fairness principle*—that society, employers, and hospitals keep them safe and ensure that they are fairly treated, whether they live, get sick, or die."[30]

Professional Decisions and Medical Futility. Professional concerns about risk exposure were coupled with proposals to minimize circumstances when providing medical treatments or interventions, particularly cardiopulmonary resuscitation, were (1) deemed to be nonbeneficial or futile for patients and also (2) heightened infection risks for physicians, critical care nurses, respiratory therapists, and other members of a code team. Some organizations allowed greater latitude to physicians to override or disregard the patient's or family's request for all life-saving measures and to unilaterally institute a do-not-resuscitate order.[31] The American Medical Association indicated that the "ethical obligations" of physicians—that is, the profession's ethical integrity for promoting public health, ensuring availability of physicians to provide care, and stewardship of limited resources—permitted withholding resuscitative interventions "without seeking patient agreement."[32]

Although the ethics of providing cardiopulmonary resuscitation procedures have historically been directed by respect for personal autonomy and familial decision-making, some professionals argued that patients and families were not always informed of or fully comprehended the generally relatively low rate of survival (25.7%) to hospital discharge of a patient who undergoes CPR. The diminished prospect of benefit to patients and the heightened risk to professionals in the COVID-19 context of providing what for many patients was most likely to be a nonbeneficial intervention led physicians to advocate for increased use of goals of care conversations with patients, forgoing CPR in certain circumstances, and protecting professional staff.[33] Insofar as LDS ecclesiastical teaching warrants refusing life-sustaining measures when dying is "inevitable" professional and organizational discussion about withholding CPR increases the importance of candid conversations between patients, their families, and their physicians.

Crisis Care and Scarce Resources. The interrelated nature of these ethical questions was magnified in public and professional discussion about developing crisis care guidance for hospitals regarding patient access to beds in intensive care units (ICU) and ventilators for respiratory assistance. In the early stages of the pandemic, concerns about high disease prevalence overwhelming the capacity of health care systems materialized in the rationing of ICU beds and respirators in Italy.[34] The Italian experience became a catalyst for an array of professional and policy guidance documents on allocation of scarce crisis care resources. The primary ethical considerations in these recommendations involved identifying the criteria or principles that should direct resource allocation in circumstances of scarcity where the critical care capacity of a hospital was overwhelmed by life-threatening patient needs for care. The values included utilitarian concerns to maximize the use of limited resources, a focus on whether health care workers or other persons who assumed essential roles in the societal pandemic response should receive priority for treatment, the principle of the moral equality of persons, and preferential concern for the "worst off" populations deemed to be at highest risk of mortality.[35]

Given the age profile of the highest at-risk population, elderly persons with compromised immune status or related health complications, it was not surprising that age-related criteria surfaced in many discussions. In my own work with the State of Oregon Ethics Task Force on Crisis Care, the most contested issue was whether, in circumstances of resource scarcity, the principle of "life cycle" should be utilized when different patients had rough prognostic equivalence about the course of their disease. The life-cycle principle in general is premised on a commitment to the moral equality of persons: health care resources should be allocated in such a manner as to ensure that each person will have an opportunity to experience a complete or full biographical life cycle. This commitment to equality entails that resource allocation would prioritize persons who have yet to experience a full life cycle, that is, to younger persons who were not in the high-risk COVID demographic. Ultimately, the varied course of the coronavirus pandemic in the United States meant that most health care systems collaborated regionally,[36] pooling resources and personnel, meaning there was minimal reliance on crisis care allocation and rationing criteria.

Health Care Disparities. The coronavirus pandemic provided an apocalyptic unveiling of the substantial inequities embedded in the structure of health care delivery in the United States, manifested most prominently in the high and disproportionate rates of mortality among many minority communities. The COVID-19 mortality rate among black Americans, for example, was twice the proportionate share of the general population. The Navajo Nation experienced rates of infection six times greater and rates of mortality five times greater than the proportionate share of the population.[37] While this prevalence disparity can be attributed to many factors, among the more significant contributors are what the Centers for Disease Control referred to as "barriers" and "inadequate access" to health care, "long-standing mistrust of the health care system," and "systemic inequalities."[38]

The COVID-19 pandemic thus brings full circle the animating question of this epilogue: the necessity for developing a health care system that provides a level of adequate health for all persons. This certainly requires equitable access to the kinds of critical care interventions that have saved the lives of many patients. More fundamentally, however, the social and ecclesiastical commitment to the moral equality of persons requires equitable access to preventive care measures that could prevent many persons from experiencing

life-threatening illness in the first place. Now more than ever, the commitment in LDS moral culture to engaged responsible citizenship and increasingly to being "good global citizens" means it is essential that the religious
community bear a prophetic witness in support of a health care system that
serves all persons rather than a sick care system that stratifies care.

Notes

1. Sarah Kliff, "University to Students on Medicaid: Buy Private Coverage or Drop Out,"
   New York Times, November 24, 2019, https://www.nytimes.com/2019/11/24/us/
   medicaid-students-brigham-young.html
2. Allie Arnell, "BYU Student Health Plan Not Compliant with Obamacare,"
   Daily Universe, September 4, 2015, https://universe.byu.edu/2015/09/04/
   byu-student-health-plan-not-compliant-with-obamacare1/
3. Sarah Kliff, "University Reverses Its Decision to Stop Accepting Medicaid," *New York
   Times*, November 26, 2019, https://www.nytimes.com/2019/11/26/us/medicaid-
   brigham-young-university.html
4. The contraceptive coverage mandate of the Affordable Care Act has been a source
   of considerable controversy and was struck down as unconstitutional by the U.S.
   Supreme Court in *Burwell v. Hobby Lobby Stores, Inc.*, 573 U.S. ___ 2014.
5. Walter Cronkite, as quoted by the Clinton Foundation, "Health Care Is Local," https://
   stories.clintonfoundation.org/healthcare-is-local-bcc165cc22eb
6. Mohammadreza Hojat et al, "The Devil Is in the Third Year: A Longitudinal Study of
   Erosion of Empathy in Medical School," *Academic Medicine* 84 (2009), 1182–1191.
7. Center for Medicare and Medicaid Services, "National Health Expenditures Fact
   Sheet: 2018," https://www.cms.gov/Research-Statistics-Data-and-Systems/Statistics-
   Trends-and-Reports/NationalHealthExpendData/NHE-Fact-Sheet
8. The Church of Jesus Christ of Latter-Day Saints, "Providing for Temporal Needs
   and Building Self-Reliance," in *General Handbook: Serving in the Church of
   Jesus Christ of Latter-day Saints*, Chapter 22, February 19, 2020, https://www.
   churchofjesuschrist.org/study/manual/general-handbook/38-church-policies-and-
   guidelines?lang=eng#title_number91
9. The Church of Jesus Christ of Latter-Day Saints, "Providing for Temporal Needs and
   Building Self-Reliance: Health," in *General Handbook: Serving in the Church of Jesus
   Christ of Latter-day Saints*, Chapter 22.1.1.1.
10. State of Utah, "Healthy Utah," December 2014, https://ccf.georgetown.edu/wp-
    content/uploads/2014/12/healthyutahplan.pdf
11. Newsroom, "Church Encourages Principled Approach to Health Care Coverage for
    Needy Utahns," December 2014, https://newsroom.churchofjesuschrist.org/article/
    church-encourages-principled-approach-health-care-coverage-needy-utahns
12. Abby Goodnough, "In Red-State Utah, a Surge Toward Obamacare," *New York Times*,
    March 3, 2017, https://www.nytimes.com/2017/03/03/health/utah-obamacare.html

13. David Goldhill, "How American Health Care Killed My Father," *The Atlantic*, September 2009, https://www.theatlantic.com/magazine/archive/2009/09/how-american-health-care-killed-my-father/307617/

14. Alastair V. Campbell, *Health as Liberation: Medicine, Theology, and the Quest for Justice* (Eugene, OR: Wipf and Stock, 1995).

15. Latter-day Saint Charities, "2019 Annual Report" (Salt Lake City: Intellectual Reserve, 2020), https://www.churchofjesuschrist.org/bc/content/shared/english/charities/pdf/2019/LDS-Charities-Annual-Report-2019_R11.pdf?lang=eng; "Humanitarian Programs," The Church of Jesus Christ of Latter-day Saints, https://www.lds.org/topics/humanitarian-service/church?lang=eng&old=true

16. Russell M. Nelson, "The Second Great Commandment," The Church of Jesus Christ of Latter-day Saints, October 2019, https://www.churchofjesuschrist.org/study/general-conference/2019/10/46nelson?lang=eng

17. *Matthew* 25:36; *Leviticus* 19:33–34.

18. Patrick Kearon, "Refuge from the Storm," *The Ensign*, May 2016, https://www.lds.org/general-conference/2016/04/refuge-from-the-storm?lang=eng. The refugee identity was underscored by a *Church News* editorial: "The [LDS pioneer] group—many who arrived unprepared for what lay ahead—were in every way refugees, running from oppression in search of a better life. Today, the Church continues to care for refugees who, like the early pioneers, find themselves in temporary camps after fleeing oppression" Viewpoint, "Help the Refugees Among Us," *Church News*, October 27, 2015, https://www.lds.org/church/news/viewpoint-help-the-refugees-among-us?lang=eng&_r=1).

19. Newsroom, "Church Members Encouraged to Assist Refugees," October 29, 2015, https://newsroom.churchofjesuschrist.org/article/church-members-encouraged-assist-refugees; Newsroom, "I Was a Stranger: Refugee Relief," March 26, 2016, https://newsroom.churchofjesuschrist.org/article/church-leaders-encourage-women-of-the-church-serve-refugees

20. Linda K. Burton, "I Was a Stranger," *The Ensign*, May 2016, https://www.lds.org/general-conference/2016/04/i-was-a-stranger?lang=eng

21. Atul Gawande, "Is Health Care a Right?," *The Atlantic*, October 2, 2017, https://www.newyorker.com/magazine/2017/10/02/is-health-care-a-right

22. Gawande, "Is Health Care a Right?"

23. Steven A. Schroeder, "We Can Do Better: Improving the Health of the American People," *New England Journal of Medicine* 357 (2007), 1221–1228.

24. Johns Hopkins University School of Medicine, Coronavirus Resource Center, https://coronavirus.jhu.edu/

25. The Church of Jesus Christ of Latter-day Saints, "Administrative Principles in Challenging Times," April 16, 2020, https://newsroom.churchofjesuschrist.org/article/administrative-principles-in-challenging-times

26. First Presidency, LDS Church, "Opportunities to Address COVID-19 Needs," April 14, 2020, https://newsroom.churchofjesuschrist.org/article/covid-19-first-presidency-letter-april-14-2020

27. Russell M. Nelson, "Opening the Heavens for Help," April 2020 General Conference, The Church of Jesus Christ of Latter-day Saints,https://www.churchofjesuschrist.org/study/general-conference/2020/04/37nelson?lang=eng

28. Church News, "Beehive Clothing Facilities Worldwide Will Produce 200,000 Gowns, 1.5 Million Masks in COVID-19 Relief Effort," *Deseret News,* May 20, 2020, https://www.thechurchnews.com/global/2020-05-20/beehive-clothing-masks-gowns-medical-covid-relief-184696

29. David Orentlicher, "The Physician's Duty to Treat During Pandemics," *American Journal of Public Health* 108 (2018): 1459–1461; Abigail Zuger, Steven H. Miles, "Physicians, AIDS, and Occupational Risks: Historic Traditions and Ethical Obligations," *Journal of American Medical Association* 258 (1987): 1924–1928.

30. Thomas Kirsch, "What Happens If Health Care Workers Stop Showing Up?," *The Atlantic,* March 24, 2020, https://www.theatlantic.com/ideas/archive/2020/03/were-failing-doctors/608662/

31. Estaban Parra, Karl Baker, "Doctors May Overrule Coronavirus' Patients End-of-Life Treatment Wishes," *Delaware News Journal,* April 13, 2020, updated April 15, 2020, https://www.delawareonline.com/story/news/2020/04/13/christianacare-doctors-can-overrule-coronavirus-patients-end-life-wishes/5120699002/

32. American Medical Association, "DNR Orders in a Public Health Crisis," April 29, 2020, https://www.ama-assn.org/delivering-care/ethics/dnr-orders-public-health-crisis

33. Daniel B. Kramer, Bernard Lo, Neal W. Dickert, "CPR in the COVID-19 Era—An Ethical Framework," *New England Journal of Medicine,* May 6, 2020, https://www.nejm.org/doi/full/10.1056/NEJMp2010758

34. Yascha Mounk, "The Triage Decisions Facing Italian Doctors," *The Atlantic,* March 11, 2020, https://www.theatlantic.com/ideas/archive/2020/03/who-gets-hospital-bed/607807/

35. Ezekiel J. Emanuel et al., "Fair Allocation of Scarce Medical Resources in the Time of COVID-19," *New England Journal of Medicine* 382 (2020): 2049–2055, https://www.nejm.org/doi/full/10.1056/NEJMsb2005114

36. Nancy Berlinger et al., "Responding to COVID-19 as a Regional Public Health Challenge," Hastings Center, April 29, 2020, http://www.thehastingscenter.org/covid19_regional_ethics_guidelines

37. Maria Godoy, "What Do Coronavirus Racial Disparities Look Like State by State?," NPR, May 30, 2020, https://www.npr.org/sections/health-shots/2020/05/30/865413079/what-do-coronavirus-racial-disparities-look-like-state-by-state

38. Centers for Disease Control and Prevention, "COVID-19 in Racial and Ethnic Minority Groups," April 22, 2020, https://www.cdc.gov/coronavirus/2019-ncov/need-extra-precautions/racial-ethnic-minorities.html

LDS Ecclesiastical Statements on Issues in Biomedical Ethics

Abortion (2020)

The Lord commanded, "Thou shalt not . . . kill, nor do anything like unto it" (Doctrine and Covenants 59:6). The Church opposes elective abortion for personal or social convenience. Members must not submit to, perform, arrange for, pay for, consent to, or encourage an abortion. The only possible exceptions are when:

1. Pregnancy resulted from forcible rape or incest.
2. A competent physician determines that the life or health of the mother is in serious jeopardy.
3. A competent physician determines that the fetus has severe defects that will not allow the baby to survive beyond birth.

Even these exceptions do not justify abortion automatically. Abortion is a most serious matter and should be considered only after the persons responsible have consulted with their bishops and received divine confirmation through prayer.

Presiding officers carefully review the circumstances if a Church member has been involved in an abortion. A membership council may be necessary if a member submits to, performs, arranges for, pays for, consents to, or encourages an abortion. However, membership council should not be considered if a member was involved in an abortion before baptism. Nor should membership councils or restrictions be considered for members who were involved in an abortion for the reasons outlined above.

Bishops refer questions on specific cases to the stake president. The stake president may direct questions to the Office of the First Presidency if necessary.

As far as has been revealed, a person may repent and be forgiven for the sin of abortion.[1]

Abortion (2010)

The Lord commanded, "Thou shalt not . . . kill, nor do anything like unto it" (Doctrine and Covenants 59:6). The Church opposes elective abortion for personal or social convenience. Members must not submit to, perform, arrange for, pay for, consent to, or encourage an abortion. The only possible exceptions are when:

1. Pregnancy resulted from forcible rape or incest.
2. A competent physician determines that the life or health of the mother is in serious jeopardy.
3. A competent physician determines that the fetus has severe defects that will not allow the baby to survive beyond birth.

Even these exceptions do not justify abortion automatically. Abortion is a most serious matter and should be considered only after the persons responsible have consulted with their bishops and received divine confirmation through prayer.

Church members who submit to, perform, arrange for, pay for, consent to, or encourage an abortion may be subject to Church discipline.

As far as has been revealed, a person may repent and be forgiven for the sin of abortion.[2]

Abortion (1991)

In view of the widespread public interest in the issue of abortion, we reaffirm that The Church of Jesus Christ of Latter-day Saints has consistently opposed elective abortion. More than a century ago, the First Presidency of the Church warned against this evil. We have repeatedly counseled people everywhere to turn from the devastating practice of abortion for personal or social convenience.

The Church recognizes that there may be rare cases in which abortion may be justified—cases involving pregnancy by incest or rape; when the life or health of the woman is adjudged by competent medical authority to be in serious jeopardy; or when the fetus is known by competent medical authority to have severe defects that will not allow the baby to survive beyond birth. But these are not automatic reasons for abortion. Even in these cases, the couple should consider an abortion only after consulting with each other, and their bishop, and receiving divine confirmation through prayer.

The practice of abortion is fundamentally contrary to the Lord's injunction, "Thou shalt not steal; neither commit adultery, nor kill, nor do anything like unto it" (Doctrine and Covenants 59:6). We urge all to preserve the sanctity of human life and thereby realize the happiness promised to those who keep the commandments of the Lord.

The Church of Jesus Christ of Latter-day Saints as an institution has not favored or opposed specific legislative proposals or public demonstrations concerning abortion.

Inasmuch as this issue is likely to arise in all states in the United States of America and in many other nations of the world in which the Church is established, it is impractical for the Church to take a position on specific legislative proposals on this important subject.

However, we continue to encourage our members as citizens to let their voices be heard in appropriate and legal ways that will evidence their belief in the sacredness of life.[3]

Abortion (1976)

The Church opposes abortion and counsels its members not to submit to, be a party to, or perform an abortion except in the rare cases where, in the opinion of competent medical counsel, the life or health of the mother is seriously endangered or where the pregnancy was caused by rape and produces serious emotional trauma in the mother. Even then it should be done only after consulting with the local bishop or branch president and after receiving divine confirmation through prayer.

Abortion is one of the most revolting and sinful practices in this day, when we are witnessing the frightening evidence of permissiveness leading to sexual immorality.

Members of the Church guilty of being parties to the sin of abortion are subject to the disciplinary action of the councils of the Church as circumstances warrant. In dealing with this serious matter, it be well to keep in mind the word of the Lord stated in the 59th Section of the Doctrine and Covenants, verse 6: "Thou shalt not steal; neither commit adultery, nor kill, nor do anything like unto it."

As far as has been revealed, the sin of abortion is one for which a person may repent and gain forgiveness.[4]

Abortion (1974)

The Church opposes abortion and counsels its members not to submit to, perform, nor abet an abortion except in the rare cases where, in the opinion of competent medical counsel, the life or good health of the mother is seriously in danger or where the pregnancy was caused by rape and produces serious emotional trauma in the mother. Even then it should be done only after consulting with the local presiding priesthood authority and after receiving divine confirmation through prayer.[5]

Abortion (1973)

In view of a recent decision of the United States Supreme Court, we feel it necessary to re-state the position of the Church on abortion in order that there be no misunderstanding of our attitude.

The Church opposes abortion and counsels its members not to submit to or perform an abortion except in the rare cases where, in the opinion of competent medical counsel, the life or good health of the mother is seriously endangered or where the pregnancy was caused by rape and produces serious emotional trauma in the mother. Even then it should be done only after counseling with the local presiding priesthood authority and after re-ceiving divine confirmation through prayer.

Abortion must be considered one of the most revolting and sinful practices in this day, when we are witnessing the frightening evidence of permissiveness leading to sexual immorality.

Members of the Church guilty of being parties to the sin of abortion must be subjected to the disciplinary action of the councils of the Church as circumstances warrant. In dealing with this serious matter, it would be well to keep in mind the word of the Lord stated in the 59th section of the Doctrine and Covenants, verse 6, "Thou shalt not steal; neither commit adultery, nor kill, nor do anything like unto it."

As to the amenability of the sin of abortion to the laws of repentance and forgiveness, we quote the following statement made by President David O. McKay and his counselors, Stephen L. Richards and J. Reuben Clark, Jr., which continues to represent the attitude and position of the Church:

"As the matter stands today, no definite statement has been made by the Lord one way or another regarding the crime of abortion. So far as is known, he has not listed it along-side the crime of the unpardonable sin and shedding of innocent human blood. That he has not done so would suggest that it is not in that class of crime and therefore that it will be amenable to the laws of repentance and forgiveness."

This quoted statement, however, should not, in any sense, be construed to minimize the seriousness of this revolting sin.[6]

Artificial Insemination (2020)

The Church strongly discourages artificial insemination using semen from anyone but the husband. However, this is a personal matter that ultimately must be left to the judgment of the husband and wife.

Artificial insemination of single sisters is not approved. Single sisters who pursue this may have some privileges of Church membership restricted.[7]

Artificial Insemination (2010)

The Church strongly discourages artificial insemination using semen from anyone but the husband. However, this is a personal matter that ultimately must be left to the judgment of the husband and wife. Responsibility for the decision rests solely upon them.

Artificial insemination of single sisters is not approved. Single sisters who deliberately refuse to follow the counsel of Church leaders in this matter are subject to Church discipline.[8]

Artificial Insemination (1985)

The Church strongly discourages artificial insemination using semen from anyone but the husband. However, this is a personal matter that ultimately must be left to the judgment of the husband and wife. Responsibility for the decision rests solely upon them.

The Church disapproves of artificial insemination of single sisters. Single sisters who deliberately refuse to follow the counsel of their priesthood leaders in this regard will be subject to disciplinary action.[9]

Artificial Insemination (1977)

The Church discourages artificial insemination with other than the semen of the husband. Artificial insemination with semen other than from the husband may produce problems related to family harmony. The Church recognizes that this is a personal matter which must ultimately be left to the determination of the husband and wife with the responsibility for their decision resting solely upon them.

A child born by means of artificial insemination after parents are sealed in the temple is born in the covenant [i.e., post-mortally sealed to the parents]. A child born by artificial insemination before parents are sealed may be sealed subsequent to the sealing of the parents.[10]

Artificial Insemination (1974)

The Church does not approve of artificial insemination with other than the semen of the husband. Artificial insemination with semen other than from the husband may produce problems related to family harmony. The Church recognizes that this is a personal matter which must ultimately be left to the determination of the husband and wife with the responsibility for their decision resting solely upon them.[11]

Birth Control (2020)

It is the privilege of married couples who are able to bear children to provide mortal bodies for the spirit children of God, whom they are then responsible to nurture and rear. The decision as to how many children to have and when to have them is extremely intimate and private. It should be left between the couple and the Lord. Church members should not judge one another in this matter.

Physical intimacy between husband and wife is intended to be beautiful and sacred. It is ordained of God for the creation of children and for the expression of love between husband and wife.[12]

Birth Control (2010)

It is the privilege of married couples who are able to bear children to provide mortal bodies for the spirit children of God, whom they are then responsible to nurture and rear. The decision as to how many children to have and when to have them is extremely intimate and private and should be left between the couple and the Lord. Church members should not judge one another in this matter.

Married couples should also understand that sexual relations within marriage are divinely approved not only for the purpose of procreation, but also as a way of expressing love and strengthening emotional and spiritual bonds between husband and wife.[13]

Birth Control (1985)

The Lord has commanded husbands and wives to multiply and replenish the earth that they might have joy in their posterity. Husbands must be considerate of their wives, who have a great responsibility not only for bearing children but also for caring for them through childhood, and should help them conserve their health and strength. Married couples should exercise self-control in all their relationships. They should seek inspiration from the Lord in meeting their marital challenges and rearing their children according to the teachings of the gospel.[14]

Birth Control (1974)

The Lord's command imposed upon all Latter-day Saints is to "multiply and replenish the earth." Where husband and wife enjoy health and vigor and are free from inheritable defects that would be entailed upon their posterity, it is contrary to the teachings of the Church artificially to curtail or prevent the birth of children. We believe that those who practice birth control will reap disappointment by and by. The Church feels husbands must be considerate of their wives who bear the greater responsibility not only of bearing children, but of caring for them throughout childhood. To this end, the mother's health and strength should be conserved and the husband's consideration for his wife is his first duty, and self-control should be a dominant factor in all their relationships.[15]

Birth Control (1969)

The First Presidency is being asked from time to time as to what the attitude of the Church is regarding birth control.

We seriously regret that there should exist a sentiment or feeling among any members of the Church to curtail the birth of their children. We have been commanded to multiply and replenish the earth that we may have joy and rejoicing in our posterity.

Where husband and wife enjoy health and vigor and are free from impurities that would be entailed upon their posterity, it is contrary to the teachings of the Church artificially to curtail or prevent the birth of children. We believe those who practice birth control will reap disappointment by and by.

However, we feel that men must be considerate of their wives who bear the greater responsibility not only of bearing children but also for caring for them through childhood. To this end, the mother's health and strength should be conserved and the husband's consideration for his wife is his first duty, and self-control a dominant factor in all their relationships.[16]

Embryonic Stem Cell Research (2005)

The First Presidency of The Church of Jesus Christ of Latter-day Saints has not taken a position regarding the use of embryonic stem cells for research purposes. The absence of a position should not be interpreted as support for or opposition to any other statement made by Church members, whether they are for or against embryonic stem cell research.[17]

Embryonic Stem Cell Research (2001)

While the First Presidency and the Quorum of the Twelve Apostles have not taken a position at this time on the newly emerging field of stem cell research, it merits cautious scrutiny. The proclaimed potential to provide cures or treatments for many serious diseases needs careful and continuing study by conscientious, qualified investigators. As with any emerging new technology, there are concerns that must be addressed. Scientific and religious viewpoints both demand that strict moral and ethical guidelines be followed.[18]

End of Life: Euthanasia (2020)

Euthanasia is defined as deliberately putting to death a person who is suffering from an incurable condition or disease. A person who participates in euthanasia, including helping someone die by suicide, violates the commandments of God.[19]

End of Life: Prolonging Life (2020)

When severe illness strikes, members should exercise faith in the Lord and seek competent medical assistance. However, when dying becomes inevitable, it should be seen as a blessing and a purposeful part of eternal existence. Members should not feel obligated to extend mortal life by means that are unreasonable. These judgments are best made by family members after receiving wise and competent medical advice and seeking divine guidance through fasting and prayer.[20]

End of Life: Assisted Suicide (2016)

Lawmakers and voters in some states, including Colorado, are seeking to legalize physician-assisted suicide. The Church maintains a firm belief in the sanctity of human life and opposes deliberately taking the life of a person even when the person may be suffering from an incurable condition or disease (see *Handbook 2: Administering the Church* [2010], 21.3.3). Life is a sacred gift and should be cherished even in difficult circumstances.

Physician-assisted suicide is permitted by law in some countries and a few states in the United States. Experience suggests that such legalization can endanger the vulnerable, erode trust in the medical profession, and cheapen human life and dignity. Moreover, the decision to end one's life carries a lasting impact far beyond the person whose life is ending.

While the Church opposes physician-assisted suicide, members should not feel obligated to extend mortal life through means that are unreasonable. Decisions in such cases are best made by family members after receiving wise and competent medical advice and seeking divine guidance through fasting and prayer (see *Handbook 2: Administering the Church* [2010], 21.3.8).

We urge Church members to let their voices be heard in opposition to measures that would legalize physician-assisted suicide.[21]

End of Life: Euthanasia and Prolonging Life (2014)

The Church of Jesus Christ of Latter-day Saints believes in the sanctity of human life, and is therefore opposed to euthanasia. Euthanasia is defined as deliberately putting to death a person who is suffering from an incurable condition or disease. Such a deliberate act ends life immediately through, for example, frequently-termed assisted suicide. Ending a life in such a manner is a violation of the commandments of God.

The Church of Jesus Christ of Latter-day Saints does not believe that allowing a person to die from natural causes by removing a patient from artificial means of life support, as in the case of a long-term illness, falls within the definition of euthanasia. When dying from such an illness or an accident becomes inevitable, it should be seen as a blessing and a purposeful part of eternal existence. Members should not feel obligated to extend mortal life by means that are unreasonable. These judgments are best made by family members after receiving wise and competent medical advice and seeking divine guidance through fasting and prayer.[22]

End of Life: Euthanasia (2010)

Euthanasia is defined as deliberately putting to death a person who is suffering from an incurable condition or disease. A person who participates in euthanasia, including assisting someone to commit suicide, violates the commandments of God.[23]

End of Life: Prolongation of Life and Right to Die (1974)

The Church does not look with favor upon any form of mercy killing. It believes in the dignity of life and that faith in the Lord and medical science should be called upon and applied to reverse conditions that are a threat to life. There comes a time when dying becomes inevitable; when it should be looked upon as a blessing, and a purposeful part of mortality.[24]

Experimentation (1974)

The Church recognizes the need for carefully conducted and controlled experimentation to substantiate the efficacy of medicine and procedures. We believe, however, that the free agency of the individual must be protected by informed consent and that a qualified group of peers should review all research to ascertain that it is needed, is appropriately designed and not harmful to the person involved.[25]

Female Genital Mutilation (2020)

The Church condemns female genital mutilation.[26]

Health Care Reform (2014)

We recognize that providing adequate health care to individuals and families throughout Utah is a complex and weighty matter. It deserves the best thinking and efforts from both the public and the private sectors. While the economic and political realities are being debated, we hope the discussion and decisions taken in this matter will be consistent with the God-given principles regarding care for the poor and the needy that in the end benefit all of His children. We reaffirm the importance for individuals and families to be as self-sufficient as their particular circumstances allow and recognize that the lack of access to health care can impair a person's ability to provide for self and family. We commend public officials for their efforts to grapple with these difficult issues and pray for their success in finding solutions that reflect the highest aspirations of society.[27]

Hypnosis (2020)

The use of hypnosis under competent, professional medical supervision for the treatment of diseases or mental disorders is a medical question to be determined by competent medical authorities. Members should not participate in hypnosis for purposes of demonstration or entertainment.[28]

Hypnosis (1974)

The Church regards the use of hypnosis under competent, professional supervision for the treatment of disease as wholly a medical question. The Church advises members against participation in hypnosis demonstrations.[29]

Intersex (2020)

In extremely rare circumstances, a baby is born with genitals that are not clearly male or female (ambiguous genitalia, sexual ambiguity, or intersex). Parents or others may have to make decisions to determine their child's sex with the guidance of competent medical professionals. Decisions about proceeding with medical or surgical intervention are often made in the newborn period. However they can be delayed unless they are medically necessary.

Special compassion and wisdom are required when youth or adults who were born with sexual ambiguity experience emotional conflict regarding the gender decisions made in infancy or childhood and the gender with which they identify.[30]

In Vitro Fertilization (2020)

The Church strongly discourages in vitro fertilization using semen from anyone but the husband or an egg from anyone but the wife. However, this is a personal matter that ultimately must be left to the judgment of the husband and wife. Responsibility for the decision rests solely upon them.[31]

In Vitro Fertilization (2010)

The Church strongly discourages in vitro fertilization using semen from anyone but the husband or an egg from anyone but the wife. However, this is a personal matter that ulti mately must be left to the judgment of the husband and wife. Responsibility for the decision rests solely upon them.[32]

In Vitro Fertilization (1989)

In vitro fertilization using semen other than that of the husband or an egg other than that of the wife is strongly discouraged. However, this is a personal matter that ultimately must be left to the judgment of the husband and wife.

A child conceived through in vitro fertilization after parents are sealed in the temple is born in the covenant [i.e., post-mortally sealed to the parents]. Such a child born before parents are sealed in the temple may be sealed to them after they are sealed.[33]

Medical and Health Practices (2020)

Members should not use medical or health practices that are ethically or legally questionable. Local leaders should advise members who have health problems to consult with competent professional practitioners who are licensed in the countries where they practice.[34]

Medical Marijuana (August 2018)

The Church joins a coalition of medical experts, public officials, and community stakeholders in calling for a safe and compassionate approach to providing medical marijuana to those in need. The Church does not object to the medicinal use of marijuana, if doctor prescribed, in dosage form, through a licensed pharmacy.[35]

Medical Marijuana (April 2018)

We commend the Utah Medical Association for its statement of March 30, 2018, cautioning that the proposed Utah marijuana initiative would compromise the health and safety of Utah communities. We respect the wise counsel of the medical doctors of Utah.

The public interest is best served when all new drugs designed to relieve suffering and illness and the procedures by which they are made available to the public undergo the scrutiny of medical scientists and official approval bodies.[36]

Organ and Tissue Donations and Transplants (2010)

The donation of organs and tissues is a selfless act that often results in great benefit to individuals with medical conditions. The decision to will or donate one's own body organs or tissue for medical purposes or the decision to authorize the transplant of organs or tissue from a deceased family member is made by the individual or the deceased member's family.

A decision to receive a donated organ should be made after receiving competent medical counsel and confirmation through prayer.[37]

Organ Transplants (1989)

Whether an individual chooses to will his own bodily organs or authorizes the transplant of organs from a deceased family member is a decision for the individual or the deceased member's family. The decision to receive a donated organ should be made with competent medical counsel and confirmation through prayer.[38]

Organ Transplants (1974)

The question of whether one should will his bodily organs to be used as transplants or for research after death must be answered from deep within the conscience of the individual involved. Those who seek counsel from the Church on this subject are encouraged to review the advantages and disadvantages of doing so, to implore the Lord for inspiration and guidance, and then to take the course of action which would give them a feeling of peace and comfort.[39]

Religion and Healing Process ("Faith Healing") (1974)

The Church believes in the same manifestations of the Spirit, including healing, that existed in the Church organized by the Savior during his earthly ministry. Through Latter-day revelation, the Lord has directed: ". . . And the elders of the Church, two or more, shall be called, and shall pray for and lay their hands upon them (the sick) in my name; and if they die they shall die unto me, and if they live they shall live unto me."[40]

Sperm Donation (2020)

The Church strongly discourages the donation of sperm.[41]

Suicide (2020)

Mortal life is a precious gift from God, and it is wrong to take one's own life. However, a person who does so may not be responsible for his or her acts. Only God can fully understand and judge the situation. The Church strongly supports suicide prevention and encourages compassion for individuals affected by suicide.[42]

Suicide (2010)

It is wrong to take a life, including one's own. However, a person who commits suicide may not be responsible for his or her acts. Only God can judge such a matter.[43]

Suicide (1989)

A person who takes his own life may not be responsible for his acts. Only God can judge such a matter. A person who has considered suicide seriously or has attempted suicide should be counseled by his bishop and may be encouraged to seek professional help. Bishops should counsel and compassionately console the family members of a person who has committed suicide.[44]

Surgical Sterilization (including Vasectomy) (2020)

The Church strongly discourages surgical sterilization as an elective form of birth control. Surgical sterilization should be considered only if (1) medical conditions seriously jeopardize life or health or (2) a person does not have the mental competence to be responsible for his or her actions. Such conditions must be determined by competent medical judgment and in accordance with law. Even then, the persons responsible for this decision should consult with each other and with their bishop and should receive divine confirmation of their decision through prayer.[45]

Surgical Sterilization (including Vasectomy) (2010)

The Church strongly discourages surgical sterilization as an elective form of birth control. Surgical sterilization should be considered only if (1) medical conditions seriously jeopardize life or health or (2) birth defects or serious trauma have rendered a person mentally incompetent and not responsible for his or her actions. Such conditions must be determined by competent medical judgment and in accordance with law. Even then, the persons responsible for this decision should consult with each other and with their bishop and should receive divine confirmation of their decision through prayer.[46]

Sterilization (1989)

The First Presidency has declared, "We seriously deplore the fact that members of the Church would voluntarily undertake measures to render themselves incapable of further procreation."

Surgical sterilization should only be considered (1) where medical conditions seriously jeopardize life or health, or (2) where birth defects or serious trauma have rendered a person mentally incompetent and not responsible for his or her actions. Such conditions must be determined by competent medical judgment and in accordance with law. Even then, the person or persons responsible for this decision should consult with each other and with their bishop (or branch president) and receive divine confirmation of their decision through prayer.[47]

Sterilization (1974)

The Lord's commandment imposed on all Latter-day Saints is to "multiply and replenish the earth." Nevertheless there may be medical conditions related to the health of the mother where sterilization could be justified. But such conditions, rare as they may be, must be determined by competent medical judgments and in accordance with laws pertaining thereto.[48]

Sterility Tests (1974)

The Church believes that having children is a blessing and privilege and, that with any abnormal condition, it is appropriate to use medical science to diagnose and restore normal function.[49]

Stillborn Children (2020)

Temple ordinances are not performed for stillborn children. However, this does not deny the possibility that a stillborn child may be part of the family in the eternities. Parents are encouraged to trust the Lord to resolve such cases in the way He knows is best. The family may record the name of a stillborn child on the family group record, followed by the word *stillborn* in parentheses.

Memorial or graveside services may be held as determined by the parents.

It is a fact that a child has life before birth. However there is no direct revelation on when the spirit enters the body.[50]

Surrogate Motherhood (2020)

The Church strongly discourages surrogate motherhood. However, this is a personal matter that ultimately must be left to the judgment of the husband and wife.[51]

Surrogate Motherhood (2010)

The Church strongly discourages surrogate motherhood. However, this is a personal matter that ultimately must be left to the judgment of the husband and wife. Responsibility for the decision rests solely upon them.[52]

Surrogate Motherhood (1989)

Surrogate motherhood is discouraged. It might cause spiritual, emotional, and other difficulties.[53]

Transgender Individuals (2020)

Transgender individuals face complex challenges. Members and nonmembers who identify as transgender—and their family and friends—should be treated with sensitivity, kindness, compassion, and an abundance of Christlike love. . . .

Gender is an essential characteristic of Heavenly Father's plan of happiness. The intended meaning of *gender* is the family proclamation is *biological sex at birth*. Some people experience feelings of incongruence between their biological sex and their gender identity. As a result, they may identify as transgender. The Church does not take a position on the causes of people identifying themselves as transgender. . . .

Church leaders counsel against medical or elective intervention for the purpose of attempting to transition to the opposite gender of a person's birth sex ("sex reassignment"). Leaders advise that taking these actions will be cause for Church membership restrictions. . . .

Some children, youth, and adults are prescribed hormone therapy by a licensed medical professional to ease gender dysphoria or reduce suicidal thoughts. Before a person begins such therapy, it is important that he or she (and the parents of a minor) understand the potential risks and benefits.[54]

Transsexual Surgery (2010)

A member who has undergone an elective transsexual operation may not receive a temple recommend.[55]

Vaccination (1978)

Reports that increasing numbers of children are not being immunized against preventable childhood disease deeply concern us. In the United States alone approximately 20 million children, 40% of those 14 years old or younger, have not been adequately immunized against polio, measles, German measles (rubella), diphtheria, pertussis (whooping cough), mumps and tetanus.

Every parent who has agonized when these diseases have maimed or brought premature death to their children would join us, we are certain, in a plea to mobilize against these deadly enemies.

Immunization is such a simple, yet vital, matter and such a small price to pay for protection against these destroying diseases.

We urge members of the Church of Jesus Christ of Latter-day Saints to protect their own children through immunization. Then they may wish to join other public-spirited citizens in efforts to eradicate ignorance and apathy that have caused the disturbingly low levels of childhood immunization.

Failure to act could subject untold thousands to preventable lifelong physical or mental impairment, including paralysis, blindness, deafness, heart damage, and mental retardation.

Immunization campaigns in the United States and other nations, if successful, will end much needless suffering and erase the potential threat of epidemics. Such efforts are deserving of our full support.[56]

Narcotics, Vaccines, Blood, etc. (1974)

The Church regards the use of these substances, as prescribed under medical supervision for the treatment or prevention of disease, as wholly a medical question.[57]

Word of Wisdom (2020)

The only official interpretation of "hot drinks" (Doctrine and Covenants 89:9) in the Word of Wisdom is the statement made by early Church leaders that the term "hot drinks" means tea and coffee.

Members should not use any substance that contains illegal drugs. Nor should members use harmful or habit-forming substances except under the care of a competent physician.[58]

Word of Wisdom (2019)

The Word of Wisdom is a law of health for the physical and spiritual benefit of God's children. It includes instruction about what foods are good for us and those substances to avoid. Over time, Church leaders have provided additional instruction on those things that are encouraged or forbidden by the Word of Wisdom, and have taught that substances that are destructive, habit-forming or addictive should be avoided.

In recent publications for Church members, Church leaders have clarified that several substances are prohibited by the Word of Wisdom, including vaping or e-cigarettes, green tea, and coffee-based products. They also have cautioned that substances such as marijuana and opioids should be used only for medicinal purposes as prescribed by a competent physician.[59]

Word of Wisdom (1989)

The only official interpretation of "hot drinks" in the Word of Wisdom (D&C 89) is the statement that the term "hot drinks" means tea and coffee. Members should not use any substance that contains illegal drugs or other harmful or habit-forming ingredients.[60]

Dietary Laws (1974)

The Church's law pertaining to proper diet and care of the body is contained in a revelation given to the Prophet Joseph Smith under date of February 27, 1833. That revelation admonishes Church members to use judgment and temperance in the use of all food and drink. It prohibits the use of alcoholic beverages, hot drinks (interpreted to mean tea and coffee) and tobacco. It also prohibits the use of all other substances which may be injurious to the body or which might be said in violation of the spirit of the revelation. It also encourages the sparing use of meats but prohibits none outright. On the affirmative side, this health code encourages the eating of all fruits and vegetables and encourages the use of whole grain.[61]

Word of Wisdom (1972)

The Word of Wisdom (section 89 of the Doctrine and Covenants) remains as to terms and specifications as found in that section. There has been no official interpretation of that Word of Wisdom except that which was given by the Brethren in the very early days of the Church when it was declared that "hot drinks" meant tea and coffee.

With reference to cola drinks, the Church has never officially taken a position on this matter, but the leaders of the Church have advised, and we do now specifically advise against the use of any drink containing harmful habit-forming drugs under circumstances that would result in acquiring the habit. Any beverage that contains ingredients harmful to the body should be avoided.[62]

Notes

1. The Church of Jesus Christ of Latter-Day Saints, "Church Policies and Guidelines, Policies on Moral Issues: Abortion," in *General Handbook: Serving in the Church of Jesus Christ of Latter-day Saints,* Chapter 38.6.1, February 19, 2020, https://www.churchofjesuschrist.org/study/manual/general-handbook/38-church-policies-and-guidelines?lang=eng#title_number91

2. The Church of Jesus Christ of Latter-Day Saints, "Policies on Moral Issues: Abortion," in *Handbook 2: Administering the Church*, Section 21.4.1, 2010, https://www.lds.org/handbook/handbook-2-administering-the-church?lang=eng

3. "Church Issues New Statement on Abortion," *Ensign*, January 11, 1991, https://www.lds.org/ensign/1991/03/news-of-the-church/church-issues-statement-on-abortion?lang=eng.

4. "Church Issues Statement on Abortion," *Ensign*, July 1976, https://www.lds.org/ensign/1976/07/news-of-the-church/church-issues-statement-on-abortion?lang=eng&clang=ase

5. "Attitudes of the Church of Jesus Christ of Latter-day Saints Towards Certain Medical Problems," in "Mormon Medical Ethical Guidelines," ed. Lester E. Bush, Jr., *Dialogue* 12 (Fall 1979), 97, https://www.dialoguejournal.com/wp-content/uploads/sbi/articles/Dialogue_V12N03_101.pdf

6. "Policies and Procedures: Statement on Abortion," *New Era*, April 1973, https://www.lds.org/new-era/1973/04/policies-and-procedures-statement-on-abortion?lang=eng

7. The Church of Jesus Christ of Latter-Day Saints, "Church Policies and Guidelines, Policies on Moral Issues: Artificial Insemination," in *General Handbook: Serving in the Church of Jesus Christ of Latter-day Saints,* Chapter 38.6.3, https://www.churchofjesuschrist.org/study/manual/general-handbook/38-church-policies-and-guidelines?lang=eng#title_number100

8. The Church of Jesus Christ of Latter-Day Saints, "Policies on Moral Issues: Artificial Insemination," in *Handbook 2: Administering the Church*, Section 21.4.3, https://www.churchofjesuschrist.org/bc/content/shared/content/english/pdf/language-materials/08702_eng.pdf?lang=eng

9. *General Handbook of Instructions* (Salt Lake City: The Church of Jesus Christ of Latter-day Saints, 1985), p. 11–3.

10. "Attitudes of the Church of Jesus Christ of Latter-day Saints Towards Certain Medical Problems," in "Mormon Medical Ethical Guidelines," ed. Lester E. Bush, Jr., *Dialogue* 12 (Fall 1979), 101.

11. "Attitudes of the Church of Jesus Christ of Latter-day Saints Towards Certain Medical Problems," in "Mormon Medical Ethical Guidelines," ed. Lester E. Bush, Jr., *Dialogue* 12 (Fall 1979), 97.

12. The Church of Jesus Christ of Latter-Day Saints, "Church Policies and Guidelines, Policies on Moral Issues: Birth Control," in *General Handbook: Serving in the Church of Jesus Christ of Latter-day Saints,* Chapter 38.6.4, https://www.churchofjesuschrist.org/study/manual/general-handbook/38-church-policies-and-guidelines?lang=eng#title_number101

13. The Church of Jesus Christ of Latter-Day Saints, "Policies on Moral Issues: Birth Control," in *Handbook 2: Administering the Church*, Section 21.4.4, https://www.churchofjesuschrist.org/bc/content/shared/content/english/pdf/language-materials/08702_eng.pdf?lang=eng

14. *General Handbook of Instructions* (Salt Lake City: The Church of Jesus Christ of Latter-day Saints, 1985), p. 11–4.

15. "Attitudes of the Church of Jesus Christ of Latter-day Saints Towards Certain Medical Problems," in "Mormon Medical Ethical Guidelines," ed. Lester E. Bush, Jr., *Dialogue* 12 (Fall 1979), 98.

16. "Attitudes of the Church of Jesus Christ of Latter day Saints Towards Certain Medical Problems," in "Mormon Medical Ethical Guidelines," ed. Lester E. Bush, Jr., *Dialogue* 12 (Fall 1979), 101–102.

17. Newsroom, "Embryonic Stem-Cell Research," https://newsroom.churchofjesuschrist. org/ldsnewsroom/eng/public-issues/embryonic-stem-cell-research

18. Lee Davidson, "No LDS Stand on Cell Research," *Deseret News*, July 6, 2001, https:// www.deseretnews.com/article/851862/No-LDS-stand-on-cell-research.html

19. The Church of Jesus Christ of Latter-Day Saints, "Church Policies and Guidelines, Medical and Health Policies: Euthanasia," in *General Handbook: Serving in the Church of Jesus Christ of Latter-day Saints,* Chapter 38.7.3, https://www. churchofjesuschrist.org/study/manual/general-handbook/38-church-policies-and-guidelines?lang=eng#title_number122

20. The Church of Jesus Christ of Latter-Day Saints, "Church Policies and Guidelines, Medical and Health Policies: Prolonging Life," in *General Handbook: Serving in the Church of Jesus Christ of Latter-day Saints,* Chapter 38.7.9, https://www. churchofjesuschrist.org/study/manual/general-handbook/38-church-policies-and-guidelines?lang=eng#title_number128

21. Office of the First Presidency, "Letter on Assisted Suicide," October 12, 2016, https:// www.churchofjesuschrist.org/church/news/first-presidency-asks-members-to-oppose-recreational-marijuana-assisted-suicide?lang=eng

22. Newsroom, "Euthanasia and Prolonging Life," November 2014, https://www. mormonnewsroom.org/official-statement/euthanasia-and-prolonging-life

23. The Church of Jesus Christ of Latter-Day Saints, "Medical and Health Policies: Euthanasia," in *Handbook 2: Administering the Church*, Section 21.3.3, https://www.churchofjesuschrist.org/bc/content/shared/content/english/pdf/ language-materials/08702_eng.pdf?lang=eng

24. "Attitudes of the Church of Jesus Christ of Latter-day Saints Towards Certain Medical Problems," in "Mormon Medical Ethical Guidelines," ed. Lester E. Bush, Jr., *Dialogue* 12 (Fall 1979), 98.

25. "Attitudes of the Church of Jesus Christ of Latter-day Saints Towards Certain Medical Problems," in "Mormon Medical Ethical Guidelines," ed. Lester E. Bush, Jr., *Dialogue* 12 (Fall 1979), 100.

26. The Church of Jesus Christ of Latter-Day Saints, "Church Policies and Guidelines, Policies on Moral Issues: Female Genital Mutilation," in *General Handbook: Serving in the Church of Jesus Christ of Latter-day Saints,* Chapter 38.6.8, https://www. churchofjesuschrist.org/study/manual/general-handbook/38-church-policies-and-guidelines?lang=eng#title_number104.

27. Newsroom, "Church Encourages Principled Approach to Health Care Coverage for Needy Utahans," December 5, 2014, https://newsroom.churchofjesuschrist.org/ar-ticle/church-encourages-principled-approach-health-care-coverage-needy-utahns

28. The Church of Jesus Christ of Latter-Day Saints, "Church Policies and Guidelines, Medical and Health Policies: Hypnosis," in *General Handbook: Serving in the Church of Jesus Christ of Latter-day Saints,* Chapter 38.7.5, https://www.churchofjesuschrist.org/study/manual/general-handbook/38-church-policies-and-guidelines?lang=eng#title_number124

29. "Attitudes of the Church of Jesus Christ of Latter-day Saints Towards Certain Medical Problems," in "Mormon Medical Ethical Guidelines," ed. Lester E. Bush, Jr., *Dialogue* 12 (Fall 1979), 100.

30. The Church of Jesus Christ of Latter-Day Saints, "Church Policies and Guidelines, Medical and Health Policies: Individuals Whose Sex at Birth Is Not Clear," in *General Handbook: Serving in the Church of Jesus Christ of Latter-day Saints,* Chapter 38.7.6, https://www.churchofjesuschrist.org/study/manual/general-handbook/38-church-policies-and-guidelines?lang=eng#title_number125

31. The Church of Jesus Christ of Latter-Day Saints, "Church Policies and Guidelines, Policies on Moral Issues: In Vitro Fertilization," in *General Handbook: Serving in the Church of Jesus Christ of Latter-day Saints,* Chapter 38.6.11, https://www.churchofjesuschrist.org/study/manual/general-handbook/38-church-policies-and-guidelines?lang=eng#title_number106.

32. The Church of Jesus Christ of Latter-Day Saints, "Policies on Moral Issues: In Vitro Fertilization," in *Handbook 2: Administering the Church*, Section 21.4.7.

33. *General Handbook of Instructions* (Salt Lake City: The Church of Jesus Christ of Latter-day Saints, 1989), p. 11–14.

34. The Church of Jesus Christ of Latter-Day Saints, "Church Policies and Guidelines, Medical and Health Policies: Medical and Health Practices," in *General Handbook: Serving in the Church of Jesus Christ of Latter-day Saints,* Chapter 38.7.7, https://www.churchofjesuschrist.org/study/manual/general-handbook/38-church-policies-and-guidelines?lang=eng#title_number126

35. Newsroom, "Coalition Seeks Safe and Compassionate Alternative to Utah's Medical Marijuana Initiative," https://newsroom.churchofjesuschrist.org/article/coalition-seeks-safe-compassionate-alternative-utah-medical-marijuana-initiative?filter=leadership?filter=leadership#coalition-statement

36. Newsroom, "First Presidency Statement on Utah Marijuana Initiative," April 20, 2018, https://newsroom.churchofjesuschrist.org/article/first-presidency-statement-on-utah-marijuana-initiative

37. The Church of Jesus Christ of Latter-Day Saints, "Church Policies and Guidelines, Medical and Health Policies: Organ and Tissue Donations and Transplants," in *General Handbook: Serving in the Church of Jesus Christ of Latter-day Saints,* Chapter 38.7.8, https://www.churchofjesuschrist.org/study/manual/general-handbook/38-church-policies-and-guidelines?lang=eng#title_number127

38. *General Handbook of Instructions* (Salt Lake City: The Church of Jesus Christ of Latter-day Saints, 1989), p. 11–16.

39. "Attitudes of the Church of Jesus Christ of Latter-day Saints Towards Certain Medical Problems," in "Mormon Medical Ethical Guidelines," ed. Lester E. Bush, Jr., *Dialogue* 12 (Fall 1979), 98.

40. "Attitudes of the Church of Jesus Christ of Latter-day Saints Towards Certain Medical Problems," in "Mormon Medical Ethical Guidelines," ed. Lester E. Bush, Jr., *Dialogue* 12 (Fall 1979), 98.

41. The Church of Jesus Christ of Latter-Day Saints, "Church Policies and Guidelines, Policies on Moral Issues: Sperm Donation," in *General Handbook: Serving in the Church of Jesus Christ of Latter-day Saints,* Chapter 38.6.7, https://www.churchofjesuschrist.org/study/manual/general-handbook/38-church-policies-and-guidelines?lang=eng#title_number217

42. The Church of Jesus Christ of Latter-Day Saints, "Church Policies and Guidelines, Policies on Moral Issues: Suicide," in *General Handbook: Serving in the Church of Jesus Christ of Latter-day Saints,* Chapter 38.6.19, https://www.churchofjesuschrist.org/study/manual/general-handbook/38-church-policies-and-guidelines?lang=eng#title_number115

43. The Church of Jesus Christ of Latter-Day Saints, "Policies on Moral Issues: Suicide," in *Handbook 2: Administering the Church,* Section 21.4.14, https://www.churchofjesuschrist.org/bc/content/shared/content/english/pdf/language-materials/08702_eng.pdf?lang=eng

44. *General Handbook of Instructions* (Salt Lake City: The Church of Jesus Christ of Latter-day Saints, 1989), p. 11–15.

45. The Church of Jesus Christ of Latter-Day Saints, "Church Policies and Guidelines, Policies on Moral Issues: Surgical Sterilization (including Vasectomy)," in *General Handbook: Serving in the Church of Jesus Christ of Latter-day Saints,* Chapter 38.6.20, https://www.churchofjesuschrist.org/study/manual/general-handbook/38-church-policies-and-guidelines?lang=eng#title_number116

46. The Church of Jesus Christ of Latter-Day Saints, "Policies on Moral Issues: Surgical Sterilization," in *Handbook 2: Administering the Church,* Section 21.4.15, https://www.churchofjesuschrist.org/bc/content/shared/content/english/pdf/language-materials/08702_eng.pdf?lang=eng

47. *General Handbook of Instructions* (Salt Lake City: The Church of Jesus Christ of Latter-day Saints, 1989), p. 11–15.

48. "Attitudes of the Church of Jesus Christ of Latter-day Saints Towards Certain Medical Problems," in "Mormon Medical Ethical Guidelines," ed. Lester E. Bush, Jr., *Dialogue* 12 (Fall 1979), 100.

49. "Attitudes of the Church of Jesus Christ of Latter-day Saints Towards Certain Medical Problems," in "Mormon Medical Ethical Guidelines," ed. Lester E. Bush, Jr., *Dialogue* 12 (Fall 1979), 99.

50. The Church of Jesus Christ of Latter-Day Saints, "Church Policies and Guidelines, Medical and Health Policies: Stillborn Children (Children Who Die Before Birth)," in *General Handbook: Serving in the Church of Jesus Christ of Latter-day Saints,* Chapter 38.7.12, https://www.churchofjesuschrist.org/study/manual/general-handbook/38-church-policies-and-guidelines?lang=eng#title_number130

51. The Church of Jesus Christ of Latter-Day Saints, "Church Policies and Guidelines, Policies on Moral Issues: Surrogate Motherhood," in *General Handbook: Serving in the Church of Jesus Christ of Latter-day Saints,* Chapter 38.6.21, https://www.

churchofjesuschrist.org/study/manual/general-handbook/38-church-policies-and-guidelines?lang=eng#title_number117

52. The Church of Jesus Christ of Latter-Day Saints, "Policies on Moral Issues: Surrogate Motherhood," in *Handbook 2: Administering the Church*, Section 21.4.16, https://www.churchofjesuschrist.org/bc/content/shared/content/english/pdf/language-materials/08702_eng.pdf?lang=eng

53. *General Handbook of Instructions* (Salt Lake City: The Church of Jesus Christ of Latter-day Saints, 1989), p. 11–15.

54. The Church of Jesus Christ of Latter-Day Saints, "Church Policies and Guidelines, Policies on Moral Issues: Transgender Individuals," in *General Handbook: Serving in the Church of Jesus Christ of Latter-day Saints,* Chapter 38.6.22, https://www.churchofjesuschrist.org/study/manual/general-handbook/38-church-policies-and-guidelines?lang=eng#title_number118

55. The Church of Jesus Christ of Latter-Day Saints, "Members Who Have Undergone Transsexual Surgery," in *Handbook 1: Administering the Church*, Section 3.4.12.

56. First Presidency Statement, "Immunize Children, Leaders Urge," May 5, 1978, https://www.lds.org/liahona/1978/07/immunize-children-leaders-urge?lang=eng

57. "Attitudes of the Church of Jesus Christ of Latter-day Saints Towards Certain Medical Problems," in "Mormon Medical Ethical Guidelines," ed. Lester E. Bush, Jr., *Dialogue* 12 (Fall 1979), 99.

58. The Church of Jesus Christ of Latter-Day Saints, "Church Policies and Guidelines, Medical and Health Policies: Word of Wisdom," in *General Handbook: Serving in the Church of Jesus Christ of Latter-day Saints,* Chapter 38.7.13, https://www.churchofjesuschrist.org/study/manual/general-handbook/38-church-policies-and-guidelines?lang=eng#title_number131

59. Newsroom, "Statement on the Word of Wisdom," August 15, 2019, https://newsroom.churchofjesuschrist.org/article/statement-word-of-wisdom-august-2019

60. *General Handbook of Instructions* (Salt Lake City: The Church of Jesus Christ of Latter-day Saints, 1989), p. 11–16.

61. "Attitudes of the Church of Jesus Christ of Latter-day Saints Towards Certain Medical Problems," in "Mormon Medical Ethical Guidelines," ed. Lester E. Bush, Jr., *Dialogue* 12 (Fall 1979), 99, 102–104.

62. "Attitudes of the Church of Jesus Christ of Latter-day Saints Towards Certain Medical Problems," in "Mormon Medical Ethical Guidelines," ed. Lester E. Bush, Jr., *Dialogue* 12 (Fall 1979), 103.

LDS Ecclesiastical Statements on Issues in Biomedical Ethics

On July 31, 2020, the Church of Jesus Christ of Latter-day Saints (LDS) published revisions and updates to its administrative guidelines, *General Handbook: Serving in the Church of Jesus Christ of Latter-day Saints*,[1] including several modifications in the "Moral Issues" section that contains policies on issues of biomedical ethics (referenced in Appendix A). I highlight here six areas of meaningful substantive modification that express a connecting theme of expanding the scope of personal and familial moral agency and accountability in bioethical decision-making.

A Moral Preface: Ecclesiastical deference to moral agency and accountability is highlighted by the insertion of a new preface to the "Moral Issues" section:

> *A few policies in this section are about matters that the Church "discourages." Church members usually do not experience membership restrictions because of their decisions about these matters. However, all people are ultimately accountable to God for their decisions.*[2]

It is noteworthy that the preface refers to practices the LDS Church "discourages"; in almost all the changed policies, the prior language was "strongly discourages."

Birth Control: The ecclesiastical statement regarding birth control now incorporates policy on surgical sterilization. This organizational modification means that policies on birth control, in which decisions are "personal and private" matters between the married couple and God, constitute the controlling guideline, with sterilization procedures situated as a subset of birth control decisions rather than as a separate category of marital moral choice. The July 2020 guidance softens the language on surgical sterilization from "strongly discourages" to "discourages," makes the method of moral choice consistent with other procreative and birth control decisions as a "personal matter" for the "judgment and prayerful consideration" of the married couple, and recommends consultation with medical professionals rather than an ecclesiastical leader:

> *The Church discourages surgical sterilization as an elective form of birth control. Surgical sterilization includes procedures such as vasectomies and tubal ligations. However, this decision is a personal matter that is ultimately left to the judgment and prayerful consideration of the husband and wife. Couples should counsel together in unity and seek the confirmation of the Spirit in making this decision.*
>
> *Surgical sterilization is sometimes needed for medical reasons. Members may benefit from counseling with medical professionals.*[3]

Human Gamete Donation: Other policies regarding procreative choices and reproductive technologies have likewise been modified. The previous policy statement on human gametes provided a cryptic one-sentence statement that "strongly" discouraged donation of sperm, with no mention of egg donation, let alone the expanding market in "selling" human gametes. The July 2020 policy is more comprehensive and provides a theological

APPENDIX

rationale for ecclesiastically "discouraging" such practices but, as with other choices in reproductive technology, defers to the moral agency of the donor:

> *The pattern of a husband and wife providing bodies for God's spirit children is divinely appointed. For this reason, the Church discourages donating sperm or eggs. However, this is a personal matter that is ultimately left to the judgment and prayerful consideration of the potential donor. The Church also discourages selling sperm or eggs.*[4]

Fertility Treatments: The July 2020 moral policies consolidate what were previously separate statements regarding artificial insemination and in vitro fertilization into a coherent category of "fertility treatments" and provide for the first time an *affirmative* theological rationale for these widespread practices: "Reproductive technology can assist a married woman and man in their righteous desire to have children." This positive rationale precedes and thereby contextualizes a posture of "discouraging" (previously, "strongly discouraging") third-party gamete donations:

> *The pattern of a husband and wife providing bodies for God's spirit children is divinely appointed. When needed, reproductive technology can assist a married woman and man in their righteous desire to have children. This technology includes artificial insemination and in vitro fertilization. The Church discourages artificial insemination or in vitro fertilization using sperm from anyone but the husband or an egg from anyone but the wife. However, this is a personal matter that is ultimately left to the judgment and prayerful consideration of a lawfully married man and woman.*[5]

Suicide: The July 2020 guidelines frame suicide as a cry for help by persons seeking "relief from physical, mental, emotional, or spiritual pain" and thereby continues the long-standing advocacy of suicide prevention coupled with an increasing emphasis on ecclesiastical ministering and professional counseling. The policy modifies the judgment that "it is wrong to take one's own life" to "it is not right for a person to take his or her own life," a phrasing that diminishes somewhat cultural and moral stigmas. The assessment "it is not right" has also been moved from the first sentence of the policy to the fifth paragraph, indicating that broader contextual concerns of counseling and compassion have priority. Furthermore, an appeal to God's omniscience and providential care has replaced prior acknowledgments of human epistemic finitude about personal responsibility for suicidal actions:

> *Mortal life is a precious gift from God—a gift that should be valued and protected. The Church strongly supports the prevention of suicide. . . . It is not right for a person to take his or her own life. However, only God is able to judge the person's thoughts, actions, and level of accountability.*[6]

Medical Marijuana: As indicated in chapter 8, the LDS Church has had an evolving perspective in the past decade on the legalization of medical marijuana. The July 2020 guidelines provide the first statement of conditional approval of medical marijuana to appear in the administrative handbook:

> *The Church opposes the use of marijuana for non-medical purposes. However, marijuana may be used for medicinal purposes when the following conditions are met:*

> - *The use is determined to be medically necessary by a licensed physician or another legally approved medical provider.*
> - *The person follows the dosage and other directions for use from the physician or other authorized medical provider.*

> *The Church does not approve of the use of marijuana in smoking or vaping forms.*[7]

Taken collectively, these modifications expand personal moral agency, reflect the principle of professional moral integrity, and represent an ecclesiastical responsibility to articulate correct principles with individuals and families assuming ultimately moral accountability.

Notes

1. Newsroom, "Significant Updates Made to Five Chapters of the General Handbook," July 31, 2020, https://newsroom.churchofjesuschrist.org/article/handbook-update-july-2020
2. The Church of Jesus Christ of Latter-day Saints, "Church Policies and Guidelines, Policies on Moral Issues," in *General Handbook: Serving in the Church of Jesus Christ of Latter-day Saints,* Chapter 38.6, July 31, 2020, https://www.churchofjesuschrist.org/study/manual/general-handbook/38-church-policies-and-guidelines?lang=eng#title_number90.
3. The Church of Jesus Christ of Latter-day Saints, "Church Policies and Guidelines, Policies on Moral Issues: Birth Control," in *General Handbook: Serving in the Church of Jesus Christ of Latter-day Saints,* Chapter 38.6.4, July 31, 2020, https://www.churchofjesuschrist.org/study/manual/general-handbook/38-church-policies-and-guidelines?lang=eng#title_number101
4. The Church of Jesus Christ of Latter-day Saints, "Church Policies and Guidelines, Policies on Moral Issues: Donation of Sperm or Egg" in *General Handbook: Serving in the Church of Jesus Christ of Latter-day Saints,* Chapter 38.6.7, July 31, 2020, https://www.churchofjesuschrist.org/study/manual/general-handbook/38-church-policies-and-guidelines?lang=eng#title_number217
5. The Church of Jesus Christ of Latter-day Saints, "Church Policies and Guidelines, Policies on Moral Issues: Fertility Treatments," in *General Handbook: Serving in the Church of Jesus Christ of Latter-day Saints,* Chapter 38.6.9, July 31, 2020, https://www.churchofjesuschrist.org/study/manual/general-handbook/38-church-policies-and-guidelines?lang=eng#title_number218
6. The Church of Jesus Christ of Latter-day Saints, "Church Policies and Guidelines, Policies on Moral Issues: Suicide," in *General Handbook: Serving in the Church of Jesus Christ of Latter-day Saints,* Chapter 38.6.19, July 31, 2020, https://www.churchofjesuschrist.org/study/manual/general-handbook/38-church-policies-and-guidelines?lang=eng#title_number115
7. The Church of Jesus Christ of Latter-day Saints, "Church Policies and Guidelines, Medical and Health Policies: Medical Marijuana," in *General Handbook: Serving in the Church of Jesus Christ of Latter-day Saints,* Chapter 38.7.8, July 31, 2020, https://www.churchofjesuschrist.org/study/manual/general-handbook/38-church-policies-and-guidelines?lang=eng#title_number219

Bibliography

ABIM Foundation, ACP–ASIM Foundation, and European Federation of Internal Medicine, "Medical Professionalism in the New Millennium: A Physician's Charter," *Annals of Internal Medicine* 136 (February 2002), 243–246, http://abimfoundation.org/what-we-do/physician-charter

Kaashif A. Ahmad et al., "Two-Year Neurodevelopmental Outcome of an Infant Born at 21 Weeks' 4 Days' Gestation," *Pediatrics* 140 (December 2017), http://pediatrics.aappublications.org/content/140/6/e20170103

Jenne Erigero Alderks, "Rediscovering the Legacy of Mormon Midwifes," *Sunstone*, April 3, 2012, https://www.sunstonemagazine.com/rediscovering-the-legacy-of-mormon-midwives/

Shana Alexander, "They Decide Who Lives, Who Dies," *Life* 53 (1962), 102–125.

American Academy of Pediatrics Committee on Fetus and Newborn, "Non-Initiation or Withdrawal of Intensive Care for High-Risk Newborns," *Pediatrics* 119 (February 2007), 401–403.

American Academy of Pediatrics, Infant Bioethics Task Force and Consultants, "Guidelines for Infant Bioethics Committees," *Pediatrics* 74 (August 1984), 306–310.

American Medical Association, "DNR Orders in a Public Health Crisis," April 29, 2020, https://www.ama-assn.org/delivering-care/ethics/dnr-orders-public-health-crisis

American Society for Reproductive Medicine, Ethics Committee, "Access to Fertility Treatment for Gays, Lesbians, and Unmarried Persons: A Committee Opinion," *Fertility and Sterility* 100 (2013), 1524–1527, https://www.asrm.org/globalassets/asrm/asrm-content/news-and-publications/ethics-committee-opinions/access_to_fertility_treatment_by_gays_lesbians_and_unmarried_persons-pdfmembers.pdf

Vanessa Amundson, "My Position Regarding Surrogacy," *A Perfectly Imperfect Perfectionist*, October 12, 2011, http://myamundson5.blogspot.com/2011/10/my-position-regarding-surrogacY.html?showComment=1330523452486#c4840420031267940597

Taylor W. Anderson, Benjamin Wood, "LDS Church Announces Opposition to Utah Medical Marijuana Initiative," *Salt Lake Tribune*, August 23, 2018, https://www.sltrib.com/news/politics/2018/08/23/lds-church-announces/

Natalie Anger, "Fierce Competition Marked Fervent Race for Cancer Gene," *New York Times*, September 20, 1994, https://www.nytimes.com/1994/09/20/science/fierce-competition-marked-fervid-race-for-cancer-gene.html

Allie Arnell, "BYU Student Health Plan Not Compliant with Obamacare," *Daily Universe*, September 4, 2015, https://universe.byu.edu/2015/09/04/byu-student-health-plan-not-compliant-with-obamacare1/

Caroline Arveseth, "Five Under Five," *Mormon Women Project*, May 19, 2010, https://www.mormonwomen.com/interview/five-under-five/

Carlos E. Asay, "God's Love for All Mankind," in *Mormons and Muslims: Spiritual Foundations and Modern Manifestations,* 2nd ed., ed. Spencer J. Palmer (Provo, UT: Brigham Young University, 2002), 51–61.

Rachel Aviv, "The Death Treatment," *The New Yorker,* June 22, 2015, http://www.newyorker.com/magazine/2015/06/22/the-death-treatment

David H. Bailey, Jeffrey M. Bradshaw, "Science and Mormonism," *Interpreter: A Journal of Latter-day Saint Faith and Scholarship* 19 (2016), 17–37.

Komal Bajaj, Susan J. Gross, "Carrier Screening: Past, Present, and Future," *Journal of Clinical Medicine* 3 (September 2014), 1033–1042.

Katie Barlow, "My Vaccination Decision," *LDS Living,* http://www.ldsliving.com/-LDSL-Blog-My-Vaccination-Decision/s/65450

David Barnard, "The Physician as Priest, Revisited," *Journal of Religion and Health* 24 (Winter 1985), 274–275.

Vincent Barry, *Bioethics in a Cultural Context* (Boston: Wadsworth, Cengage Learning, 2012).

Melvin L. Bashore et al., "Mortality on the Mormon Trail, 1847–1868," *BYU Studies Quarterly* 53 (2014), 109–123, https://byustudies.byu.edu/content/mortality-mormon-trail-1847-1868

Michelle Bayefsky, "Who Should Regulate Pre-Implantation Diagnosis in the United States?," *AMA Journal of Ethics* 20 (2018), E1160–E1167, https://journalofethics.ama-assn.org/article/who-should-regulate-preimplantation-genetic-diagnosis-united-states/2018-12

Becca, "I'm a Mormon with Infertility (Infertile in a Family-Focused Faith)," *Love Our Crazy Life,* April 21, 2018, http://www.loveourcrazylife.com/infertility-mormon-lds/

Amy Julia Becker, "The Social Construction of Selective Abortion," *The Atlantic,* January 22, 2013, https://www.theatlantic.com/sexes/archive/2013/01/the-social-construction-of-selective-abortion/267386/

Robert A. Bednarczyk et al., "Sexual Activity-Related Outcomes After Human Papillomavirus Vaccination of 11- to 12-Year-Olds," *Pediatrics* 130 (November 2012), http://pediatrics.aappublications.org/content/130/5/798

Henry K. Beecher, "Ethics and Clinical Research," *New England Journal of Medicine* 274 (June 14, 1966), 1354–1360.

Emily Beitiks, "5 Reasons Why We Need People with Disabilities in the CRISPR Debates," Center for Genetics and Society, September 7, 2016, https://www.geneticsandsociety.org/biopolitical-times/5-reasons-why-we-need-people-disabilities-crispr-debates

Sharon Belknap, "Don't Let My Baby Die," *The Ensign,* December 2001, https://www.lds.org/ensign/2001/12/dont-let-my-baby-die?lang=eng

April Young Bennett, "Five More LDS Church Discipline Policies That Affect Women Unequally," *Exponent,* February 22, 2014, https://www.the-exponent.com/five-more-lds-church-discipline-policies-that-affect-women-unequally/

April Young Bennett, "'Irrespective of Cause': A Story about Jon Huntsman, Sr., and HPV Vaccine," *Exponent,* February 3, 2018, https://www.the-exponent.com/irrespective-of-cause-a-story-about-jon-huntsman-sr-and-hpv-vaccine/

Benjamin E. Berkman, Sara Chandros Hull, "The 'Right Not to Know' in the Genomic Era: Time to Break from Tradition?," *American Journal of Bioethics* 14 (March 2014), 28–31.

Nancy Berlinger et al., "Responding to COVID-19 as a Regional Public Health Challenge," Hastings Center, April 29, 2020, http://www.thehastingscenter.org/covid19_regional_ethics_guidelines

Alicia L. Best et al., "Examining the Influence of Religious and Spiritual Beliefs on HPV Vaccine Uptake Among College Women," *Journal of Religion and Health* 58 (2019), 2196–2207.

Brian D. Birch, "A Portion of God's Light: Mormonism and Religious Pluralism," *Dialogue: A Journal of Mormon Thought* 51 (2018), 85–102.

Chris Bodenner, "When Does Abortion Become Eugenics?," *The Atlantic*, May 24, 2016, https://www.theatlantic.com/notes/all/2016/05/when-does-an-abortion-become-eugenics/483659/

Julie Bodson et al., "Religion and HPV Vaccine-Related Awareness, Knowledge, and Receipt Among Insured Women Aged 18–26 in Utah," *PLOS One* 12 (2017), 1–11, https://www.ncbi.nlm.nih.gov/pmc/articles/PMC5571930/

Bowen v. American Hospital Association, 476 US 610 (1986).

Sherry I. Brandt-Rauf et al., "Ashkenazi Jews and Breast Cancer: The Consequences of Linking Ethnic Identity to Genetic Disease," *American Journal of Public Health* 96 (November 2006), 1979–1988, https://www.ncbi.nlm.nih.gov/pmc/articles/PMC1751808/

Megan Brenan, "Americans' Strong Support for Euthanasia Persists," Gallup, May 31, 2018, https://news.gallup.com/poll/235145/americans-strong-support-euthanasia-persists.aspx

R. Lanier Britsch, "What Is the Relationship of the Church of Jesus Christ of Latter-day Saints to the Non-Christian Religions of the World?," *The Ensign*, January 1988, https://www.lds.org/ensign/1988/01/i-have-a-question/what-is-the-relationship-of-the-church-of-jesus-christ-of-latter-day-saints-to-the-non-christian-religions-of-the-world?lang=eng

Samuel Morris Brown, *In Heaven as It Is on Earth: Joseph Smith and the Early Mormon Conquest of Death* (New York: Oxford University Press, 2012).

Eric Burns, *The Smoke of the Gods: A Social History of Tobacco* (Philadelphia: Temple University Press, 2007).

Linda K. Burton, "I Was A Stranger," *The Ensign,* May 2016, https://www.lds.org/general-conference/2016/04/i-was-a-stranger?lang=eng

Burwell v. Hobby Lobby Stores, Inc., 573 U.S. 682 (2014).

Lester E. Bush, Jr., ed., "Attitudes of the Church of Jesus Christ of Latter-day Saints Towards Certain Medical Problems," in "Mormon Medical Ethical Guidelines," *Dialogue* 12 (Fall 1979), 97–106.

Lester E. Bush, Jr., "Birth Control Among the Mormons: Introduction to an Insistent Question," *Dialogue* 10 (1977), 12–44.

Lester E. Bush, Jr., *Health and Medicine Among the Latter-day Saints: Science, Sense, and Scripture* (New York: Crossroad, 1993).

Lester E. Bush, Jr., "The Mormon Tradition," in *Caring and Curing: Health and Medicine in the Western Religious Traditions,* ed. Ronald L. Numbers, Darrel W. Amundsen (Baltimore, MD: Johns Hopkins University Press, 1986), 397–420.

Lester E. Bush, Jr., "The Word of Wisdom in Early Nineteenth-Century Perspective," *Dialogue* 14 (1981), 46–65.

Richard Lyman Bushman, *Joseph Smith: Rough Stone Rolling* (New York: Vintage Books, 2005).

Ira Byock, *Dying Well: Peace and Possibilities at the End of Life* (New York: Riverhead Books, 1997).

Lisa Sowle Cahill, "Abortion: Roman Catholic Perspectives," in *Bioethics,* 4th ed., vol. 1, ed. Bruce Jennings (New York: Macmillan Reference USA, 2014), 37–41.

Shawneequa Callier, Rachel Simpson, "Genetic Diseases and the Duty to Disclose," *AMA Journal of Ethics,* August 2012, https://journalofethics.ama-assn.org/article/genetic-diseases-and-duty-disclose/2012-08

Alastair V. Campbell, *Health as Liberation: Medicine, Theology, and the Quest for Justice* (Eugene, OR: Wipf and Stock, 1995).

Courtney S. Campbell, *Bearing Witness: Religious Meanings in Bioethics* (Eugene, OR: Cascade Books, 2019).

Courtney S. Campbell, "Embodiment and Ethics: A Latter-day Saint Perspective," in *Theological Developments in Bioethics: 1990–1992,* ed. B. A. Lustig (Dordrecht, The Netherlands: Kluwer Academic, 1993), 43–67.

Courtney S. Campbell, "The Gift and the Market: Cultural Symbolic Perspectives," in *Transplanting Human Tissue: Ethics, Policy, and Practice,* ed. Stuart J. Youngner, Martha W. Anderson, Renie Schapiro (New York: Oxford University Press, 2004), 139–159.

Courtney S. Campbell, "Moral Meanings of Physician-Assisted Death for Hospice Ethics," in *Hospice Ethics: Policy and Practice in Palliative Care,* ed. Timothy W. Kirk, Bruce Jennings (New York: Oxford University Press, 2014), 223–249.

Courtney S. Campbell, "Mortal Responsibilities: Bioethics and Medically Assisted Dying," *Yale Journal of Biology and Medicine* 92 (2019), 733–739.

Courtney S. Campbell, "Prophecy and Citizenry: The Case of Human Cloning," *Sunstone* 21 (1998), 11–15.

Courtney S. Campbell, "Religious Perspectives on Human Cloning," in National Bioethics Advisory Commission, *Cloning Human Beings,* vol. 2 (Rockville, MD: NBAC, 1997), D1–D64.

Courtney S. Campbell, "Sounds of Silence: The Latter-day Saints and Medical Ethics," in *Theological Developments in Bioethics: 1988–1990,* ed. B. Andrew Lustig (Dordrecht, The Netherlands: Kluwer Academic, 1991), 23–40.

Courtney S. Campbell, "What More in the Name of God?: Theologies and Theodicies of Faith Healing," *Kennedy Institute of Ethics Journal* 20:1 (March 2010), 1–25.

Lincoln Cannon, "Mormonism Mandates Transhumanism," in *Religion and Human Enhancement: Death, Values, and Modernity,* ed. Tracy J. Trothen, Calvin Mercer (New York: Palgrave Macmillan, 2017), 49–66.

Lincoln Cannon, "What Is Mormon Transhumanism?," *Theology and Science* 13 (2015), 213–214.

Thomas Carter, *Building Zion: The Material World of Mormon Settlement* (Minneapolis: University of Minnesota Press, 2015).

Eric J. Cassell, *The Nature of Healing: The Modern Practice of Medicine* (New York: Oxford University Press, 2013).

Eric J. Cassell, *The Nature of Suffering and the Goals of Medicine,* 2nd ed. (New York: Oxford University Press, 2004).

Center for Medicare and Medicaid Services, "National Health Expenditures Fact Sheet: 2018," https://www.cms.gov/Research-Statistics-Data-and-Systems/Statistics-Trends-and-Reports/NationalHealthExpendData/NHE-Fact-Sheet

Centers for Disease Control and Prevention, *2017 Assisted Reproductive Technology Fertility Clinic Success Rates Report* (Atlanta: U.S. Dept. of Health and Human Services, 2019), ftp://ftp.cdc.gov/pub/Publications/art/ART-2017-Clinic-Report-Full.pdf

Centers for Disease Control and Prevention, "COVID-19 in Racial and Ethnic Minority Groups," April 22, 2020, https://www.cdc.gov/coronavirus/2019-ncov/need-extra-precautions/racial-ethnic-minorities.html

Centers for Disease Control and Prevention, "Healthy People 2020 Mid-Course Review: Immunization and Infectious Diseases," January 20, 2017, https://www.cdc.gov/nchs/data/hpdata2020/HP2020MCR-C23-IID.pdf

Centers for Disease Control and Prevention, "Measles Cases and Outbreaks: 2019," https://www.cdc.gov/measles/cases-outbreaks.html

Centers for Disease Control and Prevention, "More US Adolescents Up to Date on HPV Vaccination," August 23, 2018, https://www.cdc.gov/media/releases/2018/p0823-HPV-vaccination.html

Centers for Disease Control and Prevention, "National, Regional, State, and Selected Local Area Vaccination Coverage Among Adolescents Aged 13–17 Years—United States, 2017," *Morbidity and Mortality Weekly Report* 67:33 (2018), 909–917, https://www.cdc.gov/mmwr/volumes/67/wr/mm6733a1.htm#T1_down

Dawn Chan, "The Immortality Upgrade," *The New Yorker*, April 20, 2016, https://www.newyorker.com/tech/annals-of-technology/mormon-transhumanism-and-the-immortality-upgrade

Daphne Chen, "What You Don't Know About the HPV Vaccine May Hurt You," *Deseret News*, January 15, 2016, https://www.deseretnews.com/article/865645597/What-you-dont-know-about-the-HPV-vaccine-may-hurt-you.html

Daphne Chen, "With More Parents Choosing Not to Vaccinate, Utah on Brink of Losing 'Herd Immunity,'" February 9, 2016, KSL.com, https://www.ksl.com/?sid=38451248&nid=148&fm=home_page&s_cid=toppick2

James F. Childress, *Who Should Decide?: Paternalism in Health Care* (New York: Oxford University Press, 1982).

James F. Childress, Eric M. Meslin, Harold T. Shapiro, eds., *Belmont Revisited: Ethical Principles for Research with Human Subjects* (Washington, D.C.: Georgetown University Press, 2005).

Danielle Christensen, "Finding Hope: When Babies Return to Their Heavenly Home Before Ever Coming to Their Earthly One," *LDS Living*, January 16, 2020, http://www.ldsliving.com/Knowing-Heaven-How-2-Latter-day-Saint-Families-Who-Didn-t-Take-Their-Baby-Home-From-the-Hospital-Hope-for-an-Eternal-Family/s/92240?utm_source=ldsliving&utm_medium=email

Larry R. Churchill, *Rationing Health Care in America: Perceptions and Principles of Justice* (Notre Dame, IN: University of Notre Dame Press, 1987).

The Church of Jesus Christ of Latter-day Saints, "Abortion," https://www.lds.org/topics/abortion?lang=eng

The Church of Jesus Christ of Latter-day Saints, "Administrative Principles in Challenging Times, April 16, 2020, https://newsroom.churchofjesuschrist.org/article/administrative-principles-in-challenging-times

The Church of Jesus Christ of Latter-day Saints, *The Book of Mormon* (Salt Lake City: Church of Jesus Christ of Latter-day Saints, 1981).

The Church of Jesus Christ of Latter-day Saints, "Church Makes Immunizations an Official Initiative, Provides Social Mobilization," June 13, 2012, https://www.lds.org/church/news/church-makes-immunizations-an-official-initiative-provides-social-mobilization?lang=eng

The Church of Jesus Christ of Latter-day Saints, "Church Policies and Guidelines, Medical and Health Policies," in *General Handbook: Serving in the Church of*

Jesus Christ of Latter-day Saints, Chapter 38.7, February 19, 2020, https://www.churchofjesuschrist.org/study/manual/general-handbook/38-church-policies-and-guidelines?lang=eng#title_number91

The Church of Jesus Christ of Latter-day Saints, "Church Policies and Guidelines, Policies on Moral Issues," in *General Handbook: Serving in the Church of Jesus Christ of Latter-day Saints,* Chapter 38.6, February 19, 2020, https://www.churchofjesuschrist.org/study/manual/general-handbook/38-church-policies-and-guidelines?lang=eng#title_number91

The Church of Jesus Christ of Latter-day Saints, "Church Says Yes to Regulated Medical Marijuana but No to Utah Initiative," September 20, 2018, https://www.lds.org/church/news/church-says-yes-to-regulated-medical-marijuana-but-no-to-utah-initiative?lang=eng

The Church of Jesus Christ of Latter-day Saints, *The Doctrine and Covenants* (Salt Lake City: Church of Jesus Christ of Latter-day Saints, 1981).

The Church of Jesus Christ of Latter-day Saints, "The Family: A Proclamation to the World," September 1995, https://www.lds.org/topics/family-proclamation?lang=eng&old=true

The Church of Jesus Christ of Latter-day Saints, *General Handbook of Instructions* (Salt Lake City: Church of Jesus Christ of Latter-day Saints, 1989).

The Church of Jesus Christ of Latter-day Saints, *The Holy Bible* (Salt Lake City: Church of Jesus Christ of Latter-day Saints, 1979).

The Church of Jesus Christ of Latter-day Saints, "Humanitarian Programs," https://www.lds.org/topics/humanitarian-service/church?lang=eng&old=true

The Church of Jesus Christ of Latter-Day Saints, "Medical and Health Practices," in *Handbook 2: Administering the Church*, Section 21.3; https://www.lds.org/handbook/handbook-2-administering-the-church?lang=eng

The Church of Jesus Christ of Latter-Day Saints, "Members with Disabilities," in *Handbook 2: Administering the Church*, Section 21.1.26, https://www.lds.org/handbook/handbook-2-administering-the-church?lang=eng.

The Church of Jesus Christ of Latter-Day Saints, "Members Who Have Undergone Transsexual Surgery," in *Handbook 1: Administering the Church*, Section 3.4.12, 2010.

The Church of Jesus Christ of Latter-day Saints, *The Pearl of Great Price* (Salt Lake City: Church of Jesus Christ of Latter-day Saints, 1981).

The Church of Jesus Christ of Latter-Day Saints, "Policies on Moral Issues," in *Handbook 2: Administering the Church*, Section 21.4, 2010, https://www.lds.org/handbook/handbook-2-administering-the-church?lang=eng

The Church of Jesus Christ of Latter-Day Saints, "Providing for Temporal Needs and Building Self-Reliance: Health," in *General Handbook: Serving in the Church of Jesus Christ of Latter-day Saints,* Chapter 22.1.1.1, February 19, 2020, https://www.churchofjesuschrist.org/study/manual/general-handbook/38-church-policies-and-guidelines?lang=eng#title_number91

The Church of Jesus Christ of Latter-day Saints, *Teachings of the Presidents of the Church: Joseph Smith* (Salt Lake City: Intellectual Reserve, 2011).

The Church of Jesus Christ of Latter-day Saints, Office of the First Presidency, "Letter on Assisted Suicide," October 13, 2016, https://www.churchofjesuschrist.org/church/news/first-presidency-asks-members-to-oppose-recreational-marijuana-assisted-suicide?lang=eng

Church News, "Beehive Clothing Facilities Worldwide Will Produce 200,000 Gowns, 1.5 Million Masks in COVID-19 Relief Effort," *Deseret News*, May 20, 2020, https://www.thechurchnews.com/global/2020-05-20/beehive-clothing-masks-gowns-medical-covid-relief-184696

"Church Opposes Abortion Bill," *Deseret News*, January 23, 1969.

Jan Cienski, "Mormons May Be Key in Stem Cell Debate," *World-Wide Religious News*, August 4, 2001, https://wwrn.org/articles/5900/

Drew Clark, "The Mormon Stem Cell Choir," *Slate*, August 3, 2001, https://slate.com/news-and-politics/2001/08/the-mormon-stem-cell-choir.html

Stephen L. Clark, "A Doctor Looks at Amniocentesis," *The Ensign*, April 1985, https://www.lds.org/study/ensign/1985/04/random-sampler/a-doctor-looks-at-amniocentesis?lang=eng

Kristin Clift, "Glimpses of Eternity: Sampled Mormon Understandings of Disability, Genetic Testing, and Reproductive Choice in New Zealand," M.A. thesis, University of Otago, New Zealand, 2012, https://ourarchive.otago.ac.nz/bitstream/handle/10523/3935/CliftKristin2012MA.pdf?sequence=1&isAllowed=y

The Clinton Foundation, "Health Care Is Local," https://stories.clintonfoundation.org/healthcare-is-local-bcc165cc22eb

I. Glenn Cohen, "Selling Bone Marrow: *Flynn v. Holder*," *New England Journal of Medicine* 366 (January 26, 2012), 296–297.

Committee on Human Gene Editing: Scientific, Medical, and Ethical Considerations, *Human Genome Editing: Science, Ethics, and Governance* (Washington, D.C.: National Academies Press, 2017), https://www.nap.edu/catalog/24623/human-genome-editing-science-ethics-and-governance

Quentin L. Cook, "Restoring Morality and Religious Freedom," *The Ensign*, September 2012, https://www.lds.org/ensign/2012/09/restoring-morality-and-religious-freedom?lang=eng

S. Creighton, E. W. Almqvist, D. MacGregor, et al., "Predictive, Pre-Natal and Diagnostic Genetic Testing for Huntington's Disease: The Experience in Canada from 1987 to 2000," *Clinical Genetics* 63 (2003), 462–475.

Cruzan v. Director, Missouri Department of Health, 497 U.S. 261 (1990).

Kelsey Dallas, "Faith Leaders, Including LDS Presiding Bishop, Call on Obama, Congress, to Reject Controversial Religious Liberty Report," *Deseret News*, October 12, 2016, https://www.deseret.com/2016/10/12/20598089/faith-leaders-including-lds-presiding-bishop-call-on-obama-congress-to-reject-controversial-religiou#president-barack-obama-gestures-while-saying-its-important to-think-about-the-why-when-choosing-a-goal-or-profession-as-he-speaks-to-young-african-leaders-initiative-event-at-the-omni-shoreham-hotel-wednesday-aug-3-2016-in-washington

Kelsey Dallas, "Finding God in One of Science's Biggest Debates: Gene Editing," *Deseret News*, January 13, 2016, https://www.deseretnews.com/article/865645387/Finding-God-in-one-of-sciences-biggest-debates-2-Genetic-editing.html

Kelsey Dallas, "The Morality of Playing God with Your Baby's DNA," *Deseret News*, March 28, 2015, https://www.deseretnews.com/article/865666153/The-morality-of-playing-God-with-Your-babys-DNA.html

Marcy Darnovsky, "Do Not Open the Door to Editing Genes in Future Humans," *National Geographic*, August 2016, https://www.nationalgeographic.com/magazine/2016/08/human-gene-editing-pro-con-opinions/

Lee Davidson, "No LDS Stand on Cell Research," *Deseret News,* July 6, 2001, https://www.deseretnews.com/article/851862/No-LDS-stand-on-cell-research.html

Dena S. Davis, *Genetic Dilemmas: Reproductive Technology, Parental Choices, and Children's Futures* (New York: Oxford University Press, 2010).

Jonathan Decker, "Is Birth Control Against the Commandments?," *LDS Living,* November 30, 2018, http://www.ldsliving.com/Ask-a-Latter-day-Saint-Therapist-Is-Birth-Control-Against-the-Commandments/s/89844?utm_source=ldsliving&utm_medium=email

Sherri L. Dew, "Are We Not All Mothers?," *The Ensign,* October 2001, https://www.lds.org/general-conference/2001/10/are-we-not-all-mothers?lang=eng

Robert T. Divett, *Medicine and the Mormons: An Introduction to the History of Latter-day Saint Health Care,* 2nd ed. (Charleston, SC: CreateSpace, 2010).

DNA Learning Center, "Mark Skolnick," https://www.dnalc.org/view/15718-Mark-Skolnick.html

Jennifer Dobner, "Genetics and U," *Continuum,* Spring 2014, https://continuum.utah.edu/features/genetics-and-u/

Donate Life America, "National Donor Sabbath," https://www.donatelife.net/nds/

Elliot N. Dorff, "Testimony: National Bioethics Advisory Commission," *Ethical Issues in Human Stem Cell Research: Religious Perspectives,* vol. 3 (Rockville, MD: Government Printing Office, 2000), C3–C5, https://bioethicsarchive.georgetown.edu/nbac/stemcell3.pdf

Donald B. Doty, "Prolonging Life," in *Encyclopedia of Mormonism,* ed. Daniel H. Ludlow (New York: Macmillan, 1992), 1159–1160, https://eom.byu.edu/index.php/Prolonging_Life

Ross Douthat, "Romney's Mormon Story," *New York Times,* August 11, 2012, https://www.nytimes.com/2012/08/12/opinion/sunday/douthat-romneys-mormon-story.html

Daniel L. Dreisbach, Mark David Hall, eds., *The Sacred Rights of Conscience: Selected Readings on Religious Liberty and Church-State Relations in the American Founding* (Indianapolis: Liberty Fund, 2009).

Melanie Drolet, Elodie Benard, Norma Perez, Marc Brisson, "HPV Vaccination Programmes: Updated Systematic Review and Data Analysis," *The Lancet,* June 26, 2019, https://doi.org/10.1016/S0140-6736(19)30298-3

Raymond S. Duff, A. G. M. Campbell, "Moral and Ethical Dilemmas in the Special-Care Nursery," *New England Journal of Medicine* 289 (October 25, 1973), 890–894, https://www.nejm.org/doi/full/10.1056/NEJM197310252891705

Paul H. Dunn, Richard M. Eyre, *The Birth We Call Death* (Salt Lake City: Aspen Books, 1999).

W. Cole Durham, Jr., "Church and State," in *Encyclopedia of Mormonism,* ed. Daniel H. Ludlow (New York: Macmillan, 1992), 282–283; https://contentdm.lib.byu.edu/digital/collection/EoM/id/3541

Homer S. Ellsworth, "Birth Control," in *Encyclopedia of Mormonism,* ed. Daniel H. Ludlow (New York: Macmillan, 1992), 116–117; https://contentdm.lib.bYu.edu/digital/collection/EoM/id/5515

Ezekiel J. Emanuel et al., "Fair Allocation of Scarce Medical Resources in the Time of COVID-19," *New England Journal of Medicine* 382 (2020): 2049–2055, https://www.nejm.org/doi/full/10.1056/NEJMsb2005114

James E. Enstrom, "Health Practices and Cancer Mortality Among Active California Mormons," *Journal of the National Cancer Institute* 81 (1989), 1809–1810.

James E. Enstrom, Lester Brenslow, "Lifestyle and Reduced Mortality Among Active California Mormons, 1980–2004," *Preventive Medicine* 46 (2008), 133–136.

Ethics Committee of the American Society for Reproductive Medicine, "Use of Reproductive Technology for Sex Selection for Non-Medical Reasons," *Fertility & Sterility* 10 (June 2015), 1418–1422.

John H. Evans, *The History and Future of Bioethics: A Sociological View* (New York: Oxford University Press, 2012).

Van Evans, Daniel W. Curtis, and Ram A. Cnaan, "Volunteering Among Latter-Day Saints," *Journal for the Scientific Study of Religion* 52 (December 2013), 827–841.

Ashley Evanson, "Mormons and Science," *LDS Living*, September 15, 2008, http://www.ldsliving.com/Mormons-and-Science/s/4696

Anne Fadiman, *The Spirit Catches You and You Fall Down* (New York: Farrar, Straus and Giroux, 1997).

John E. Ferguson III, Benjamin R. Knoll, Jana Riess, "The Word of Wisdom in Contemporary American Mormonism: Perceptions and Practice," *Dialogue* 51(2018), 39–77.

Leland A. Fetzer, "Tolstoy and Mormonism," *Dialogue: A Journal of Mormon Thought* 6 (1972), 13–29, https://www.dialoguejournal.com/wp-content/uploads/2010/05/Dialogue_V06N01_15.pdf

Peter G. Filene, *In the Arms of Others: A Cultural History of the Right-to-Die in America* (Chicago: Ivan R. Dee, 1998).

L. B. Finer, M. R. Zolna, "Declines in Unintended Pregnancy in the United States, 2008–2011," *New England Journal of Medicine* 374 (2016), 843–852.

First Presidency, LDS Church, "Opportunities to Address COVID-19 Needs," April 14, 2020, https://newsroom.churchofjesuschrist.org/article/covid-19-first-presidency-letter-april-14-2020

First Presidency, LDS Church, "The Origin of Man," *The Ensign*, February 2002, https://www.lds.org/ensign/2002/02/the-origin-of-man?lang=eng

First Presidency, LDS Church, "Policies and Procedures: Statement on Abortion," *New Era* 3, April 1973, https://www.lds.org/new-era/1973/04/policies-and-procedures-statement-on-abortion?lang=eng

Kelsey Foreman, "The Church Gets Involved in Re-Writing Utah's Medical Marijuana Laws," *Utah Business*, September 3, 2019, https://www.utahbusiness.com/prop2-medical-marijuana/

Daniel Fox, H. M. Leichter, "Rationing Care in Oregon: The New Accountability," *Health Affairs* 10 (Summer 1991), 7–27.

Renee C. Fox, *Essays in Medical Sociology: Journeys in the Field*, 2nd ed. (New Brunswick, NJ: Transaction Books, 1988).

Renee C. Fox, "'It's the Same, but Different': A Sociological Perspective on the Case of the Utah Artificial Heart," in *After Barney Clark: Reflections on the Utah Artificial Heart Program*, ed. Margery W. Shaw (Austin: University of Texas Press, 1984), 68–90.

Renee C. Fox, Judith P. Swazey, *The Courage to Fail: A Social View of Organ Transplants and Dialysis*, 2nd ed. (New York: Routledge, 2002).

Renee C. Fox, Judith P. Swazey, *Spare Parts: Organ Replacement in American Society* (New York: Oxford University Press, 1992).

Russell Arben Fox, "About the McKay Quote," March 11, 2004, *Times and Seasons*, https://www.timesandseasons.org/harchive/2004/03/about-the-mckay-quote/

Arthur W. Frank, *The Wounded Storyteller: Body, Ethics and Illness*, 2nd ed. (Chicago: University of Chicago Press, 2013).

Cary Funk et al., "U.S. Public Opinion on the Future Use of Gene Editing," Pew Research Center, July 26, 2016, http://www.pewinternet.org/2016/07/26/u-s-public-opinion-on-the-future-use-of-gene-editing/

Cary Funk et al., "U.S. Public Wary of Biomedical Technologies to 'Enhance' Human Abilities," Pew Research Center, July 26, 2016, http://www.pewinternet.org/2016/07/26/u-s-public-wary-of-biomedical-technologies-to-enhance-human-abilities/

Melvin K. Gardner, "Elder Russell M. Nelson," *The Ensign*, June 1984, https://www.lds.org/ensign/1984/06/elder-russell-m-nelson-applying-divine-laws?lang=eng

Atul Gawande, *Being Mortal: Medicine and What Matters in the End* (New York: Henry Holt and Company, 2014).

Atul Gawande, "Is Health Care a Right?," *The Atlantic*, October 2, 2017, https://www.newyorker.com/magazine/2017/10/02/is-health-care-a-right

Robert Gehrke, "LDS Church's Stance on Medical Marijuana Doesn't Make Sense," *Salt Lake Tribune*, April 13, 2018, https://www.sltrib.com/opinion/2018/04/12/gehrke-lds-churchs-stance-on-medical-marijuana-doesnt-make-sense-why-is-it-ok-for-a-nevada-mormon-but-not-a-utah-mormon/

"The Genetics of Genealogy," *The Liahona*, August 1979, https://www.lds.org/liahona/1979/08/discovery/the-genetics-of-genealogy?lang=eng&clang=ara

Genetics Science Learning Center, "Genetics in Utah," https://learn.genetics.utah.edu/content/science/utah/

Fatemeh Ghorbani et al., "Causes of Family Refusal for Organ Donation," *Transplantation Proceedings* 43 (March 2011), 405–406.

Nancy Gibbs, "Defusing the War Over the 'Promiscuity' Vaccine," *Time*, June 21, 2006, http://content.time.com/time/nation/article/0,8599,1206813,00.html

L. Kay Gillespie, "Death and Dying," in *Encyclopedia of Mormonism*, ed. Daniel H. Ludlow (New York: Macmillan, 1992), 364–366, https://eom.byu.edu/index.php/Death_and_Dying

Terryl L. Givens, Philip L. Barlow, *The Oxford Handbook of Mormonism* (New York: Oxford University Press, 2015).

Hannah C. Glass et al., "Outcomes for Extremely Premature Infants," *Anesthesia and Analgesia* 120 (June 2015), 1337–1351.

Maria Godoy, "What Do Coronavirus Racial Disparities Look Like State by State?," *NPR*, May 30, 2020, https://www.npr.org/sections/health-shots/2020/05/30/865413079/what-do-coronavirus-racial-disparities-look-like-state-by-state

David Goldhill, "How American Health Care Killed My Father," *The Atlantic*, September 2009, https://www.theatlantic.com/magazine/archive/2009/09/how-american-health-care-killed-my-father/307617/

Abby Goodnough, "In Red-State Utah, A Surge Toward Obamacare," *New York Times*, March 3, 2017, https://www.nytimes.com/2017/03/03/health/utah-obamacare.html

Laurie Goodstein, "Utah Passes Antidiscrimination Bill Backed by Mormon Leaders," *New York Times*, March 12, 2015, https://www.nytimes.com/2015/03/12/us/politics/utah-passes-antidiscrimination-bill-backed-by-mormon-leaders.html

Neil M. Gorsuch, *The Future of Assisted Suicide and Euthanasia* (Princeton, NJ: Princeton University Press, 2006).

Government of Canada, "Medical Assistance in Dying," https://www.canada.ca/en/health-canada/services/medical-assistance-dying.html

Jennifer Graham, "Behind the Numbers: How to Make Sense of Utah's Abortion Rate for Married Women," *Deseret News*, June 14, 2017, https://www.deseretnews.com/article/865681905/Behind-the-numbers-How-to-make-sense-of-Utahs-abortion-rate-for-married-women.html

Jennifer Graham, "Can Your Child Wait Until College to Get the HPV Vaccine?," *Deseret News*, https://www.deseretnews.com/article/865678797/Can-your-child-wait-until-college-to-get-the-HPV-vaccine.html

Emma Green, "The Mormon Church Tries to Create a Little More Space for LGBTQ Families," *The Atlantic*, April 7, 2019, https://www.theatlantic.com/family/archive/2019/04/lgbtq-mormons-latter-day-saints-apostasy-child-baptism/586630/

Kent Greenawalt, *Religious Convictions and Political Choice* (New York: Oxford University Press, 1988).

Vanessa Grigoriadis, "IVF Coverage Is the Benefit Everyone Wants," *New York Times*, January 30, 2019, https://www.nytimes.com/2019/01/30/style/ivf-coverage.html

Matthew J. Grow, Richard E. Turley, Jr., Steven C. Harper, Scott A. Hales, *Saints: The Story of the Church of Jesus Christ of Latter-day Saints*, vol. 1, *The Standard of Truth: 1815–1846* (Salt Lake City: Church of Jesus Christ of Latter-day Saints, 2018).

Guttmacher Institute, "Induced Abortion in the United States," September 2019, https://www.guttmacher.org/fact-sheet/induced-abortion-united-states

Steven I. Hadju, Manjunath S. Vadmal, "A Note from History: The Use of Tobacco," *Annals of Clinical & Laboratory Science* 40 (2010), 178–181.

Michael L. Hadley, ed., *The Spiritual Roots of Restorative Justice* (New York: SUNY Press, 2001).

Joan Halifax, *Being with Dying: Cultivating Compassion and Fearlessness in the Presence of Death* (Boston: Shambhala, 2009).

Daniel E. Hall, Keith G. Meador, and Harold G. Koenig, "Measuring Religiousness in Health Research: Review and Critique," *Journal of Religion and Health* 47 (2008), 134–163.

Erin Hallstrom, "When Mother's Day is Hard: A 40-Year-Old Single Women's Thoughts, *LDS Living*, May 10, 2017, https://www.ldsliving.com/When-Mother-s-Day-Is-Hard-A-40-Year-Old-Single-Woman-s-Thoughts/s/85290

Arla Halpin, "Finding Answers for Family Caregiving," *The Ensign*, June 2018, https://www.lds.org/ensign/2018/06/finding-answers-for-family-caregiving?lang=eng

Heidi A. Hamann et al., "Interpersonal Responses Among Sibling Dyads Tested for BRCA1/BRCA 2 Genetic Mutations," *Health Psychology* 27 (2008), 100–108.

Steven C. Harper, *Setting the Record Straight: The Word of Wisdom* (Orem, UT: Millennium Press, 2007).

John Harris, *Enhancing Evolution: The Ethical Case for Making Better People* (Princeton, NJ: Princeton University Press, 2007).

Hastings Center, "Belonging: On Disability, Technology, and Community," December 3, 2019, https://www.thehastingscenter.org/the-art-of-flourishing-events-series/

J. B. Haws, *The Mormon Image in the American Mind: Fifty Years of Public Perception* (New York: Oxford University Press, 2013).

Erika Check Hayden, "Should You Edit Your Children's Genes?," *Nature* 530, February 23, 2016, https://www.nature.com/news/should-you-edit-your-children-s-genes-1.19432

"Healing," The Church of Jesus Christ of Latter-day Saints, https://www.churchofjesuschrist.org/study/history/topics/healing?lang=eng

Health Resources and Services Administration, "General FAQ: Bone Marrow Transplantation," https://bloodcell.transplant.hrsa.gov/about/general_faqs/index. html#1990%20number%20tx%20inUS

Gordon Bitner Hinckley, Ezra T. Benson, et al., *Speaking Out on Moral Issues* (West Valley City, UT: Bookcraft, 1998).

Mohammadreza Hojat et al, "The Devil Is in the Third Year: A Longitudinal Study of Erosion of Empathy in Medical School," *Academic Medicine* 84 (2009), 1182–1191.

Jeffrey R. Holland, "The Inconvenient Messiah," *The Ensign,* February 1984, https://www. lds.org/ensign/1984/02/the-inconvenient-messiah?lang=eng&_r=1

C. Christopher Hook, "Transhumanism and Posthumanism," in *Bioethics,* 4th ed., vol. 6, ed. Bruce Jennings (New York: Macmillan Reference, 2014), 3096–3102.

Catherine Hough-Telford, "Vaccine Delays, Refusals, and Patient Dismissals: A Survey of Pediatricians," *Pediatrics* 138 (September 2016), http://pediatrics.aappublications.org/ content/pediatrics/138/3/e20162127.full.pdf.

Humanity+, "Transhumanist Declaration," https://humanityplus.org/philosophy/ transhumanist-declaration/

James Hunter, *Culture Wars: The Struggle to Define America* (New York: Basic Books, 1991).

"Immunization—A Reminder," *The Ensign,* July 1985, https://www.lds.org/ensign/1985/ 07/random-sampler/immunizations-a-reminder?lang=eng

"Immunize Children, Leaders Urge," *The Liahona,* July 1978, https://www.lds.org/ liahona/1978/07/immunize-children-leaders-urge?lang=eng

Independent Lens, "Made in Boise: The Complex World of Surrogacy," http://www.pbs. org/independentlens/films/made-in-boise/

In re Quinlan, 70 N.J. 10, 355 A.2d 647 (NJ 1976).

Jeanne B. Inouye, "Stillborn Children," in *Encyclopedia of Mormonism,* ed. Daniel H. Ludlow (New York: Macmillan, 1992), 1419, https://eom.byu.edu/index.php/ Stillborn_Children

Jacobson v. Massachusetts, 197 U.S. 11 (1905).

Andrew Jameton, "Dilemmas of Moral Distress: Moral Responsibility and Nursing Practice," *AWHONN's Clinical Issues in Perinatal and Women's Health* 4 (1993), 542–551.

Andrew Jameton, "What Moral Distress in Nursing History Could Suggest About the Future of Health Care," *AMA Journal of Ethics,* June 2017, https://journalofethics.ama-assn.org/article/what-moral-distress-nursing-history-could-suggest-about-future-health-care/2017-06

Annie Janvier et al., "Moral Distress in the Neonatal Intensive Care Unit: Caregiver's Experience," *Journal of Perinatology* 27 (2007), 203–208, https://www.nature.com/articles/7211658.pdf?origin=ppub

Thomas Jefferson, "Letter to Danbury Baptists," in *The Sacred Rights of Conscience: Selected Readings on Religious Liberty and Church-State Relations in the American Founding,* ed. Daniel L. Dreisbach, Mark David Hall (Indianapolis: Liberty Fund, 2009), 528.

Thomas Jefferson, "Virginia Statute for Religious Freedom," in *The Sacred Rights of Conscience: Selected Readings on Religious Liberty and Church-State Relations in the American Founding,* ed. Daniel L. Dreisbach, Mark David Hall (Indianapolis: Liberty Fund, 2009), 250–251.

Melinda E. Jennings, "Our Stillborn Baby," *The Ensign,* February 2006, https://www.lds. org/ensign/2006/02/our-stillborn-baby?lang=eng

John Paul II, "*Evangelium Vitae* (The Gospel of Life)," March 1995, http://w2.vatican.va/content/john-paul-ii/en/encyclicals/documents/hf_jp-ii_enc_25031995_evangelium-vitae.html

Johns Hopkins University School of Medicine, Coronavirus Resource Center, https://coronavirus.jhu.edu/

Kirk Johnson, "By Accident, Utah Is Proving an Ideal Genealogical Laboratory," *New York Times,* July 31, 2004, https://www.nytimes.com/2004/07/31/us/by-accident-utah-is-proving-an-ideal-genetic-laboratory.html

Scott A. Johnson, *The Word of Wisdom: Discovering the LDS Code of Health* (Springville, UT: Cedar Fort Press, 2013).

Crystal Nicole Jones, "Our Blessing of Infertility," *The Ensign,* April 2011, https://www.lds.org/ensign/2011/04/faith-and-infertility-expanded/crystal-nicole-jones?lang=eng

James H. Jones, *Bad Blood: The Tuskegee Syphilis Experiment,* 2nd ed. (New York: Free Press, 1993).

Robert P. Jones, *Liberalism's Troubled Search for Equality: Religion and Cultural Bias in the Oregon Physician-Assisted Suicide Debates* (Notre Dame, IN: University of Notre Dame Press, 2007).

Albert R. Jonsen, *The New Medicine and the Old Ethics* (Cambridge, MA: Harvard University Press, 1990), 17–37.

Albert R. Jonsen et al., "Critical Issues in Newborn Intensive Care: A Conference Report and Policy Proposal," *Pediatrics* 55 (1975), 756–765.

Kaiser Health News, "End of Life Care: A Challenge in Terms of Costs and Quality," June 4, 2013, https://khn.org/morning-breakout/end-of-life-care-17/

Leon R. Kass, "Implications of Prenatal Diagnosis for the Quality of, and Right to, Human Life," in *Biomedical Ethics and the* Law, ed. James M. Humber, Robert F. Almeder (New York: Plenum Press, 1976), 313–327.

Patrick Kearon, "Refuge from the Storm," *The Ensign,* May 2016, https://www.lds.org/general-conference/2016/04/refuge-from-the-storm?lang=eng

Caroline Kee, "Here Is What An Abortion at 22 Weeks Is Actually Like," *Buzzfeed,* October 20, 2016, https://www.buzzfeed.com/carolinekee/woman-shares-late-term-abortion-story

Joanne Kenen, "Mitt's Son Has Twins Via Surrogate," *Politico,* May 4, 2012, https://www.politico.com/story/2012/05/romneys-son-has-twin-boys-through-surrogate-075939

Nephi K. Kezerian, "Sick, Blessing the," in *Encyclopedia of Mormonism,* ed. Daniel H. Ludlow (New York: Macmillan, 1992), 1308–1309, https://contentdm.lib.byu.edu/digital/collection/EoM/id/4194

Spencer W. Kimball, "President Kimball Speaks Out on Administration to the Sick," *New Era,* October 1981, https://www.lds.org/new-era/1981/10/president-kimball-speaks-out-on-administration-to-the-sick?lang=eng

Thomas Kirsch, "What Happens If Health Care Workers Stop Showing Up?," *The Atlantic,* March 24, 2020, https://www.theatlantic.com/ideas/archive/2020/03/were-failing-doctors/608662/

Sarah Kliff, "University Reverses Its Decision to Stop Accepting Medicaid," *New York Times,* November 26, 2019, https://www.nytimes.com/2019/11/26/us/medicaid-brigham-young-university.html

Sarah Kliff, "University to Students on Medicaid: Buy Private Coverage or Drop Out," *New York Times,* November 24, 2019, https://www.nytimes.com/2019/11/24/us/medicaid-students-brigham-young.html

Daniel B. Kramer, Bernard Lo, Neal W. Dickert, "CPR in the COVID-19 Era—An Ethical Framework," *New England Journal of Medicine*, May 6, 2020, https://www.nejm.org/doi/full/10.1056/NEJMp2010758

Carol Kuruvilla, "Mormon Mom Has an Abortion Story That Donald Trump Needs to Hear," *Huffington Post*, October 21, 2016, https://www.huffingtonpost.com/entry/mormon-mom-abortion-donald-trump_us_580a565ce4b000d0b1566cf0

Jonathan Lambert, "Measles Cases Mount in Pacific Northwest Outbreak," *NPR: Public Health*, February 8, 2019, https://www.npr.org/sections/health-shots/2019/02/08/692665531/measles-cases-mount-in-pacific-northwest-outbreak

Mitchell Landsberg, "Romney's Conservative Roots Lie in Mormon Faith," *Los Angeles Times,* September 21, 2012, https://www.latimes.com/politics/la-xpm-2012-sep-21-la-na-mormon-conservatism-20120921-story.html

John D. Lantos, "Hooked on Neonatology," *Health Affairs* 20 (September/October 2001), https://www.healthaffairs.org/doi/full/10.1377/hlthaff.20.5.233

John D. Lantos, William L. Meadows, *Neonatal Bioethics: The Moral Challenges of Medical Innovation* (Baltimore: Johns Hopkins University Press, 2006), 74.

Latter-day Saint Charities, "2019 Annual Report" (Salt Lake City: Intellectual Reserve,2020), https://www.churchofjesuschrist.org/bc/content/shared/english/charities/pdf/2019/LDS-Charities-Annual-Report-2019_R11.pdf?lang=eng

"Leader of Mormon Church Looks to Future," *Sunstone* 20:2 (1997), 72.

Anne Leahy, "Mormonism Dysembodied: Placing LDS Theology in Conversation with Disability," *element* 5 (2009), 29–42.

Heidi Ledford, "CRISPR Fixes Embryo Error," *Nature* 548 (August 3, 2017), 13–14.

Bruce Y. Lee, "Should Kids Be Required to Get the HPV Vaccine?," *Forbes*, February 13, 2018, https://www.forbes.com/sites/brucelee/2018/02/13/hpv-vaccine-should-kids-be-required-to-get-the-vaccine/#26b8f914a6df

Don Lefevre, "LDS Church Public Affairs Office: Statement on Human Cloning," March 21, 1997.

Tamar Lewin, "Industry's Growth Leads to Leftover Embryos, and Painful Choices," *New York Times*, June 17, 2015, https://www.nytimes.com/2015/06/18/us/embryos-egg-donors-difficult-issues.html

Barbara Lockhart, "The Body: A Burden or a Blessing?," *The Ensign*, February 1985, https://www.lds.org/ensign/1985/02/the-body-a-burden-or-a-blessing?lang=eng

Ben Lockhart, "The Church of Jesus Christ of Latter-day Saints Joins Utah Coalition," *Deseret News*, August 23, 2018, https://www.deseretnews.com/article/900029171/mormon-church-joins-utah-coalition-saying-no-to-marijuana-ini

Ben Lockhart, "Report: More Utah Youth Getting HPV Vaccine for Cancer Prevention," *Deseret News*, September 22, 2018, https://www.deseretnews.com/article/900033147/report-more-utah-youth-getting-hpv-vaccine-for-cancer-prevention.html

Ben Lockhart, "13 Things Voters Should Know About Utah's Medical Marijuana Initiative," *Deseret News*, September 23, 2018, https://www.deseretnews.com/article/900033184/13-things-voters-should-know-about-utahs-medical-marijuana-initiative.html

Joseph Lynn Lyon, "Word of Wisdom," in *Encyclopedia of Mormonism*, ed. Daniel H. Ludlow (New York: Macmillan, 1992), 991–992, https://contentdm.lib.byu.edu/digital/collection/EoM/id/4353

Val D. MacMurray, Kim Ventura, "Decision Models in Bioethics," in *Perspectives in Mormon Ethics: Personal, Societal, Legal, and Medical*, ed. Donald G. Hill, Jr. (Salt Lake City: Publishers Press, 1983), 253–284.

Mary Madabhushi, "I Feel Like I've Been Here Before," *Mormon Women Project,* November 13, 2017, https://www.mormonwomen.com/2017/11/feel-like-ive/

James Madison, "Memorial and Remonstrance against Religious Assessments," in *The Sacred Rights of Conscience: Selected Readings on Religious Liberty and Church-State Relations in the American Founding,* ed. Daniel L. Dreisbach, Mark David Hall (Indianapolis: Liberty Fund, 2009), 309–313.

Nancy Madsen-Wilkerson, "When One Needs Care, Two Need Help," *The Ensign,* March 2016, https://www.lds.org/ensign/2016/03/when-one-needs-care-two-need-help?lang=eng

Marijuana Policy Project, "Summary of Utah's Medical Cannabis Law," https://www.mpp.org/states/utah/summary-of-utahs-medical-cannabis-law/

Dustin Marlan, "Beyond Cannabis: Psychedelic Decriminalization and Social Justice," *Lewis and Clark Law Review* 23 (2019), 851–892.

Elaine S. Marshall et al., "'This Is a Spiritual Experience': Perspectives of Latter-day Saint Families Living with a Child with Disabilities," *Qualitative Health Research* 13 (2003), 57–76.

Elaine S. Marshall, "The Power of God to Heal," in *Joseph & Hyrum: Leading as One,* ed. Mark E. Mendenhall et al. (Salt Lake City: Deseret Book, 2010), 165–184.

Adam Martin, "You Can Now Sell Your Bone Marrow for $3,000," *The Atlantic,* December 2, 2011, https://www.theatlantic.com/national/archive/2011/12/you-can-now-sell-your-bone-marrow-3000/334788/

Emily Martin, *The Woman in the Body: A Cultural Analysis of Reproduction* (Boston: Beacon Press, 2001).

David Masci, "Human Enhancement: The Scientific and Ethical Dimensions of Striving for Perfection," Pew Research Center, July 26, 2016, http://www.pewinternet.org/essay/human-enhancement-the-scientific-and-ethical-dimensions-of-striving-for-perfection/

Patrick Q. Mason, "The Possibilities of Mormon Peacebuilding," *Dialogue: A Journal of Mormon Thought* 37 (2004), 12–45.

Patrick Q. Mason, ed., *Directions for Mormon Studies in the Twenty-First Century* (Salt Lake City: University of Utah Press, 2016).

R. Matesanz et al., "How Spain Reached 40 Deceased Organ Donors per Million Population," January 9, 2017, https://onlinelibrary.wiley.com/doi/full/10.1111/ajt.14104

Thomas H. Maugh II, "The Man Who Makes the Pieces of the Puzzle Fit," *Los Angeles Times,* March 20, 1994, http://articles.latimes.com/1994-09-20/news/ls-40953_1_breast-cancer-gene

Heather May, "What Science Says About Mormonism's Health Code," *Salt Lake Tribune,* October 6, 2012, http://archive.sltrib.com/article.php?id=54897327&itype=CMSID

William F. May, *The Physician's Covenant: Images of the Healer in Medical Ethics* (Louisville, KY: Westminster John Knox Press, 2001).

William F. May, "Religious Justifications for Donating Body Parts," *Hastings Center Report* 15 (February 1985), 38–42.

William F. May, *Testing the Medical Covenant: Active Euthanasia and Health Care Reform* (Grand Rapids: Eerdmans, 1996).

The Mayo Clinic, "Amniocentesis," https://www.mayoclinic.org/tests-procedures/amniocentesis/about/pac-20392914

Matthew McBride, James Goldberg, eds., *Revelations in Context: The Stories Behind the Sections of the Doctrine and Covenants* (Salt Lake City: Intellectual Reserve, 2016).

Bruce R. McConkie, "The Salvation of Little Children," *The Ensign,* April 1977, https://www.lds.org/ensign/1977/04/the-salvation-of-little-children?lang=eng

Mark L. McConkie, R. Wayne Boss, "'I Teach Them Correct Principles and They Govern Themselves': The Leadership Genius of the Mormon Prophet Abstract," *International Journal of Public Administration* 28 (2005), 437–463.

J. E. McCulloch, *Home: The Savior of Civilization* (Washington, D.C.: Southern Cooperative League, 1924).

Laurie McGinley, "HPV-Related Cancers Are Rising. So Are Vaccine Rates, Just Not Fast Enough," *Washington Post,* August 23, 2018, https://www.washingtonpost.com/news/to-your-health/wp/2018/08/23/hpv-related-cancer-rates-are-rising-so-are-vaccine-rates-just-not-fast-enough/?utm_term=.26a38bef2943

Chephra McKee, Kristen Bohannon, "Exploring the Reasons Behind Parental Refusals of Vaccines," *Journal of Pediatric Pharmacology and Therapeutics* 21 (March–April 2016), 104–109, https://www.ncbi.nlm.nih.gov/pmc/articles/PMC4869767/

Alan Meisel, *The Right to Die* (New York: John Wiley and Sons, 1989), 436–437.

Bettina Meiser, "Psychological Impact of Genetic Testing for Huntington's Disease: An Update of the Literature," *Journal of Neurology, Neurosurgery, and Psychiatry* 69 (2000), 574–578.

Melissa Merrill, "Faith and Infertility," *The Ensign,* April 2011, https://www.lds.org/ensign/2011/04/faith-and-infertility?lang=eng

Ray A. Merrill, "Tobacco Smoking and Cancer in Utah," *Dialogue: A Journal of Mormon Thought* 35 (2002), 73–77.

Ray A. Merrill, Hala N. Mandanat, Joseph L. Lyon, "Active Religion and Health in Utah," *Dialogue: A Journal of Mormon Thought* 35 (2002), 78–90.

Christopher Meyers, "Cruel Choices: Autonomy and Critical-Care Decision-Making," *Bioethics* 18 (2004), 104–119.

John Stuart Mill, "On Liberty," in *The English Philosophers: From Bacon to Mill,* ed. Edwin A. Burtt (New York: Modern Library, 1939), 949–1041.

Adam S. Miller, "Suffering, Agency, and Redemption: Mormonism and Transhumanism," in *Transhumanism and the Body: The World Religions Speak,* ed. Calvin Mercer, Derek F. Maher (New York: Palgrave Macmillan, 2014), 121–136.

David Gibbes Miller, Rebecca Dresser, Scott Y. H. Kim, "Advance Euthanasia Directives: A Controversial Case and Its Ethical Implications," *Journal of Medical Ethics* 45 (2019), 84–89.

Millman Research Report, "2017 U.S. Organ and Tissue Transplant Cost Estimates and Discussion," August 2017, http://us.milliman.com/uploadedFiles/insight/2017/2017-Transplant-Report.pdf

Richard A. Mimer, "Blood Transfusions," in *Encyclopedia of Mormonism,* ed. Daniel H. Ludlow (New York: Macmillan Reference, 1992), 131–132, https://contentdm.lib.byu.edu/digital/collection/EoM/id/5527

Wayne A. Mineer, "Organ Transplants and Donation," in *Encyclopedia of Mormonism,* ed. Daniel H. Ludlow (New York: Macmillan Reference, 1992), 1051–1052, https://contentdm.lib.byu.edu/digital/collection/EoM/id/4024

James C. Mohr, *Abortion in America: The Origins and Evolution of National Policy* (New York: Oxford University Press, 1978).

Thomas S. Monson, "Your Jericho Road," *The Ensign*, May 1977, https://www.lds.org/general-conference/1977/04/your-jericho-road?lang=eng

Jonathan D. Moreno, "The End of the Great Bioethics Compromise," *Hastings Center Report* 35 (January–February 2005), 14–15.

Mormon Transhumanist Association, "Mormon Transhumanist Affirmation," https://transfigurism.org/mormon-transhumanist-affirmation

Mormon Transhumanist Association, "The Basics of Mormon Transhumanism," https://transfigurism.org/primers/1

Mormon Transhumanist Association, "What Is Mormon Transhumanism?," https://transfigurism.org/primer-mormon-transhumanism

Mormon Women for Ethical Government, *The Little Purple Book: MWEG Essentials* (Salt Lake City: By Common Consent Press, 2018).

"Mother in Heaven," https://www.churchofjesuschrist.org/study/manual/gospel-topics-essays/mother-in-heaven?lang=eng

Kristen Moulton, "Large Families Make Utah Fertile Ground for Genetics Researchers," *Los Angeles Times*, October 25, 1998, http://articles.latimes.com/1998/oct/25/local/me-36126

Yascha Mounk, "The Triage Decisions Facing Italian Doctors," *The Atlantic*, March 11, 2020, https://www.theatlantic.com/ideas/archive/2020/03/who-gets-hospital-bed/607807/

Jonathan Muraskas, Kayhan Parsi, "The Cost of Saving the Tiniest Lives: NICUs Versus Prevention," *AMA Journal of Ethics* 10 (October 2008), 655–658, https://journalofethics.ama-assn.org/article/cost-saving-tiniest-lives-nicus-versus-prevention/2008-10

Thomas H. Murray, "Gifts of the Body and the Needs of Strangers," *Hastings Center Report* 17 (April 1987), 30–38.

National Academy of Medicine, *Physician Assisted Death: Scanning the Landscape* (Washington, D.C.: National Academies Press, 2018).

National Bioethics Advisory Commission, *Ethical Issues in Human Stem Cell Research*, 1999, https://bioethicsarchive.georgetown.edu/nbac/execsumm.pdf

National Commission for the Protection of Human Subjects of Biomedical and Behavioral Research, "The Belmont Report," April 1979, https://www.hhs.gov/ohrp/regulations-and-policy/belmont-report/read-the-belmont-report/index.html

National Institute on Drug Abuse, "Overdose Death Rates," August 2018, https://www.drugabuse.gov/related-topics/trends-statistics/overdose-death-rates.

Jaime L. Natoli et al., "Pre-Natal Diagnosis of Down Syndrome: A Systematic Review of Termination Rates (1995–2011)," *Prenatal Diagnosis*, March 14, 2012, https://obgyn.onlinelibrary.wiley.com/doi/full/10.1002/pd.2910

Geoffrey S. Nelson, "Mormons and Compulsory Vaccination," *MormonPress*, March 30, 2015, http://www.mormonpress.com/mormon_vaccination;

Russell M. Nelson, "Abortion: An Assault on the Defenseless," *The Ensign*, October 2008, https://www.lds.org/ensign/2008/10/abortion-an-assault-on-the-defenseless?lang=eng

Russell M. Nelson, *The Gateway We Call Death* (Salt Lake City: Deseret Book, 1995).

Russell M. Nelson, "Opening the Heavens for Help," April 2020 General Conference, https://www.churchofjesuschrist.org/study/general-conference/2020/04/37nelson?lang=eng

Russell M. Nelson, "Reverence for Life," *The Ensign*, November 1985, https://www.lds.org/general-conference/1985/04/reverence-for-life?lang=eng

Russell M. Nelson, "The Second Great Commandment," October 2019, https://www. churchofjesuschrist.org/study/general-conference/2019/10/46nelson?lang=eng

Russell M. Nelson, "We Are Children of God," *The Ensign*, November 1998, https://www.churchofjesuschrist.org/study/general-conference/1998/10/ we-are-children-of-god?lang=eng

Linda King Newell, "A Gift Given, A Gift Taken: Washing, Anointing, and Blessing the Sick Among Mormon Women," *Sunstone* 6 (1981), 16–25.

Quincy D. Newell, Eric F. Mason, *New Perspectives in Mormon Studies: Creating and Crossing Boundaries* (Norman: University of Oklahoma Press, 2013).

Jerreld L. Newquist, ed., *Gospel Truth: Discourses and Writings of President George Q. Cannon*, vol. 2 (Salt Lake City: Deseret Book, 1974).

Newsroom, "Abortion," https://newsroom.churchofjesuschrist.org/official-statement/ abortion

Newsroom, "Church Encourages Principled Approach to Health Care Coverage for Needy Utahns," December 5, 2014, https://newsroom.churchofjesuschrist.org/article/ church-encourages-principled-approach-health-care-coverage-needy-utahns

Newsroom, "Church Expresses Support for 'Fairness for All' Approach," May 13, 2019, https://newsroom.churchofjesuschrist.org/article/church-expresses-support-fairness-for-all-approach

Newsroom, "Church Family History Records Lead to Groundbreaking Genetic Research," July 18, 2008, https://newsroom.churchofjesuschrist.org/article/church-family-history-records-lead-to-groundbreaking-genetic-research

Newsroom, "The Church of Jesus Christ Supports the Federal Fairness for All Act," December 6, 2019, https://newsroom.churchofjesuschrist.org/article/federal-fairness-for-all-support-december-2019

Newsroom, "Church Leaders Counsel Members After Supreme Court Same-Sex Marriage Decision," June 30, 2015, https://newsroom.churchofjesuschrist.org/article/ top-church-leaders-counsel-members-after-supreme-court-same-sex-marriage-decision

Newsroom, "Church Members Encouraged to Assist Refugees," October 29, 2015, https://newsroom.churchofjesuschrist.org/article/church-members-encouraged-assist-refugees

Newsroom, "Church Supports Principles of *Utah Compact* on Immigration," November 11, 2010, https://newsroom.churchofjesuschrist.org/article/church-supports-principles-of-utah-compact-on-immigration

Newsroom, "The Divine Institution of Marriage," July 15, 2019, https://newsroom. churchofjesuschrist.org/article/the-divine-institution-of-marriage

Newsroom, "Deferred Action for Childhood Arrivals (DACA) Statement," January 26, 2018, https://newsroom.churchofjesuschrist.org/article/daca-statement-january-2018

Newsroom, "Embryonic Stem Cell Research," October 10, 2018, https://newsroom. churchofjesuschrist.org/official-statement/embryonic-stem-cell-research

Newsroom, "Euthanasia and Prolonging Life," January 10, 2017, https://newsroom. churchofjesuschrist.org/official-statement/euthanasia-and-prolonging-life

Newsroom, "First Presidency Statement on Utah Marijuana Initiative," https:// newsroom.churchofjesuschrist.org/article/first-presidency-statement-on-utah-marijuana-initiative

Newsroom, "Health Practices," August 1, 2018, https://newsroom.churchofjesuschrist. org/article/health-practices

Newsroom, "I Was a Stranger: Refugee Relief," March 26, 2016, https://newsroom.churchofjesuschrist.org/article/church-leaders-encourage-women-of-the-church-serve-refugees
https://www.mormonnewsroom.org/article/church-leaders-encourage-women of the church-serve-refugees

Newsroom, "Major Medical Findings Aided by Church History Records," February 5, 2019   https://newsroom.churchofjesuschrist.org/ldsnewsroom/eng/news-releases-stories/major-medical-findings-aided-by-church-family-history-records

Newsroom, "The Mormon Ethic of Community," October 16, 2012, https://newsroom.churchofjesuschrist.org/article/the-mormon-ethic-of-community

Newsroom, "Mormon Physicians Pioneer Research in Genetics," February 5, 2019, https://newsroom.churchofjesuschrist.org/ldsnewsroom/eng/news-releases-stories/mormon-physicians-pioneer-research-in-genetics

Newsroom, "Spreading the Word of Immunization," https://www.churchofjesuschrist.org/media/image/information-statistics-graphics-lds-church-aa8cc67?lang=eng

Newsroom, "Statement on the Word of Wisdom," August 15, 2019, https://newsroom.churchofjesuschrist.org/article/statement-word-of-wisdom-august-2019

Newsroom, "Supreme Court Decision Will Not Alter Doctrine on Marriage," June 26, 2015, https://newsroom.churchofjesuschrist.org/article/supreme-court-decision-will-not-alter-doctrine-on-marriage

Newsroom, "Utah Medical Marijuana Initiative," May 11, 2018, https://newsroom.churchofjesuschrist.org/article/marijuana-analysis

Newsroom Canada, "Euthanasia and Prolonging Life," February 10, 2018, https://canada.lds.org/euthanasia-and-prolonging-life

Reinhold Niebuhr, *An Interpretation of Christian Ethics* (New York: Seabury, 1979).

Nikolas T. Nikas, Dorinda C. Bordlee, Madeline Moriera, "Determination of Death and the Dead Donor Rule: A Survey of the Current Law on Brain Death," *Journal of Medicine and Philosophy* 41:3 (June 2016), 237–256.

Nicole Nixon, "Patient Advocates May Sue over LDS Church Involvement in Utah Medical Marijuana Laws," *NPR Utah*, November 15, 2018, http://www.kuer.org/post/patient-advocates-may-sue-over-lds-church-involvement-utah-medical-marijuana-laws#stream/0

NOLO, "Death with Dignity in Utah," January 10, 2020, https://www.nolo.com/legal-encyclopedia/death-with-dignity-utah.html

Karen Norrgard, "Ethics of Genetic Testing: Medical Insurance and Genetic Discrimination," *Nature Education* 1 (2008), 90.

Dallin H. Oaks, "Good, Better, Best," The Church of Jesus Christ of Latter-day Saints, October 2007, https://www.churchofjesuschrist.org/study/general-conference/2007/10/good-better-best?lang=eng

Dallin H. Oaks, "The Gospel Culture," *The Ensign*, March 2012, https://www.lds.org/ensign/2012/03/the-gospel-culture?lang=eng&_r=1

Dallin H. Oaks, "The Great Plan of Happiness," The Church of Jesus Christ of Latter-day Saints, October 1993, https://www.lds.org/general-conference/1993/10/the-great-plan-of-happiness?lang=eng

Dallin H. Oaks, "Healing the Sick," *The Ensign*, May 2010, https://www.lds.org/general-conference/2010/04/healing-the-sick?lang=eng

Dallin H. Oaks, "Religious Values and Public Policy," *The Ensign*, October, 1992, https://www.lds.org/ensign/1992/10/religious-values-and-public-policy?lang=eng&_r=1

Dallin H. Oaks, "Truth and the Plan," The Church of Jesus Christ of Latter-day Saints, https://www.lds.org/general-conference/2018/10/truth-and-the-plan?lang=eng

Dallin H. Oaks, "Weightier Matters," *The Ensign*, January 2001, https://www.lds.org/en-sign/2001/01/weightier-matters?lang=eng.

Jacqueline K. Olive et al., "The State of the Antivaccine Movement in the United States: A Focused Examination of Nonmedical Exemptions in States and Counties," *PLOS Medicine*, June 12, 2018, https://doi.org/10.1371/journal.pmed.1002578;

Oregon Public Broadcasting, "A Question of Genes: Inherited Risks," https://repository.library.georgetown.edu/handle/10822/549317

David Orentlicher, "The Physician's Duty to Treat During Pandemics," *American Journal of Public Health* 108 (2018), 1459–1461.

Organ Procurement and Transplantation Network, "Ethical Principles in the Allocation of Human Organs," June 2015, https://optn.transplant.hrsa.gov/resources/ethics/ethical-principles-in-the-allocation-of-human-organs/

Organ Procurement and Transplantation Network, "An Evaluation of the Ethics of Presumed Consent," https://optn.transplant.hrsa.gov/resources/ethics/an-evaluation-of-the-ethics-of-presumed-consent/

Organ Procurement and Transplantation Network, "National Data," https://optn.transplant.hrsa.gov/data/view-data-reports/national-data/

Boyd K. Packer, "Our Moral Environment," The Church of Jesus Christ of Latter-day Saints, April 1992, https://www.lds.org/general-conference/1992/04/our-moral-environment?lang=eng

Tyler Pager, "'Monkey, Rat and Pig DNA': How Misinformation Is Driving the Measles Outbreak Among Ultra-Orthodox Jews," *New York Times,* April 10, 2019, https://www.nytimes.com/2019/04/09/nyregion/jews-measles-vaccination.html

Benjamin E. Park, "Salvation Through a Tabernacle: Joseph Smith, Parley P. Pratt, and Early Mormon Theologies of Embodiment," *Dialogue: A Journal of Mormon Thought* 43 (2010), 1–44.

Estaban Parra, Karl Baker, "Doctors May Overrule Coronavirus' Patients End-of-Life Treatment Wishes," *Delaware News Journal*, https://www.delawareonline.com/story/news/2020/04/13/christianacare-doctors-can-overrule-coronavirus-patients-end-life-wishes/5120699002/

Ardis E. Parshall, "Even with Proven Smallpox Vaccines, 19th-Century Utahns Balked," *Salt Lake Tribune*, December 2, 2011, http://archive.sltrib.com/article.php?id=53038477&itype=CMSID

Erich Robert Paul, "Science and Religion," in *The Encyclopedia of Mormonism*, ed. Daniel H. Ludlow (New York: Macmillan, 1992), 1270–1272, https://contentdm.lib.byu.edu/digital/collection/EoM/id/4169

David L. Paulsen, Martin Pulido, "'A Mother There': A Survey of Historical Teachings about Mother in Heaven," *BYU Studies Quarterly* 50 (2011), 70–97.

PBS, "Frontline: The Vaccine War," April 27, 2010, https://www.pbs.org/wgbh/frontline/film/vaccines/

Edmund D. Pellegrino, "Doctors Must Not Kill," in *Euthanasia: The Good of the Patient, The Good of Society*, ed. Robert I. Misbin (Frederick, MD: University Publishing Group, 1991), 27–42.

Joshua J. Perkey, "Loss and Childlessness: Finding Hope amid the Pain," *The Ensign*, October 2012, https://www.lds.org/ensign/2012/10/loss-and-childlessness-finding-hope-amid-the-pain?lang=eng

Paul H. Peterson, Ronald W. Walker, "Brigham Young's Word of Wisdom Legacy," *BYU Studies Quarterly* 42 (2003), 29–64.

Taylor Petrey, "A Mormon Leader Signals New Openness on Transgender Issues," *Slate*, February 13, 2015, https://slate.com/human-interest/2015/02/mormons-and-transgender-elder-dallin-h-oaks-says-the-lds-church-is-open-to-rethinking-its-teachings.html

Taylor Petrey, "Mormons and the Politics of Mother's Day," *Patheos*, May 12, 2013, https://www.patheos.com/blogs/peculiarpeople/2013/05/mormons-and-the-politics-of-mothers-day/

Pew Research Center, "Reactions to the 'Mormon Moment,'" January 12, 2012, https://www.pewforum.org/2012/01/12/mormons-in-america-mormon-moment/

Mary Pflum, "Egg Freezing 'Startups' Have Wall Street Talking—and Traditional Fertility Doctors Worried,"*NBCNews,*March4,2019,https://www.nbcnews.com/health/features/egg-freezing-startups-have-wall-street-talking-traditional-fertility-doctors-n978526

Michael Pollan, "The Trip Treatment," *The New Yorker*, February 9, 2015, https://www.newyorker.com/magazine/2015/02/09/trip-treatment

Allison Pond, "Myths About Religion and Organ Donation Cause Hesitation," *Deseret News*, August 5, 2001, https://www.deseretnews.com/article/700168701/Myths-about-religion-and-organ-donation-cause-hesitation.html

Thaddeus Mason Pope, "Medical Aid in Dying: When Legal Safeguards Become Burdensome Obstacles," *ASCO Post*, December 25, 2017, https://www.ascopost.com/issues/december-25-2017/medical-aid-in-dying-when-legal-safeguards-become-burdensome-obstacles/

Allen Porter, "Bioethics and Transhumanism," *Journal of Medicine and Philosophy* 42 (June 2017), 237–260.

President's Council on Bioethics, "Organ Transplantation: Ethical Dilemmas and Policy Choices," https://bioethicsarchive.georgetown.edu/pcbe/background/org_transplant.html

Melissa Proctor, "Bodies, Babies, and Birth Control," *Dialogue: A Journal of Mormon Thought* 36 (2003), 159–175.

Jonathan D. Quick, Heidi Larson, "The Vaccine-Autism Myth Started 20 Years Ago. Here's Why It Still Endures Today," *Time*, February 28, 2018, http://time.com/5175704/andrew-wakefield-vaccine-autism/

James Rachels, "Active and Passive Euthanasia," *New England Journal of Medicine* 292 (January 9, 1975), 78–80.

Luke Ramseth, "LDS Church Issues Statement Opposing Medical Marijuana Ballot Initiative," *Salt Lake Tribune*, April 12, 2018, https://www.sltrib.com/news/health/2018/04/10/lds-church-issues-statement-opposing-medical-marijuana-ballot-initiative-which-a-majority-of-utah-voters-supports/

Paul Ramsey, *The Patient as Person: Explorations in Medical Ethics* (New Haven, CT: Yale University Press, 1970).

Abbas Rana, Elizabeth Louise Godfrey, "Outcomes in Solid Organ Transplantation: Success and Stagnation," *Texas Heart Institute Journal* 48 (2019), 75–76.

Rayna Rapp, "Refusing Prenatal Diagnosis: The Meanings of Bioscience in a Multicultural World," *Science, Technology, and Human Values* 23 (Winter 1998), 45–70.

Sumathi Reddy, "Fertility Clinics Let You Select Your Baby's Sex," *Wall Street Journal*, August 17, 2015, http://www.wsj.com/article_email/fertility-clinics-let-you-select-your-babys-sex-1439833091-lMYQjAxMTI1MDE1ODkxMjgzWj

Lindsey Redfern, "5 Things Couples Dealing with Infertility in Your Ward Wish You Knew," *LDS Living,* May 28, 2019, http://www.ldsliving.com/5-Things-Couples-Dealing-With-Infertility-In-Your-Ward-Wish-You-Knew/s/80393

T. R. Reid, "How We Spend $3,400,000,000,000," *The Atlantic,* June 15, 2017, https://www.theatlantic.com/health/archive/2017/06/how-we-spend-3400000000000/530355/

"Religious Freedom," The Church of Jesus Christ of Latter-day Saints, https://www.lds.org/religious-freedom?lang=eng

Dale G. Renlund, "Family History and Temple Work: Sealing and Healing," *The Ensign,* May 2018, https://www.lds.org/general-conference/2018/04/family-history-and-temple-work-sealing-and-healing?lang=eng

Dale G. Renlund, "Preserving the Heart's Mighty Change," *The Ensign,* November 2011, https://www.lds.org/general-conference/2009/10/preserving-the-hearts-mighty-change?lang=eng

David B. Resnik, "Genetics and Personal Responsibility for Health," *New Genetics and Society* 33 (June 30, 2014), 113–125.

Susan M. Reverby, "More Than Fact and Fiction: Cultural Memory and the Tuskegee Syphilis Study," *Hastings Center Report* 31 (September–October 2001), 22–28.

Jana Riess, "How Welcoming Are Mormons to People with Disabilities?," *Religion News Service,* October 17, 2016, https://religionnews.com/2016/10/17/how-welcoming-are-mormons-to-people-with-disabilities-at-new-hartford-temple-some-room-for-improvement/

Jana Riess, "The Incredible Shrinking Mormon American Family," *Religion News Service,* June 15, 2019, https://religionnews.com/2019/06/15/the-incredible-shrinking-mormon-american-family/

Jana Riess, *The Next Mormons: How Millennials Are Changing the LDS Church* (New York: Oxford University Press, 2019).

John A. Robertson, *Children of Choice: Freedom and the New Reproductive Technologies* (Princeton, NJ: Princeton University Press, 1994).

Roe v. Wade, 410 U.S. 113 (1973).

William J. Rorabaugh, "Alcohol in America," *OAH Magazine of History* 6 (1991), 17–19.

William J. Rorabaugh, *The Alcoholic Republic: An American Tradition* (New York: Oxford University Press, 1979).

Richard Rorty, *Philosophy and Social Hope* (New York: Penguin Books, 1999).

Aaron Saguil, Karen Phelps, "The Spiritual Assessment," *American Family Physician* 86 (2012), 546–550.

Cecil O. Samuelson, Jr., "Medical Practices," in *Encyclopedia of Mormonism,* ed. Daniel H. Ludlow (New York: Macmillan, 1992), 875, https://contentdm.lib.byu.edu/digital/collection/EoM/id/3912

Cecil O. Samuelson, Jr., "I Am Thinking of Donating Some Organs for Transplantation. Am I Wrong in Wanting to Do So?" *Ensign,* February 1988, https://www.lds.org/ensign/1988/02/i-have-a-question/am-i-wrong-in-wanting-to-donate-organs-for-transplantation?lang=eng&clang=ase

Michael Sandel, *The Case Against Perfection* (Cambridge, MA: The Belknap Press of Harvard University Press, 2007).

Kory Scadden, "Caring for Loved Ones in Their Final Hours," *LDS Living,* http://www.ldsliving.com/Caring-for-Loved-Ones-in-Their-Final-Hours/s/75170

Gilbert W. Scharffs, "The Case Against Easier Abortion Laws," *The Ensign*, August 1972, https://www.lds.org/ensign/1972/08/the-case-against-easier-abortion-laws?lang=eng

David Schenck, Larry R. Churchill, *Healers: Extraordinary Clinicians at Work* (New York: Oxford University Press, 2012).

Lawrence J. Schneiderman, Nancy S. Jecker, Albert R. Jonsen, "Medical Futility: Response to Critiques," *Annals of Internal Medicine* 125 (1996), 669–674.

Bryan Schott, "Majority of Utahns Support Right-to-Die Legislation," *UtahPolicy.com*, December 10, 2015, https://utahpolicy.com/index.php/features/today-at-utah-policy/7910-poll-majority-of-utahns-support-right-to-die-legislation

Steven A. Schroeder, "We Can Do Better: Improving the Health of the American People," *New England Journal of Medicine* 357 (2007), 1221–1228.

Gene A. Sessions, Craig J. Oberg, eds., *The Search for Harmony: Essays on Science and Mormonism* (Salt Lake City: Signature Books, 1993).

Anthony Shaw, "Dilemmas of 'Informed Consent' in Children," *New England Journal of Medicine* 289 (October 25, 1973), 885–890.

Rachel Sheffield, "Finding Peace from Stories of Infertility in the Bible," *The Ensign*, September 2018, https://www.lds.org/ensign/2018/09/finding-peace-from-stories-of-infertility-in-the-bible?lang=eng

Mark Lawrence Shrad, "Does Down Syndrome Justify Abortion?," *New York Times*, September 4, 2015, https://www.nytimes.com/2015/09/04/opinion/does-down-syndrome-justify-abortion.html

Connor Simpson, "Google Wants to Cheat Death," *The Atlantic*, September 18, 2013, https://www.theatlantic.com/technology/archive/2013/09/google-wants-cheat-death/310943/

Devyn Smith, "The Human Genome Project, Modern Biology, and Mormonism: A Viable Marriage?," *Dialogue: A Journal of Mormon Thought* 35 (2002), 61–71.

Gregory A. Smith, "A Growing Share of Americans Say It's Not Necessary to Believe in God to Be Moral," Pew Research Center, October 16, 2017, http://www.pewresearch.org/fact-tank/2017/10/16/a-growing-share-of-americans-say-its-not-necessary-to-believe-in-god-to-be-moral/

Joseph Fielding Smith, ed., *Teachings of the Prophet Joseph Smith* (Salt Lake City: Deseret Book, 1977).

Leah M. Smith et al., "Effect of Human Papillomavirus (HPV) Vaccine on Clinical Indicators of Sexual Behavior Among Adolescent Girls: The Ontario Grade 8 HPV Vaccine Cohort Study," *CMAJ* 187:2 (February 03, 2015), E74–E81, http://www.cmaj.ca/content/187/2/E74

Morgan Smith, "Lawsuit Drops Claim Mormon Church Swayed Medical Pot Changes," AP News, May 3, 2019, https://apnews.com/bac195211378427080568acd03f14954

Matthew Soerens, Jenny Yang, Leith Anderson, *Welcoming the Stranger: Justice, Compassion, and Truth in the Immigration Debate* (Downers Grove, IL: IVP Books, 2016).

A. D. Sorenson, "Zion," in *Encyclopedia of Mormonism*, ed. Daniel H. Ludlow (New York: Macmillan, 1992), 1625–1626, https://contentdm.lib.byu.edu/digital/collection/EoM/id/4373

Michael Specter, "How the DNA Revolution Is Changing Us," *National Geographic*, August 2016, https://www.nationalgeographic.com/magazine/2016/08/dna-crispr-gene-editing-science-ethics/

Carolynn R. Spencer, "Learning to Cope with Infertility," *The Ensign*, June 2012, https://www.lds.org/ensign/2012/06/learning-to-cope-with-infertility?lang=eng&_r=1

Carol Mason Spicer, "Nature and Role of Codes and Other Ethics Directives," *Encyclopedia of Bioethics*, 3rd ed., vol. 5, ed. Stephen G. Post (New York: Macmillan Reference USA, 2004), 2621–2629.

Peggy Fletcher Stack, "Surprise! The LDS Church Can Be Seen as More 'Pro-Choice' Than "Pro-Life' on Abortion," *Salt Lake Tribune*, June 2, 2019, https://www.sltrib.com/religion/2019/06/01/surprise-lds-church-can/

Delbert L. Stapley, "Why Are Medical Doctors So Important If the Priesthood Has the Power to Cure Diseases?," *New Era*, March 1971, https://www.lds.org/new-era/1971/03/q-and-a-questions-and-answers/why-are-medical-doctors-so-important-if-the-priesthood-has-the-power-to-cure-diseases?lang=eng

Jonathan A. Stapley, "'Pouring in Oil': The Development of the Modern Mormon Healing Ritual," in *By Our Rites of Worship: Latter-day Saint Views on Ritual in Scripture, History, and Practice*, ed. Daniel L. Belnap (Provo, UT: Religious Studies Center; Salt Lake City: Deseret Book, 2013), 283–316.

Jonathan A. Stapley, Kristine Wright, "Female Ritual Healing in Mormonism," *Journal of Mormon History* 37 (2011), 1–85.

Jonathan A. Stapley, Kristine Wright, "The Forms and the Power: The Development of Mormon Ritual Healing to 1847," *Journal of Mormon History,* 35 (2009), 42–87.

"Statement of The Church of Jesus Christ of Latter-day Saints on the Question of Physician-Assisted Suicide," May 29, 1997.

State of Utah, "Healthy Utah," December 2014, https://ccf.georgetown.edu/wp-content/uploads/2014/12/healthyutahplan.pdf

Rachel Hunt Steenblik, "A Mormon Ethic of Food," *Dialogue: A Journal of Mormon Thought* 48 (2015), 59–74.

William T. Stephenson, "Cancer, Nutrition, and the Word of Wisdom," *The Ensign*, July 2008, https://www.lds.org/ensign/2008/07/cancer-nutrition-and-the-word-of-wisdom-one-doctors-observations?lang=eng

Barbara Stoll et al., "Trends in Care Practices, Morbidity, and Mortality of Extremely Preterm Neonates, 1993–2012," *Journal of the American Medical Association* 314 (2015), 139–151, https://jamanetwork.com/journals/jama/fullarticle/2434683

Jason Swensen, "The Church Has Partnered with the Red Cross Nearly 100 Years," *Church News*, March 26, 2015, https://www.lds.org/church/news/the-church-has-partnered-with-the-red-cross-nearly-100-years-?lang=eng

Jason Swensen, "Church Says Yes to Regulated Medical Marijuana but No to Utah Initiative," September 20, 2018, https://www.lds.org/church/news/church-says-yes-to-regulated-medical-marijuana-but-no-to-utah-initiative?lang=eng

Stephen C. Taysom, *Shakers, Mormons, and Religious Worlds: Conflicting Visions, Contested Boundaries* (Bloomington: Indiana University Press, 2017).

Melanie Thernstrom, *The Pain Chronicles: Cures, Myths, Mysteries, Prayers, Diaries, Brain Scans, Healing, and the Science of Suffering* (New York: Farrar, Straus and Giroux, 2010).

William M. Timmins, "On Death and Dying," *The Ensign*, April 1989, https://www.lds.org/ensign/1989/04/on-death-and-dying?lang=eng

Alexis de Tocqueville, *Democracy in America*, ed. J. P. Mayer (New York: Doubleday & Company, 1969).

"Toward a Twenty-First-Century *Jacobson v. Massachusetts*," *Harvard Law Review* 121 (2008), 1820–1842.

Pamela Mahoney Tsigdonis, "The Big IVF Add-On Racket," *New York Times*, December 13, 2019, https://www.nytimes.com/2019/12/12/opinion/ivf-add-ons.html?smid= nytcore-ios-share

United Network for Organ Sharing, "Data," Accessed November 10, 2017, https://unos. org/data/

United Network for Organ Sharing, "Transplant Costs," Accessed November 10, 2017, https://transplantliving.org/financing-a-transplant/transplant-costs/

United Network for Organ Sharing, "Transplant Trends," Accessed November 10, 2017, https://unos.org/data/transplant-trends/

United States Census Bureau, "Health Insurance Coverage in the United States: 2018," November 8, 2019, https://www.census.gov/library/publications/2019/demo/p60-267. html

United States Commission on Civil Rights, *Peaceful Coexistence: Reconciling Nondiscrimination Principles with Civil Liberties* (Washington, D.C.: USCCR, 2016).

United States Department of Health and Human Services, "Organ Donation Statistics," Accessed November 10, 2017, https://www.organdonor.gov/statistics-stories/statistics.html

Utah State Legislature, "HB 93: End of Life Prescription Provisions," https://le.utah.gov/ ~2020/bills/static/HB0093.html

"Vaccines: Poll," *LDS Living*, Accessed September 25, 2017, http://www.ldsliving.com/- Poll-Vaccines/s/65438

Wouter van Beek, "Covenants," in *The Encyclopedia of Mormonism*, ed. Daniel H. Ludlow (New York: Macmillan, 1992), 331–333, https://contentdm.lib.byu.edu/digital/collec-tion/EoM/id/5640

"Vaping, Coffee, Tea, and Marijuana," *New Era*, August 2019, https://www. churchofjesuschrist.org/study/new-era/2019/08/vaping-coffee-tea-and-marijuana? lang=eng

Viewpoint, "Help the Refugees Among Us," *Church News*, October 27, 2015, https://www. lds.org/church/news/viewpoint-help-the-refugees-among-us?lang=eng&_r=1

Danielle B. Wagner, "Abortion: How President Nelson and Other Church Leaders' Teachings Provide Hope, Compassion, and Direction," *LDS Living*, January 28, 2019, http://www.ldsliving.com/Abortion-How-President-Nelson-and-Other-Church-Leaders-Teachings-Provide-Hope-Compassion-and-Direction/s/90173

Danielle B. Wagner, "14 Myths and Truths We Know About Our Heavenly Mother," *LDS Living*, May 10, 2018, https://www.ldsliving.com/14-Myths-and-Truths-We-Know-About-Our-Heavenly-Mother/s/88439

Tad Walsh, "LDS Leaders Ask Mormons to Oppose Legalization of Assisted Suicide, Recreational Marijuana," *Deseret News*, October 13, 2016, https://www.deseretnews. com/article/865664777/LDS-leaders-ask-Mormons-to-oppose-legalization-of-assisted-suicide-recreational-marijuana.html

LeRoy Walters, "Religion and the Renaissance of Medical Ethics in the United States: 1965–1975," in *Theology and Bioethics: Exploring the Foundations and Frontiers*, ed. Earl E. Shelp (Dordrecht, The Netherlands: D. Reidel, 1985), 3–16.

Lynn D. Wardle, "Teaching Correct Principles: The Experience of the Church of Jesus Christ of Latter-day Saints Responding to Widespread Social Acceptance of Elective Abortion," *BYU Studies Quarterly* 53 (2014), 107–140.

C. Terry Warner, "Agency," in *The Encyclopedia of Mormonism*, ed. Daniel H. Ludlow (New York: Macmillan, 1992), 26–27, https://contentdm.lib.byu.edu/digital/collec-tion/EoM/id/5455

SarahJaneWeaver,"Mother,DaughterReachOuttoStrangers,"*ChurchNews*,February2,2006, https://www.thechurchnews.com/archive/2006-02-04/mother-daughter-reach-out-to-strangers-28305

Emily Webster, "Health Department Recommends HPV Vaccine," *Daily Universe*, January 23, 2008, https://universe.byu.edu/2008/01/23/health-department-recommends-hpv-vaccine/

Tiffanie Wen, "Why Don't More People Want to Donate Their Organs?," *The Atlantic*, November 10, 2014, https://www.theatlantic.com/health/archive/2014/11/why-dont-people-want-to-donate-their-organs/382297/

J. Kael Weston, "The Mormon Church vs. Pot," *New York Times*, November 6, 2018, https://www.nytimes.com/2018/11/06/opinion/mormon-church-medical-marijuana-utah-referendum.html

Michael White, "The End at the Beginning," *Ochsner Journal* 11:4 (Winter 2011), 309–316, https://www.ncbi.nlm.nih.gov/pmc/articles/PMC3241062/

John A. Widtsoe, ed., *Discourses of Brigham Young* (Salt Lake City: Deseret Book, 1975).

Stephen Wilkinson, "The Sale of Human Organs," *Stanford Encyclopedia of Philosophy*, 2015, https://plato.stanford.edu/entries/organs-sale/

Richard N. Williams, "Knowledge," in *The Encyclopedia of Mormonism*, ed. Daniel H. Ludlow (New York: Macmillan, 1992), 799–800, https://contentdm.lib.byu.edu/digital/collection/EoM/id/3856

"Willowbrook Hepatitis Experiments," https://science.education.nih.gov/supplements/webversions/bioethics/guide/pdf/Master_5-4.pdf

Jim Woolf, "4 of 5 LDS Senators Taking Stand in Support of Stem Cell Research," *Salt Lake Tribune*, July 19, 2001, http://www.sltrib.com/07192001/nation_w/114784.htm

Walker Wright, "LDS Moral Commitments on Immigration Are Grounded in History and Scripture," *Deseret News*, June 19, 2018, https://www.deseretnews.com/article/900022114/op-ed-lds-moral-commitments-on-immigration-are-grounded-in-history-and-scripture.html

Matthew Wynia, "The Birth of Medical Professionalism: Professionalism and the Role of Professional Associations," in *Healing as Vocation: A Medical Professional Primer*, ed. Kayhan Parsi, Myles N. Sheehan (New York: Rowman and Littlefield, 2006), 23–34.

Ed Yeates, "In Vitro Procedure Weeds Out Genetic Defects," *Deseret News*, June 13, 2012, https://www.deseretnews.com/article/865557436/In-vitro-procedure-weeds-out-genetic-defects.html

Sherry Young, "HPV Vaccinations Are a Choice to Ponder Seriously," *Deseret News*, July 7, 2008, https://www.deseretnews.com/article/700240635/HPV-vaccinations-are-a-choice-to-ponder-seriously.html

Danish Zaidi, "Influences of Religion and Spirituality in Medicine," *AMA Journal of Ethics* 20 (2018), E609–E612, https://journalofethics.ama-assn.org/article/influences-religion-and-spirituality-medicine/2018-07

Sarah Zhang, "What Mormon Family Trees Tell Us About Cancer," *The Atlantic*, June 23, 2017, https://www.theatlantic.com/science/archive/2017/06/mormon-genetic-testing/530781/

Steven R. Zimmerman, "The Mormon Health Traditions: An Evolving View of Modern Medicine," *Journal of Religion and Health* 32 (Fall 1993), 189–196.

Abigail Zuger, Steven H. Miles, "Physicians, AIDS, and Occupational Risks: Historic Traditions and Ethical Obligations," *Journal of American Medical Association* 258 (1987), 1924–1928.

Index

For the benefit of digital users, indexed terms that span two pages (e.g., 52–53) may, on occasion, appear on only one of those pages.